Development economics between markets and institutions

Development economics between markets and institutions

Incentives for growth, food security and sustainable use of the environment

edited by:
Erwin Bulte
Ruerd Ruben

Mansholt publication series - Volume 4

Wageningen Academic
P u b l i s h e r s

ISBN 978-90-8686-047-0

ISSN 1871-9309

First published, 2007

Wageningen Academic Publishers
The Netherlands, 2007

Mansholt Publication Series

The Mansholt Publication Series (MPS) contains peer-reviewed textbooks, conference proceedings and thematic publications focussing on social changes and control processes in rural areas and (agri)food chains, and the institutional contexts in which these changes and processes take place. MPS provides a platform for researchers and educators who would like to increase the quality, status and (international) exposure of their teaching materials or of their research output.

MPS is supported by the Mansholt Graduate School which also appoints the members of the editorial board responsible for the quality and content of the overall series. MPS is published and marketed internationally by Wageningen Academic Publishers.

The Mansholt Publication Series editors are:

Prof. Wim Heijman
Prof. Kees de Hoog
Dr. Arjen E.J. Wals
Prof. Leontien Visser

Preface

Development economics is a scientific discipline with a strongly growing importance in international discussions on sustainable agricultural development. This collection of articles in a special edition of the Mansholt Graduate School Series, under the challenging title *'Development economics between markets and institutions'* provides a concise overview of recent academic and policy debates regarding the contributions of agriculture for reducing poverty, enhancing food security and guaranteeing sustainable use of natural resources. Many contributions also point to the complex interfaces of agricultural producers in between market forces and public, private or voluntary institutions.

The academic development of the discipline of development economics has shown considerable progress in several areas. Studies on macro-economic development and international trade are increasingly related with analyses regarding micro-economic incentives and producers' responses. Moreover, linkages between development economics and technical sciences strongly increased, particularly concerning issues such as sustainable land use and supply chain integration. Finally, strategic discussions on rural development pathways recognize the key importance of adequately dovetailing public policies with private sector engagement.

For a long time and natural/technical scientists had different perspectives on necessary developments in agriculture in developing countries. Of course the solutions for the issues are seldom found in one of the disciplinary domains. However, only few scientists develop the seniority to think across disciplinary borders and actively try to integrate insights from the different disciplines. Only a very few are able to stimulate others to do so and co-operate in large interdisciplinary programs. Prof Arie Kuyvenhoven is one of those scientific leaders.

The development economics group at Wageningen University, during the last 20 years under the leadership of Prof. Dr Arie Kuyvenhoven, received a strong international reputation for its contributions to international academic debates in these areas. The articles included in this volume and prepared by a large number of colleagues – including seven former Wageningen UR PhD students - acknowledge the wide appreciation for his work within our academic community.

Because of his leadership skills he was appointed as director of Mansholt Graduate school from 2003 to 2007. He strongly focused on enhancement of the quality of the science and firmly stimulated a strong publication culture among the scientists. That had serious impact as can be seen from the evaluations of the scientists in the mid term review of MGS. On behalf of Wageningen University I would like to indicate my appreciation for Prof. Arie Kuyvenhoven's contributions as a scientist, as head of the chair group and as director of MGS and thank him for this.

Martin J. Kropff
Rector Magnificus, Wageningen University and Research Centre

Acknowledgements

The editors of this volume would like to acknowledge the support of Dr. Luuk Knippenberg (CIDIN/Radbout University) and Dr. Jack Peerlings (MGS) for the review of the contributed articles, and assistance of Dr. Marrit van den Berg and Ir. Ellen Mangnus (Development Economics Group) in preparing the index.

Prof. Erwin Bulte
Prof. Ruerd Ruben

Contents

Sustainable management of natural resources

Strategies for enhancing food security

Markets and the role of the state

Introduction

Erwin Bulte and Ruerd Ruben

Development economics is a rapidly evolving scientific discipline, where new and exciting issues are greedily incorporated into the research agenda and innovative approaches are warmly welcomed. In addition to being dynamic, the field is also unusually broad. It is impossible to do justice to all developments in a single volume. Fortunately, that is not the aim of this book.

This collection of articles is published as a festschrift or *Liber Amoricum* for professor Arie Kuyvenhoven. His retirement from the Development Economics Group at Wageningen University (The Netherlands) is a great opportunity to take stock of recent developments and insights in a subset of areas in development economics. Taken together, the chapters in this volume present an overview of the state of the art in many important fields. It is certainly possible to identify common themes among the various contributions, connecting them to a rather coherent piece of work. Examples of such overarching themes are agricultural development, and linkages between the rural sector and the rest of the economy. However, a possibly more important common theme is that all contributions have been prepared by friends and colleagues of professor Kuyvenhoven.

While any classification of articles is necessarily somewhat ad hoc and arbitrary, we have taken the liberty to group the 23 chapters in this book under 5 headers. These are:
1. prospects for (rural) poverty alleviation;
2. options for sustainable management of natural resources;
3. strategies for enhancing food security;
4. markets and the role of trade; and
5. institutions and the role of governance.

It is obvious that these headers define overlapping research areas, and that most of the papers in this volume could have been placed under multiple labels. For example, the close connection between institutions and sustainable resource management is evident, as is the intimate relationship between poverty alleviation and food security. With that caveat in mind, and also reminding that various fields important for development are omitted altogether for space reasons, we will now proceed by briefly introducing the various contributions.

1. Rural poverty alleviation

Studies regarding the causes and consequences of chronic poverty increasingly point attention to the importance of understanding rural households' resource allocation decisions within a framework of market distortions, insecurity and risk. The four contributions addressing policies towards poverty alleviation consequently focus on options for strengthening spatial, sectoral and political linkages towards a more inclusive strategy of development.

The contribution by Von Braun focuses on poverty reduction policies within the framework of urbanization dynamics, spatial differentiation and rural stagnation. These new contextual conditions tend to modify the opportunities for creating dynamic rural-urban linkages in developing countries. The article reviews the evolution of economic thought around spatial allocation of economic activities and the potential role of rural-urban linkages for enhancing the effectiveness of economic policies towards technology, infrastructure and market integration. Recent progress in regionally disaggregated analysis and economic modelling provides better insights in the potential contributions of migration flows and trade networks for scaling-up innovation and communication and enhancing socio-economic integration of poor people in remote areas.

In a similar vein, Escobal analyses the impact of transaction costs for rural households' welfare, devoting attention to the causes and implications of the lack of spatial and vertical integration of product and factor markets. Based on an innovative procedure for a disaggregated analysis of the transport, information, negotiation and monitoring costs of farmers incorporated in the potato market of Peru, the study shows that public infrastructure is vital for lowering transaction costs for reaching market outlets and establishing transaction in those markets. Lowering transaction cost enables specialization and division of labour and hence is a driving force for improving efficiency and income generating opportunities for the rural poor.

Market failures are also key for the understanding of household resource use decisions. The contribution by Barrett analyses the spill-over effects of limited access to (formal or informal) credit or insurance markets for farmers' behaviour in other (asset, factor and product) markets, demonstrating that households rationally exploit other market and non-market resource allocation mechanisms to resolve, at least partly, their financing problems. These 'displaced distortions' of financing constraints commonly manifest themselves in allocative inefficiency that may lead researchers and policymakers to mistakenly direct interventions towards the symptoms manifest in other markets rather than towards the root financial markets failures.

The political conditions for sustainable development and peace are addressed in the contribution by Pronk. Searching for a new paradigm in international policy making, the mutuality of peace and development is increasingly recognised. However, in many developing countries globalisation and exclusion are leading to persistent feelings of injustice, neglect and alienation. These may become a source of increased instability, insecurity, conflict and violence. Globalisation has also resulted into ecological distortions, sharpening of inequalities, a greater conflict potential and a weakening of the capacity of the polity to deal with these concerns. It is, therefore, imperative to strive for sustainability as an inclusive concept that reinforces mutual thrust that justice will be maintained and secured for all people, without any discrimination, thus providing ultimate guarantees of mutual security.

2. Natural resources

Van Rheenen and Olofinbiyi discuss some of the recent challenges that face land use policies. These include: climate change, urbanisation, production of biofuels, and the renewed attention on the multiple potential roles of agriculture ("multifunctionality"). While the primary role of agriculture (in developing countries and elsewhere) is to feed the people, too much attention to that service will diminish other valuable services provided by agriculture. There may be trade-offs between short-term and long-term objectives of land use policies. The chapter advocates the improvement of tools and methods of analysis for land policies, in order to enhance the way that the poor can adapt to (and possibly mitigate) global developments.

So-called less-favoured areas are areas "neglected by man and nature." This implies that both agro-ecological conditions for agriculture as well as the degree of market development are modest. Hazell and co-authors discuss development strategies for such less-favoured areas, paying special attention to technological innovations and institutional change. However, as argued by the authors, the diversity of less-favoured areas implies that it is unlikely that one set of policies will be uniformly successful. Instead, tailor-made solutions are called for.

A key issue when it comes to promoting development in less favoured areas is adoption and diffusion of low-input agricultural technologies among smallholder farmers. Arellanes and Lee explore the adoption of one such technology, namely minimum tillage (*labranza minima*) among resource-poor agricultural households in central Honduras. The study identifies candidates for minimum tillage adoption, and relates adoption decisions to household characteristics and prior land use decisions. Interestingly, income does not appear to be a determinant of adoption, suggesting that minimum tillage may be a suitable low-input technology for resource-poor households.

Kruseman discusses two important approaches to studying the relationship between household behaviour and agricultural resource degradation in developing countries. Bioeconomic programming and econometric analysis present two complementary avenues to capture both behavioural and material-system relations in a single model. Both approaches are illustrated using African case studies.

3. Food security

Economic policies for enhancing food security traditionally address options for guaranteeing physical, social and economic access to sufficient, safe and nutritious food that meets the dietary needs and food preferences for an active and healthy life. New attention is given to the political conditions for pursuing food security in fragile states and the governance dimensions of food aid programs. Moreover, more differentiated food security strategies tailored towards the existing resource and environmental conditions, are likely to better respond to available options for overcoming critical economic or institutional constraints.

Pinstrup-Andersen outlines the serious ethical, economic and stability problems for both developing and developed countries if poverty, hunger and malnutrition prevail and the Millennium Development Goals and the World Food Summit goals are not converted into concrete plans, goals, strategies and actions. Whereas poverty, hunger and lack of social justice contribute to armed conflict and terrorism, the probability of armed conflict is much larger in the poorest developing countries. Many failed or fragile states are caught in a poverty-conflict trap. Therefore, development strategies for low-income developing countries should pursue the dual goal of eliminating poverty, hunger and inequality and reducing the probability of armed conflict and terrorism. While the main responsibility for action rests with each developing country, donor countries should provide appropriate support in the form of fair trade policies and developing assistance.

Keyzer and Van Wesenbeeck refer to some biblical notions to outline the importance of three critical governance dimensions for guaranteeing food security: (1) income entitlements, (2) supply of goods, and (3) adequate funding of food security programs. While the assurance of food security is considered a core task of government, secure funding for the delivery of food aid is a commonly neglected element in food security management. Based on a spatially explicit model for optimal food aid provision in Sub-Saharan Africa, it is demonstrated that foreign donors that intervene during food crises should program an adequate exit strategy for when the emergency is over, so as to ensure that local authorities can take over these tasks, eventually relying on foreign financial support and technical assistance during a transitional period.

Focussing specifically on West Africa, Benoit-Cattin outlines the particular challenges for promoting food security in a highly diverse geographic, agro-ecological, historical and socio-economic context. Given the differential character of local food security problems, 'one size fits all' policy options are not suitable. Instead, different combinations of conceivable technological options for improving the productivity and resilience of production systems, combined with institutional options for establishing 'contestable markets' and mutual learning are required in the post structural adjustment context. In addition, new attention should be given to the critical role of public infrastructure investments and service provision to guarantee a more integrated performance of national food systems.

The implication of land use changes for grain self-sufficiency and environmental sustainability in China are discussed by Tan, Heerink and Futian Qu. Feeding the large Chinese population with limited land implies that available resources are used intensively. Grain self-sufficiency is under threat due to continued conversion of arable land for construction land or other uses. Further increases in grain productivity thus have to rely on even higher levels of input use that may further threaten the quality of the natural resource base (i.e. soil degradation, water scarcity, water pollution, etc.). The chapter uses provincial land use data to show that the centre of grain production is gradually moving northwards towards more fragile and water scarce areas. Ecological restoration programs are concentrated in regions with relatively low land productivity, while land conversion for construction usually takes place in areas with relatively high land productivity. Since the stock of potentially cultivable land is almost exhausted, China's grain self-sufficiency policy can only be maintained in the near future by preserving the available stock of arable land and increasing its productivity in a sustainable way.

4. Markets and trade

Turning to the markets and trade theme, we start off with a country-specific evaluation of agricultural trade policies. Policy makers Kenya (and elsewhere) wrestle with the dilemma that they want to give proper incentives to agricultural producers, but also wish to provide low-cost food to urban populations. The study by Karanja focuses on protection in the sugar market to illustrate a few key issues. The chapter evaluates price and welfare impacts on producers and consumers. The 1990s in Kenya saw a policy change from the promotion of competition to policies aimed mainly at protection of local producers. In spite of the high protection for Kenyan sugar farmers, there is intense competitive pressure from regional and international markets. Trade policies are ineffective (possibly counterproductive), and out of tandem with recent international developments.

Jansen and Morley analyse the consequences of the Central America-Dominican Republic-United States Free Trade Agreement (CAFTA) for the textile sector (the apparel value chain) – a key economic sector in terms of export earnings and employment for the region. CAFTA has increased the scope of trade relations with the U.S., and offers potential to spur growth in the sector (promising jobs and increased integration). These issues are explored with general equilibrium models. However, the competitiveness of firms is unclear in light of enhanced competition from Asia. The study proposes several policy implications to promote opportunities for the textile sector in a globalizing world.

Ruben, focuses on the importance of transaction costs – information, negotiation and monitoring costs – that farmers incur when using the market. While such costs will be high in early phases of market development, they may fall over time as contractual relations develop. To make the shift from wholesale markets towards preferred supplier regimes requires that savings in transaction costs are sufficiently large. In this light, the evolution of fresh produce procurement channels by supermarkets in developing countries is analyzed. One insight of the chapter is that the nature of supplier-buyer relationships changes from one phases to another in the evolution to integral chain care, and that small farmers may require assistance to become engaged in such procurement regimes.

The chapter by Roseboom revisits the sugar trade, but zooms in on the reform of EU trade policies. Sugar industries in so-called Sugar Protocol (SP) countries will lose as their export revenues will decline sharply. Other changes affecting the sugar industry are the Everything-But-Arms (EBA) agreement and the increase use of sugarcane (or sugarcane waste) as biofuel. Biofuel promises greater gains for SP countries in the short run, and in the long run it may be the case that energy production becomes the core business (with sugar as a by-product). These issues are explored and discussed in relation to price and subsidy developments.

5. Governance and institutions

From trade and transaction costs to governance and institutions is a small step. Van Huylenbroek and Espinel examine three case studies (cattle trading in Uganda, water management in the Peninsula of Santa Elena, and biodiversity conservation in relation to indigenous knowledge) to highlight the importance of institutions and governance structures, and in particular describe how the asymmetry of information embedded in markets hinders smallholders to fully benefit from the values incorporated in their property rights.

Swinnen changes the scenery and takes us to transition countries in eastern Europe. He describes and analyses the evolution in the agricultural sector (in terms of farm size, technology and agricultural institutions) in response to reforms and restructuring processes. One insight is that a great diversity of farm structures across countries has developed. This diversity reflects initial conditions and external constraints, but also policy differences, and could have permanent implications for development of the rural sector in transition economies (due to "lock-in" effects). Swinnen also revisits the issue of vertical integration, and discusses the evidence regarding exclusion of small farmers from integrated value chains.

Moll explores the potential contribution of rural microfinance institutions for promoting rural development and integration of rural economies into national and international markets. Barriers to the flourishing of the financial sector in the countryside are identified, followed by a discussion of the implications for the design of microfinance institutions (and financial products offered). The chapter also discusses the incorporation of microfinance institutions into national and international financial systems.

Besides credit, it is obvious that land is a major input in agricultural production. Holden describes the functioning of land markets in Africa, and discusses the importance of emerging land markets for economic growth and income distribution. A review of the evidence suggests there are complex interactions between land markets, the distribution of resources, and the characteristics of non-land factor markets (including credit and insurance markets). Implications for small farmers and the distribution of assets is also discussed. While land rental markets are often beneficial for the poor (offering them greater flexibility to meet their needs), there is also some evidence that land sales markets might accentuate inequalities through distress sales of the poor.

Pingali and Raney contrast two waves of agricultural technology development – the Green Revolution (making improved germplasm freely available as a public good) and the more recent Gene Revolution (dominated by profit-maximising multinationals involved in transgenic innovations). In light of large asymmetries between developed and developing countries, it is not evident that poor farmers will be able to benefit from the Gene Revolution to the same extent as they were able to capture technology spill-overs associated with the Green revolution. Similarly, inequalities within countries could be magnified as a result of the Gene Revolution. The chapter advocates a strong public sector – working cooperatively with multinationals – to ensure that the poor are able to benefit from the second wave of agricultural technologies.

Damania and Bulte revisit the issue of interlinkages between institutions and income. In light of potentially powerful feedback effects (institutions affect income, but income may also affect institutions) it has become common to instrument for institutions in growth regressions. This chapter follows the reverse route, and uses a panel approach to instrument for income in a regression aiming to explain institutional quality. Rainfall is used as an instrument for income in sub-Saharan Africa where rainfed agriculture continues to be important. The main result is that there is no evidence of a significant direct effect of income (shocks) on the quality of governance.

6. Conclusions and outlook

Taken together, the 23 chapters that make up this *Liber Amoricum* paint a varied picture. The book includes qualitative as well as quantitative contributions, and identifies various challenges and opportunities for agricultural development in developing and transition countries. The chapters in this volume provide a concise overview of recent debates in development economics regarding the role of state and market institutions for economic growth, food security and poverty alleviation, and the sustainable use of natural resources.

Not surprisingly, these issues have also been at the core of the professional work of professor Arie Kuyvenhoven, whose research has contributed to all these topics. After graduating from the Netherlands School of Economics in Rotterdam (1971), he started his career as a visiting lecturer at the University of Lagos, Nigeria. Upon his return to The Netherlands, he accepted a position at Erasmus University Rotterdam, and subsequently became engaged in different positions with the Netherlands Economic Institute (NEI), ultimately as Statutory Director. In 1987 he was appointed as chair of the Development Economics Group at Wageningen University. Finally, in 2003 he became Scientific Director of the Mansholt Graduate School of Social Sciences in Wageningen.

During his professional career, Professor Kuyvenhoven maintained a wide network of international collaboration. He served as member of the National Advisory Council for Development (NAR), as member on several advisory committees of the Netherlands Organization for Scientific Research (NWO, WOTRO, MaGW), as board member of the Centre for World Food Studies (SOW) at Free University Amsterdam, and as board member of the ESCAP CGPRT Centre, Bogor (Indonesia). In 1998 he became Member and Vice-Chair of the Board of Trustees of the International Food Policy Research Institute (IFPRI) at Washington, D.C. (U.S.A). In addition, he has been a visiting professor at Cornell University, Ithaca (New York) and at the University of California, Davis (USA), and was awarded as Honorary Professor at Nanjing Agricultural University (China).

Many colleagues and former graduates considered it a real honour and pleasure to contribute to this special volume. This illustrates that professional collaboration with Arie Kuyvenhoven, on many occasions, turned out to become personal friendship. His warm personality and open relationships with colleagues and students became the cornerstone for the growing international reputation of the development economics group in Wageningen. He deserves wide credit and our warm appreciation.

Prospects for (rural) poverty alleviation

Policies for pro-poor rural-urban linkages[1]

Joachim von Braun

Abstract

Policy must address market failures in urbanization dynamics and in rural stagnation. New contextual conditions are changing the opportunities for rural- urban linkages in developing countries. The chapter reviews the evolution of economic thought around spatial allocation of economic activity and rural - urban linkages and focuses on the roles of technology, infrastructure and market policies to foster optimal rural - urban linkages.

1. Introduction

Traditionally, development policy, strategy and research have adopted a simplified conceptualization of rural and urban areas, with 'rural' referring to the remote 'farmland' and 'urban' to 'crowded cities.' This view has facilitated the isolated treatment of issues affecting each space, and has, as a result, failed to acknowledge the important poverty reducing inter-linkages that exist between the two spaces and the many variants of them. Moreover, new contextual and exogenous conditions are changing the opportunities for rural-urban linkages as well as the intensification of these linkages. Elements of the changing conditions include (a) increasing trade and capital flows which have prompt the fast changing agriculture and food system with urban consumers increasingly influencing the nature and level of the interactions of the various stakeholders in the agri-food chain; (b) the information revolution with an increasing number of rural communities benefiting from enhanced access to communications technologies carrying relevant content of information and facilitating new market institutions and services; and (c) increasingly decentralized governance structures across the developing world with national governments and policymakers, as well as private investors involved in regional development and inter-regional competitiveness.

While urbanization is part of a healthy economic development process, its unguided shape and speed, often combined with market and other institutional failures, can result in adverse effects for people and the environment. The urban share of poverty is increasing: in 1993, the urban share of the poor was around 19 percent, but by 2002 it had increased to almost 25 percent (Ravallion, *et al.,* 2007). Policy must address market failures in urbanization dynamics and in rural stagnation, i.e. labour market, services- and goods market failures

---

[1] Assistance by Tewodaj Mengistu and comments from Maximo Torero, and Shenggen Fan (all at IFPRI) on an earlier draft are gratefully acknowledged. The chapter draws on a keynote address at the Ethiopian Economics Association in June 2007 by the author.

due to ill-guided expectations, information gaps and missing markets (e.g. finance), and the negative environmental externalities sometimes engendered by urbanization.

Against this broad problematic the more narrowly stated question addressed in this chapter is: how can rural-urban linkages be improved to accelerate inclusive growth, expand employment, and serve the poor? These linkages do not just 'exist', they must be optimally invested in to help reduce transaction costs related to the linkages of diverse types and stimulate positive externalities and spillover effects.

The chapter starts with a brief synthesis of theoretical and conceptual frameworks of rural-urban linkages so as to identify points of entry for policy. Thereafter, trends in rural - urban linkages are reviewed. Next, policies facilitating rural - urban linkages and some criteria for setting policy priorities are identified. Finally areas for related economic research needs are mentioned.

2. Concepts and framework

In recent years, there has been some resurgence in the academic and policy fields of the importance of 'geography' to the understanding of economic transactions. This section briefly reviews the evolution of the theoretical literature on spatial allocation of economic activity and also the driving forces behind spatial differentiation.

2.1 The evolution of theory and concepts on rural-urban linkages

A well known early analysis of spatial allocation of economic activity was undertaken by J.H. von Thünen in 1826. Using a model of agricultural land use, he showed how market processes determined land use in different geographical locations, and more specifically how land use is a function of transport costs to markets and the farmer's land rent. His model generated four concentric rings of agricultural activity around a central city, with dairy and intensive farming closest to the city, followed by timber and firewood in the second circle, grain production in the third, and finally, ranching and livestock activities in the fourth circle. Urban demand is a key driver of spatial allocation of economic activities already in this basic model of marginal returns to assets and labour.

Another breakthrough was by Walter Christaller (1933), who developed the Central Place Theory to explain how urban settlements are formed and are spaced out relative to each other. The main premise of Christaller's theory was that 'if the centralization of mass around a nucleus is an elementary form of order, then the same centralistic principle can be equated in urban settlements' (Agarwal, 2007). His model, which was later refined by Lösch (1954), predicted an urban hierarchy of human settlements around hexagonal shapes (the hexagon being the most efficient way to travel between the settlements), with varying sizes of centres. The size of the centre is determined by the type of goods and services it provides, whereby larger settlements (fewer in numbers) provide 'higher order' goods and services (which require a large market both in terms of income and population, and are therefore more specialized), and smaller settlements provide 'lower order' goods and services. In this framework, since some of the demand for the goods produced in the centres (such as manufacturing) comes from peripheries, production is tied with agricultural land distribution (Krugman, 1991).

These conceptual frameworks not only define rural-urban linkages but also urban-urban linkages between centres of differing scale related to economies of scale in sub-sectors of the economy. However, the early models were based on strong assumptions such as homogeneous spaces, uniform consumer preferences, and proportionality of transport costs to distance. Therefore, their applicability to real settings is limited. Nevertheless, they do clarify the gradual nature of the differentiation between urban and rural areas: in reality, and as expected in the theory of economic geography mentioned above, the 'very rural' and the 'very urban' co-exist along a continuum with many in-between stages varying from small towns to peri-urban areas (Figure 1).

Moreover, a dynamic set of flows exists between these various spaces, creating interdependencies between them. In general, two types of flows can be distinguished. The first type is *spatial*, which includes flows of people, goods, money (in the form of remittances for example), information and waste. In bio-physical perspectives, flows of water, bio-mass products, and nutrients, are relevant. The second type is *sectoral*, which includes flows of agricultural products going to urban and peri-urban areas, and goods from the urban manufacturing areas going to more rural areas (Tacoli, 1998).

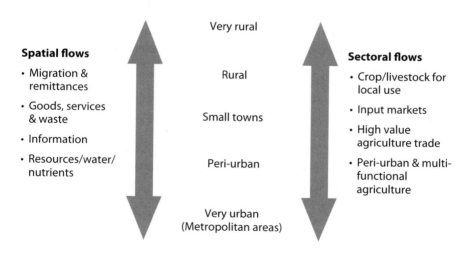

Figure 1. The stylized rural-urban continuum.

2.2 Forces behind spatial differentiation

Since the early 1990s, various economic studies explaining spatial differentiation have emerged. Generally, two types of arguments can be distinguished; the first one focuses on geographical endowments determining comparative or absolute advantages, while the second one focuses on the existence of linkages (backward and forward) that cause agglomeration of certain activities (Venables, 1996). An example of the first type of argument comes from Gallup and Sachs (1998), who focus on inherent geographical differences such as climate and location to explain spatial differentiation. According to them, people tend to converge to locations that are conducive to growth, and as such, some locations are at an absolute disadvantage over others. Through an econometric analysis, they show that landlocked (because they are subject to higher transportation costs) and tropical regions (because they are subject to higher levels of disease) are the most disadvantaged. The second type of argument is represented by Krugman's (1991) 'new economics of geography,' whose central assertion is 'the world economy can engage in a process of self organization in which locations with seemingly identical potential end up playing very different economic roles' (Krugman, 1998). Building on Christaller and Lösch, he finds that while some of the demand for the goods produced in centres (for e.g. from the manufacturing sector) comes from the periphery, demand also comes from the manufacturing sector itself because of backward linkages to other manufacturing industries. This adds a 'circular causation' dimension in that manufacturing would concentrate in a location where the market is large, and at the same time, the market would be large where manufacturing is concentrated. Krugman also points to forward linkages as sources of agglomeration of economic activities, as other things equal, people prefer to live and produce near the concentrated manufacturing sector because it would be cheaper to buy goods in this central location where it is produced.

Additionally, Krugman adds a dynamic perspective incorporating changes in economic parameters, such as transportation costs, share of non-agricultural products in expenditure, and economies of scale to his model. He finds that populations start concentrating in central areas and regions start diverging when the non-agricultural expenditure passes a critical threshold and there are large economies of scale to production and lower transportation costs (Krugman, 1991). However, focusing on transport and communication costs, Venables (1996) finds that spatial agglomeration of economic activity happens only at intermediate price levels. He finds that in the other two extreme situations where transport and communication costs are either very high or very low, economic activity tends to be dispersed, but for different reasons; in the case of high costs, economic activity is dispersed because firms need to be closer to consumers, and in the reverse case of low costs, firms would consider other issues besides costs of transport and communications when deciding on location.

Against the backdrop of the above reviewed conceptual and theoretical evolutions, in the following section some mega trends of change in the rural urban linkages and their main policy determinants are discussed.

2.3 Trends and domains of policies facilitating urban-rural linkages

The level of interaction between urban and rural areas has intensified, with both positive and negative implications. A large divide between rural and urban areas in developing countries still exists and as a result, major inequalities persist between the two spaces.

2.3.1 Flows between rural and urban space

Driven by technological progress, improvements in infrastructure, and liberalization and creation of markets, globalization has meant rapid rural transformation across the developing world. As such, the interactions connecting the various spaces between the 'very rural' and the 'very urban' have become deeper. The evolution of spatial and sectoral flows is explored next.[2]

Spatial flows

Urbanization is happening at accelerated speeds across the developing world; from 1994 to 2004, the annual population growth rate in urban areas for developing countries as a whole was on average 2.6 percent, as opposed to 0.6 percent for rural population growth. The average annual rate of urban population growth is even higher for sub-Saharan Africa at 4.4 percent (as opposed to 1.5 percent for rural population growth; World Bank, 2006). This trend is expected to continue over the coming decades with urban populations in less developed regions surpassing rural populations by 2020.[3]

The dynamics of migration are complex. The economic thinking behind migration decisions dates back from the Harris-Todaro model (1970) which asserts that an individual's decision to migrate is based on the differences in expected earnings in the formal urban sector and the expected earnings in the village. Since then, economic theory has evolved to focus on families as a unit of analysis, as opposed to just the individual. Stark and Bloom (1985) found that migration is a household decision and families invest in migrant(s) in return for future receipts of remittances. As such, migration is a source of income diversification for households, and is also circular in that it entails continued interaction between migrant(s) and their families, who remain in the area of origin.

---

[2] The following sub-section does not comprehensively address all rural-urban flows. It is therefore important to keep in mind that a number of other types of flows exist, for instance, through nutrition and health (including animal health/human health interactions, and adaptation in nutrition behaviors in rural areas which are becoming more urban).

[3] In 2020, the urban and rural populations in less developed regions are forecasted to reach 3.3 billion and 3.1 billion respectively. The numbers in 2005 were 2.3 and 3 billion respectively (UNDESA 2006).

More broadly though migration is determined by push and pull factors; push factors e.g. include droughts, land scarcity, low wages or absence of wage labour in out-migration areas, and pull factors include better job opportunities and/or possibility of higher income in destination areas (Von Braun, 2005). Moreover, migration is not always permanent, there is a strong component of seasonal migration in developing countries (especially in Asia), whereby people are 'pulled' into urban areas as a result of strong growth in manufacturing and industry. One of the advantages of such migrations, beyond increased earnings, is that the availability of these urban jobs is not tied to the agricultural season, which entails that people can work both in rural and urban areas (Deshingkar, 2005). Seasonal migration can be welfare enhancing; for instance a recent study in Vietnam found that seasonal migration resulted in an annual increase of about 5 percent of household expenditure, and a 3 percentage point decrease in the poverty headcount (De Brauw and Harigaya, 2007).

One of the main outcomes of increased migration (including international migration) linkages is growing remittance receipts in many developing countries. Such remittances can play a very important role in supplementing incomes in receiving households. Additionally, the increase in purchasing power of receiving households can stimulate the local economy, and in the particular case of rural areas, increased remittance receipts can stimulate the rural non-farm economy (Thanh *et al.*, 2005).

Much of the *global* change in labour allocation is related to intersectoral shifts due to enhanced growth in other sectors (manufacturing, industry and services) in urban areas. Over the next 15 years, the economically active population is projected to increase from 3 billion to 3.5 billion, and while farm employment may go down probably by about 300 million, employment in services and industry, both in urban and rural areas, are estimated to employ an addition 400 million people each (see Table 1).

The intensification of urban-rural linkages with respect to environmental flows has occurred as a result of increased urban demands for rural resources such as rural land, water and

Table 1. Global employment 2005 and 2020 (in billions).

	Farm	Services and industry- rural areas	Services and industry-urban areas	Total
2005	0.9	0.6	1.5	3.0
2020	0.6	1.0	1.9	3.5
Change 2005-2020	-0.3	+0.4	+0.4	+0.5

Source: author's estimates of sector shares based on ILO economically active population projections.

air. The most visible change is associated with the physical expansion of urban areas, as urbanization has lead to the extension of urban space onto rural space to accommodate the increasing population and level of activity. This has increased the demand for land around cities to build residences, industries, transport corridors such as roads and highways, as well as for the disposal urban waste (both industrial and household waste) (Mc Granahan *et al.*, 2004).

In many low and middle income countries, the growth of cities has sometimes produced densely populated and impoverished squatter settlements in peri-urban areas, with little access to adequate shelter, sanitation and other types of services. People residing in these urban sprawls are prone to diseases as they often do not have access to safe water and sanitation. Additionally, they have to cope with high levels of pollution with the manufacturing food processing and urban building activities release their chemical waste into the atmosphere, soil and waterways in the nearby peri-urban areas (McGregor *et al.*, 2006).

Sectoral flows

Based on the works of Hirschman (1958), in the 1970s it was widely believed in the developing world that the agricultural sector had weak linkages to the rest of the economy, and that as a result the policy focus should be on promoting industrialization rather than agricultural productivity. However, the failures of development strategies based on import-substituting industrialization (for e.g. in many Latin American and African countries) and the successes of countries that pursued agricultural-led growth (for e.g. China and later on India) have demonstrated that agricultural productivity growth is essential to launching an economy-wide growth, especially in predominately agrarian societies. Indeed, agriculture growth engenders both backward linkages in the form of increased demand for farm inputs such as fertilizers and farm equipment, and forward linkages, as increased farm households' income translates into increased demand for consumption goods and services (Mellor, 1995; Hazell and Röell, 1983; Diao *et al.*, 2007; De Ferranti *et al.*, 2005). These linkages can subsequently lead to rural transformation, with an expansion of the rural non-farm economy (RNFE) and better linkages with the rest of the economy, with increasing sectoral and spatial flows between rural and urban areas.

In the last couple of decades, significant gains in agricultural labour productivity have been achieved; agricultural value added per worker in low and middle income countries increased by approximately 43 percent from 1985 to 2003, going from US$ 405 to US$580[4] (World Bank, 2006). Additionally, globalization of agriculture has led to enhanced agricultural trade and the commercialization of traditional agricultural processes. And, as a part of the general trend toward the liberalization of markets, many developing countries have reduced the level of government intervention (for e.g. elimination of price controls on agricultural commodities, elimination of export taxes, privatization and/or dissolution of

---

[4] The data are in 2000 constant dollars.

state-owned enterprises, reduction of subsidies, etc.) in the agricultural sector. As a result of these reforms, some countries, especially those where reforms were fully implemented, experienced an increase in the number of traders and higher levels of competition, as well as reduced marketing margins, although in some cases, marketing margins remained high due to inadequate transport infrastructure and high levels of uncertainty. Furthermore, the reforms enabled a reduction in price instability as well as lower price spreads between different markets (Kuyvenhoven *et al.*, 2000; see Kherallah *et al.*, 2002 for an overview of market reforms specifically in sub-Saharan Africa).

An additional outcome of the increased opportunities brought about by globalization and the liberalization of the 1980s and 1990s, is the expansion of the rural non-farm employment (RNFE). Indeed, the RNFE now accounts for approximately 25 percent of full time employment and about 30-40 percent of rural household income in developing countries (Haggblade *et al.*, 2007b).

Another feature of the globalization of agriculture is that consumer preferences across the globe have become major driving forces of agricultural production systems. They spend about $4 trillion on food and beverages (Von Braun, 2005). The retail industry caters to them, while the food processing and trading industry supplies the retail sector and procures from the farm sector, which in turn is supplied by agriculture input industries. Rising consumer incomes and urbanizing lifestyles have increased the demand for high value agricultural products, which include meat and fish, fruits and vegetables, and dairy products. In order to meet these demands, agricultural producers in developing countries are diversifying towards these products.

2.3.2 The rural-urban divide

Despite the increasing levels of rural-urban interactions, major rural-urban disparities continue to exist across the developing world; as a result of adverse terms of trade between agricultural and non-agricultural and prices, as well as urban-biases in government spending in health, education, and physical infrastructure across the developing world, major inequalities between urban and rural areas persist, not only in terms of income, but also in asset endowment and human development (Eastwood and Lipton, 2004). Thus, while inequality exists separately within the rural and urban spheres, the largest differences are between urban and rural areas; most of the poor, as much as seventy-five percent (Ravallion *et al.*, 2007), live in rural areas and depend on agriculture and related trade, services and processing activities for their livelihoods. Additionally, in many countries, rural inhabitants do not have the same level of access to social services and infrastructure as their urban counterparts, further perpetuating the existing inequalities.

Even more concerning, the rural-urban divide seems to be increasing in parts of the developing world (Eastwood and Lipton, 2004). China and India provide illustrations of this increasing divide. Both countries have experienced sustained economic growth over the last decade (on average 9.5 percent annually for China, and 6.25 percent for India between 1994 and 2004), and have achieved major success in poverty reduction. However, economic growth and poverty reduction have been distributed unevenly. In both countries the bulk of the poor still live in rural areas, and are concentrated in certain regions: in China, the majority of the poor are concentrated in the interior or the country, and in India, half of the poor are concentrated in just three states; Uttar Pradesh, Bihar and Madhya Pradesh. Moreover, spatial inequality in both countries seems to be increasing. In China, the difference in the average monthly per capita income between urban and rural areas almost doubled between 1994 and 2004 going from US$ 99 to US$ 161, and the percentage of total inequality (measured using general entropy) due to inequality between Inland and Coastal areas increased from 6.5 percent in 1990 to 11.6 percent in 2004. Similarly, in India, the difference between average monthly per capita income in urban and rural areas went from US$21 to over US$27 and the percentage of total inequality due to inequality between North and South regions increased from 2.6 percent in 1990 to 15.9 percent in 2003 (Gajwani *et al.*, 2006).

2.3.3 Domains of policy facilitating urban rural linkages

This sub-section reviews how actions in three major areas alluded to in the above mentioned conceptual discussion – R&D and technology, infrastructure, and market institutions – can further stimulate rural-urban linkages to promote growth, employment creation and poverty reduction.

R&D and technology

Technologies work through factor and output markets, processing, and consumption linkages. Here, two types of technologies that have had a substantial impact on rural growth and poverty reduction are explored: innovations coming out of agricultural research and development (R&D) and improvements in information and communications technologies (ICTs).

Science and technology are fundamental for rural-urban linkages and in this context, agricultural research is essential. The Green Revolution experience, especially in Asia, has shown that agricultural R&D can result in technological breakthroughs that enable considerable improvement in agricultural productivity resulting in agricultural growth, which in turn can translate into substantial rural development and poverty reduction. The Green Revolution technologies, which include high-yielding varieties complemented with irrigation and intensive fertilizer use, were developed in response to the threat of widespread hunger in Asia in the early 1960s emanating from high population growth and the related overuse of water and land resources. The results were that these new technologies enabled not only higher production of food to meet the growing demand (and thus permitted many countries to achieve food security at the national level), but also had spillover effects in terms of rural development, as the rural non-farm sectors through consumption and production linkages (Haggblade *et al.*, 2007b).

Technological innovations in agricultural production continue to be significant sources of agricultural productivity growth, which can in turn translate into rural growth and poverty reduction. For instance, it is estimated that in Asia and Latin America, for each additional dollar of income generated in agriculture, between $0.6 and $0.9, and between $ 0.4 and $ 0.6 respectively, of income is generated in the local rural non-farm economy (Haggblade *et al.*, 2007b). The multipliers are lower for sub-Saharan Africa because of agro-climate conditions, and the lack of infrastructure and sound policies.

Nevertheless, while agricultural performance has been relatively high in recent years, there is cause for concern. Eighty developing countries today spent only a total of about $1.4 billion on agricultural R&D, which represents only 6 percent of global expenditures. Furthermore, together, the agricultural R&D expenditures of China and India represent 22 percent of the total developing country investments (Pardey *et al.*, 2006). This suggests that many small and medium sized developing countries are not investing enough in agricultural R&D. In effect, as the positive spillovers from technologies previously derived from industrialized countries' R&D dry out, and because the capacity for even intelligent borrowing and adaptation of technology is constrained by the low level of expenditures in small and medium sized developing countries, many of these countries may have only limited access to growth enhancing technologies. As a consequence, rural-urban linkages become weaker.

ICTs enable the reduction of transaction costs and open up markets possibilities for rural inhabitants, which can result in additional network externalities. Indeed, at the macro-level, tele-density is positively associated with growth. Wavermann *et al.* (2004) found that that 10 more mobile phones per 100 people increased GDP by 0.6 percent. But, there seems to be a minimum threshold of about 15 percent coverage to achieve the strongest growth effects (Torero and Von Braun, 2006). Africa and South Asia are far below these critical thresholds, especially in rural areas.

At the micro level, the welfare gains from having access to ICTs are fairly large, as the alternatives (sending a messenger or letter) are much costlier and more time consuming. In effect, the welfare gains from a telephone call range between US$ 1.62 to US$ 1.91 for instance in Bangladesh and Peru. As such, the willingness to pay for access to telephones is also relatively high, and typically exceeds the actual prevailing tariff rates (Torero and Von Braun, 2006).

Infrastructure
Infrastructure works as a bridge between the rural and urban worlds, and between the agricultural sectors and others sectors of the economy. The large returns in terms of growth and poverty reduction of investments in rural roads are well known. For example, in China, the returns to investments in rural roads in terms of national income are over three times that of investments in urban roads - for every Yuan invested in rural roads, the return are around 6 Yuan, versus 1.55 Yuan for urban roads. The returns in terms of poverty reduction are

equally better for rural road investments - 5.67 persons per 10,000 Yuan invested versus 0.31 for urban roads (Fan and Chan-Kang, 2005). As such, in the last decade, many developing have invested heavily in rural roads.

What is less well understood, however, is the interaction effect between different types of infrastructure investments. One study looking at the complementarities of investments in multiple types of rural infrastructure in Bangladesh found that the effect on household welfare impacts can be more than the sum of its individual impacts, and even multiplicative in some cases (Chowdhury and Torero, 2006; see also Escobal and Torero, 2005 for Peru case). This suggests that simultaneous provision of different types of infrastructure rather than individually could potentially strengthen the welfare and poverty reducing gains of rural infrastructure provision. Thus, optimal investment in rural-urban linkages entails complex bundles of infrastructure components.

Market institutions

In developing countries, particularly the low-income ones, market failures such as deficiencies in information markets and lack of regulation and legal enforcement mechanisms persist and restrict the level of trade between different spaces (Gabre-Madhin, 2001). As such, developing adequate market institutions and strengthening them are essential to facilitating spatial flows, and for that, an improved knowledge of the nature of transaction costs incurred by traders is important. Nevertheless, basic necessary market conditions and institutions include removing trading constraints, along with forming/strengthening institutions that can collect and disseminate good agricultural price and supply (both domestic and international) information, as well as institutions that provide effective legal systems. Additionally, improved access to financial risk management tools such as rural credit and insurance are vital to support the development of markets in rural areas (Kuyvenhoven *et al.*, 2000).

As globalization deepens, the consumer-driven agri-food chain is becoming more integrated, incorporating more small farmers through arrangements such as producer-marketing cooperatives which facilitate horizontal cooperation, and contract farming, which facilitate vertical cooperation. These types of arrangements can have substantial positive income benefits for poor rural households, by helping to reduce transaction costs and by reducing the variability in prices of agricultural products. Additionally, such arrangements give farmers better access to markets for their produce, and to technological innovations in agricultural inputs. However, such arrangements are usually information intensive, and require adequate legal frameworks and organizational capacity. Thus, in order to maximize the potential gains for farmers from integration within the globalized agri-food system, it is necessary to provide access to communications technologies, as well as training/capacity building for farmers. An opportunity here is in commodity exchanges appropriately adapted to the infrastructure and institutional environments of low income countries.

Joachim von Braun

3. Conclusions and research implications

The need for new attention to the spatial dimensions of development and to rural-urban linkages for inclusive growth is underlined in this chapter. The nature of such 'attention' by policy and advisory communities include:

a. Distinguishing the various types of dynamic flows that exist between rural und urban spaces.
b. Review of transactions costs of all economic activities between rural and urban areas with a perspective to their optimal reductions.
c. Focus on the non-trivial positive and negative externalities of spatial allocation and concentration of economic activities of different types.

The chapter has emphasized three needs of public policy actions that are crucial for enabling better rural-urban linkages:

1. scaling up of innovation in agriculture and the whole value chains;
2. scaling up of transport and communication infrastructure toward optimal densities;
3. development of market institutions, including in labour markets that enable the participation of remote areas and the poor in the national economy.

These policy issues pose new challenges for research. Much progress has been made in regionally disaggregated analysis and economic modelling. Integration of spatial analysis through for example the use of GIS technologies can be useful to the visualization and understanding of realities. Still, many of the economic models on which the rural-urban framework is based are simplistic and do not take into consideration spatial realities such as agro-ecological conditions. Blending aggregate modelling with information systems that capture local knowledge is a challenge. It is needed however for policy research to facilitate relevant guidance to policies that strengthen rural-urban linkages.

References

Chowdhury, S. and M. Torero, 2006. Urban-rural linkages in Bangladesh: Impact of infrastructure and the food value chain on livelihoods and migration of landless households, women and girls in the Northwestern region. Washington, DC: International Food Policy Research Institute.
Deshingkar, P., 2005. How rural is rural? ODI Opinions No. 52. London: Overseas Development Institute.
De Ferranti, D., G. Perry, W. Foster, D. Lederman and A. Valdés, 2005. Beyond the city. The rural contribution to development. World Bank Latin American and Caribbean Studies. The World Bank, Washington DC, USA.
De Brauw, A. and T. Harigaya, 2007. Seasonal migration and improving living standards in Vietnam. American Journal of Agricultural Economics 89 (2): 430-447.

Diao, X., B. Fekadu, A.S. Taffesse, K. Wamisho and B. Yu, 2007. Agricultural growth linkages in Ethiopia: Estimates using fixed and flexible price models. IFPRI Discussion Paper No. 00695. Washington, DC: International Food Policy Research Institute.

Eastwood, R. and M. Lipton, 2004. Rural and urban income inequality and poverty: Does convergence between sectors offset divergence within them? In: Cornia, G. A. (ed) Inequality, growth and poverty in an era of liberalization and globalization. Oxford University Press, Oxford, UK.

Escobal, J. and M. Torero, 2005. Measuring the impact of asset complementarities. The case of rural Peru. Cuadernos de Economía Vol. 42 (Mayo): 137-164.

Fan, S. and C. Chan-Kang, 2005. Road development, economic growth and poverty reduction in China. IFPRI Research Report No.138. Washington, DC: International Food Policy Research Institute.

Gabre-Madhin, E., 2001. Market institutions, transaction costs, and social capital in the Ethiopian grain market. Research Report 124. Washington, DC: International Food Policy Research Institute.

Gajwani, K., R. Kanbur and X. Zhang, 2006. Comparing the evolution of spatial inequality in China and India: A fifty-year perspective. Development Strategy and Governance Division Discussion Paper No. 44. Washington, DC: International Food Policy Research Institute.

Haggblade, S., P.B.R. Hazell and T. Reardon, 2007b. Transforming the rural non-farm economy. Johns Hopkins University Press for IFPRI, Baltimore MD, USA.

Harris J. and M. Todaro,1970. Migration, unemployment and development: A two-sector analysis. American Economic Review. 60(1):126-42.

Hazell, P.B. and A. Röell, 1983. Rural growth linkages: Household expenditure patterns in Malaysia and Nigeria. Research Report No. 41. Washington, DC: International Food Policy Research Institute.

Kherallah, M., C. Delgado, E. Gabre-Madhin, N. Minot and M. Johnson, 2002. Reforming agricultural markets in Africa. Johns Hopkins University Press for IFPRI, Baltimore MD, USA.

Krugman, P., 1991. Increasing returns and economic geography. Journal of Political Economy 99 (3): 483-499.

Krugman, P., 1998. The role of geography in development. In: B. Pleskovic and J. Stiglitz (eds.) Annual World Bank Conference on Development Economics 1998. The World Bank, Washington DC, USA.

Kuyvenhoven, A., H. Moll and A. van Tilburg, 2000. Agricultural markets beyond liberalization: issues, analysis, and findings. In: A. van Tilburg, H.A.J. Moll and A. Kuyvenhoven (eds.) Agricultural markets beyond liberalization. Kluwer Academic Publishers, Norwell MA, USA.

McGranahan, G., D. Satterthwaite and C. Tacoli, 2004. Rural-urban chance, boundary problems and environmental burdens. Working paper series on rural-urban interactions and livelihoods strategies. Working paper No.10. London: International Institute for Environment and Development.

McGregor, D., D. Simon and D. Thompson (eds.), 2006. The peri-urban interface: Approaches to sustainable natural and human resource use. Earthscan, London, UK.

Mellor, J.W. (ed.), 1995. Agriculture on the road to industrialization. Johns Hopkins University Press for IFPRI, Baltimore, MD, USA.

Pardey, P.G., J.M. Alston and R.R. Piggott, 2006. Shifting ground: Agricultural R&D worldwide. Issue Brief No. 46. Washington DC: International Food Policy Research Institute.

Ravallion, M., S. Chen, and P. Sangraula, 2007. The urbanization of global poverty. Background paper prepared for the World Bank's World Development Report 2008. Washington, DC: The World Bank.

Stark, O. and D.E. Bloom, 1985. The New Economics of Labour Migration. American Economic Review 75(2):173-178.

Tacoli, C., 1998. Rural-urban interactions: A guide to the literature. Environment and urbanization 10 (1): 147-166.

Thanh, H.X., D.N. Anh and C. Tacoli, 2005. Livelihood diversification and rural-urban linkages in Vietnam's Red River Delta. Food Consumption and Nutrition Division Discussion Paper No. 193. Washington, DC: International Food Policy Research Institute.

Torero, M. and J. von Braun, 2006. Information and communication technologies for development and poverty reduction. Washington, DC: International Food Policy Research Institute.

UNDESA (United Nations Department of Economic and Social Affairs), 2006. World Population Prospects: The 2006 Revision and World Urbanization Prospects: The 2005 Revision. Online access http://esa.un.org/unpp [last access 05.22.2007].

Venables, A.,1996. Trade Policy, Cumulative Causation, and Industrial Development. Journal of Development Economics 49(l):179-98.

Virchow, D. and J. von Braun (eds.), 2000. Villages in the future. Crops, jobs and livelihood. Springer-Verlag, Berlin, Heidelberg, Germany.

Von Braun, J., 2005. Agricultural Economics and Distributional Effects. In: Reshaping Agriculture's Contributions to Society; ed. David Colman and Nick Vink, pp. 1-20. Blackwell Publishing Ltd for IAAE: Malden MA, USA.

World Bank, 2006. World Development Report 2006: Equity and Development. Oxford University Press, New York NY, USA.

The role of public infrastructure in lowering transaction costs

Javier A. Escobal

Abstract

This chapter analyses the impact of transaction costs for rural households' welfare, devoting attention to the causes and implications of the lack of spatial and vertical integration of product and factor markets. Based on an innovative procedure for a disaggregated analysis of the transport, information, negotiation and monitoring costs of farmers incorporated in the potato market of Peru, the study shows that public infrastructure is vital for lowering transaction costs for reaching market outlets and establishing transactions in those markets. Lowering transaction cost enables specialization and division of labour and hence is a driving force for improving efficiency and income generating opportunities for the rural poor

1. Introduction

When attempting to evaluate the impact of specific policies on rural households, the specialized literature commonly assumes a complete integration of product and factor markets and factors on the part of rural households. However, empirical evidence suggests that rural markets tend to be thin, underdeveloped or even nonexistent. The dearth of markets is due to the limited economic development or to obstacles to their development.

In this context, the response of farmers, for example, to an increase in prices on the international, national, regional or local markets, has commonly been overestimated. This lack of knowledge of the microeconomic determinants of farmer integration with product factors markets and has multiple implications. The most important include those associated with the implementation of pricing policies, which attempt to have a homogeneous and almost instantaneous impact on agricultural supply and/or production, something which does not occur (to the surprise of those who promote such policies). De Janvry *et al.* (1989) showed how, in different contexts, the erroneous modeling of how rural households make decisions could lead to the overestimation of price elasticities of agricultural supply. Typically, this overestimation originates from mistakenly assuming that decisions on consumption and production are separable. Udry (1995) cites the work of Fafchamps, Rosenzweig, Foster and Rosenzweig, and that of Jacoby (the case of the Peruvian highlands) to demonstrate how imperfections in the labor market condition the non-separability of production and consumption decisions.

In the case of Peru, the topic of the market integration of farmers has received little attention. Recent studies carried out by GRADE in the framework of the Economic Research Consortium have examined the issue of agricultural trade and market integration. Escobal and Agreda (1998), using time series data of 12 agricultural goods in 12 Peruvian cities, showed that markets for agricultural products in Peru are reasonably integrated (from a spatial point of view). It also demonstrated that access to public goods and services is a determinant factor in explaining the speed at which consumer price information is disseminated to different cities around the country. Results also showed that in the long term, there is a complete transmission between wholesale and farmgate prices for some staple crops (i.e. potato or onion).

Although these results demonstrate that agricultural markets in Peru are reasonably spatially and vertically integrated in the long term, they also show important deviations in the short term. Additionally, the results obtained to date reveal little about the level of efficiency in which these markets actually operate. Finally, these results do not respond to the question of why certain producers choose to integrate into the market as net-sellers while others choose to remain subsistence farmers. Information on how access to assets in general and to public good and services in particular influences the way in which farmers integrate into markets can be used to design alternative policies to promote farmers' more successful market integration.

This study posits that there are high household-specific transaction costs, which limit the capacity of poor farmers to integrate into agricultural markets. The fact that many rural households do not participate in certain agricultural markets due to the existence of transaction costs has been documented in the economics literature. Notwithstanding, the relationship between these costs and marketing strategies has received little attention. Moreover, the relationship between access to public infrastructure and lower transaction costs has not been documented at all. Lowering transaction costs may be one of the most effective ways of integrating the poor into a market economy, allowing them to grasp the benefits that come with the division of labour and specialization that market relations promote.

Additionally, an important criticism of the literature on transaction costs is that theoretical development has not been accompanied by successful measurement of transaction costs. This chapter will attempt to partially fill this gap, proposing a methodology to estimate these costs and applying it to the case of potato farmers of the Tayacaja Province, in the Huancavelica Department, in the Andes of Peru.

This chapter is divided into four sections, besides this introduction. Section 2 defines transaction costs and the activities related to those costs. It also proposes a microeconomic model that associates transaction costs with the marketing option each rural household chooses. Additionally, it suggests an alternative to directly estimate transaction costs. Section 3 describes the study zone, presents the sample frame used to evaluate transaction costs in the Peruvian potato market and presents the main results of the study. Finally, Section 4 lay out the main conclusions and policy implications. This section also suggests future lines of research associated both with transaction costs and with the database that this study has generated.

2. The role of public infrastructure in a costly exchange environment: conceptual framework

Transaction cost theory develops from the work of Ronald Coase in his 1937 article 'The nature of the firm'. He argues in that article that market exchange was not costless and underlined the importance of transaction costs in the organization of firms and other contractual arrangements. Transaction arrangements evolve so as to minimize their implicit costs given the social, political and economic environment that prevails.

North (1990) defined transaction costs as the costs of measuring what is traded as well as the costs of monitoring compliance with agreements. In general, there are no precise definitions of these costs, but they are recognized as being the costs associated with establishing contracts, monitoring them and ensuring their compliance. Transaction cost economics, unlike traditional neoclassical economic theory, recognizes that trade activity does not occur in a frictionless economic environment. According to Eggertsson (1990), transaction costs originate from one or more of the following activities:

- The search for price and quality information for the goods or inputs to be traded, as well as the search for buyers and/or potential sellers (including relevant information about their conduct).
- The negotiation necessary to identify the relative negotiating power of buyers and sellers.
- The establishment of contractual agreements.
- The monitoring of parties to the contract to verify their compliance.
- The costs associated with fulfillment of the agreement, as well as penalties originating from non-compliance of the contractual relationship.
- The protection of property rights before third parties.

Transaction costs can be classified in three groups: information, negotiation and monitoring costs. Information costs occur before the transaction is made and include the costs of obtaining information on prices and products, as well as the costs associated with identifying commercial counterparts. Negotiation costs are costs associated with the development of the transaction and usually include commissions, the act of negotiating specific transaction conditions and the costs associated with the drawing up of contracts (whether formal or informal). Monitoring costs occur after the transaction is made and are usually associated with the costs of assuring that product quality and payments are as agreed upon.

According to Hobbs (1997) a critical element of transaction costs economics is that, *ceteris paribus*, vertical coordination among the different production, process and distribution stages will be carried out in the most transaction-cost-efficient manner.[5] The empirical literature

---

[5] Note that when we refer to a household that makes production and consumption decisions, we are actually considering an economic agent integrated vertically that produces for self-consumption to minimize transaction costs.

on transaction costs is based mainly on the strategy proposed by Williamson (1979). In this strategy, the need to directly evaluate transaction costs associated with different trade relationships is 'evaded' by reformulating arguments associated with the transaction cost economics literature in terms of the effects that certain observable attributes would have on the differential costs of implementing, or not implementing, a market transaction.

Formally, if we establish that between two possible transactions (T1 and T2) the one with lower transaction costs (TC) will occur, we would have:

$$T^* = T^1, \text{if } TC^1 < TC^2$$
$$ = T^2, \text{if } TC^1 > TC^2 \tag{1}$$

Although TC^1 and TC^2 are not directly observable, it is enough to observe vector X, which represents observable attributes that affect transaction costs:

$$TC^1 = \beta_1 X + \varepsilon_1$$
$$TC^2 = \beta_2 X + \varepsilon_2 \tag{2}$$

Empirically, the probability of observing T1 would be equivalent to:

$$Prob\ (TC^1 > TC^2) = Prob\ (e_1 - e_2 > (\beta_2 - \beta_1)\, X) \tag{3}$$

Although we will initially follow Williamson's strategy for evaluating the determinants of whether or not a farmer will participate in a particular product market, we will also attempt to determine a way to directly estimate transaction costs.

As mentioned, an important criticism of the literature on transaction costs is that its theoretical development has not been accompanied by successful measurement of transaction costs. We must remember that transaction costs, like any other cost in economic theory, are opportunity costs. As such, they can be estimated. One possibility would be to evaluate the time spent in their 'production', to later place a value on this time according to an hourly wage. However, this alternative would require a detailed recounting of all activities undertaken, as well as their duration. Another alternative would be to estimate (econometrically) how much each activity associated with these transaction costs contributes to determining the price the farmer receives.[6]

---

[6] This can be done using the 'hedonic price' technique. See Section 2.3.

3. Market integration and transaction costs

3.1 Review of literature

The fact that the existence of transaction costs keeps many rural households from participating in certain agricultural markets has been documented in the economics literature by De Janvry *et al.* (1991). Transaction costs drive wedges between purchasing and selling prices of a household, based on the concept of non-tradable goods taken from international trade theory. However, the literature has not used the same concept to determine why one household opts for a particular sales market for its product while another does not. Although risk considerations obviously could determine that a household will diversify the markets for its product, the transaction costs associated with each household and the differential transaction costs between markets would also help explain the 'mix' of destinations a farmer chooses.

We have slightly modified the methodology proposed by De Janvry *et al.* (1995) in two aspects to account for the direct measurement of transaction cost. First we are modeling the decision of selling at the farmgate or selling at market. We believe that the decision of a household to participate in a certain agricultural market depends on that household's position of supply and demand relative to the range of prices created as a result of the difference existing between effective buying and selling prices on that market. This range originates from a group of transaction costs, some of which are specific to the household, while others are related to the environment or region in which the household is located and still others are associated with the specific market of destination.

In this context, a particular market 'fails' when a household is faced with a large difference between the price at which a product or input could be bought and the price at which it could be sold. Given the wide margin between these two prices, it may be better for the household not trading the product or input on that market. While this decision occurs in all markets to which the household is associated, the household will prefer to remain self-subsistent for that crop.[7] Generally, households can be classified in different categories according to the 'mix' of markets in which they have decided to participate.

The second modification, which will be described in more detail in the next section, is the introduction of a hedonic price function to account for the transaction costs differences. If p is the effective price that determines production and consumption decisions, each household faces the following:

---

[7] In this case, the shadow or subjective price of the household (that which equals its supply and demand) falls within the margin: it is higher than the price the farmer would receive if he had sold the product, for which reason he decides not to sell; and is lower than what it would cost him to buy the product, for which reason he decides not to buy it.

Supply of agricultural product	$q = q\,(p\,,z^q)$	(4)
Demand of agricultural product in market j	$C^j = C^j\,(p^j\,,z^{dj})$	(5)
Idiosyncratic transmission of prices in market j	$p^{sj} = p^{sj}\,(z^{pj})$	(6)
Transaction costs in market j	$TC^j = TC^j\,(z^{ij})$	(7)

where z^q, z^{dj}, z^{pj} and z^{ij} are exogenous variables that affect supply, demand, sales price and transaction costs, respectively. Thus, for the retailers of a product in market j, the effective price at level of each household would be:

$$p^j = p^{sj}\,(z^{pj}) - TC^j\,(z^{ij}) \tag{8}$$

In this framework, the condition of being a retailer of potato in market j would be:

$$q[p^{sj}\,(z^{pj}) - TC^j\,(z^{ij})\,,z^q\,] - c\,[p^{sj}\,(z^{pj}) - TC^j\,(z^i)\,,z^d\,] > 0$$
$$o \quad I\,(z^q\,,z^{dj}\,,z^{pj}\,,z^{ij}) > 0 \tag{9}$$

This model can be estimated using the following probit equation:

$$Prob\,(Net\ Seller\ in\ market\ j) = Prob\,[I\,(z^q\,,z^{dj}\,,z^{pj}\,,z^{ij}) > 0] \tag{10}$$

The expanded model can make estimates based on either a probit or logit specification or a multivariate probit or logit, depending on whether we are dealing with two or more destinations. If we use the participation of sales in each market as the base and take into account that the endogenous variable is between values 0 and 1, the valid estimation method would be a Two-Limit Tobit Model. In our case, we are attempting to simulate a strategy associated with the decision to sell at the farmgate or elsewhere so we will try to capture this decision using a probit model.

3.2 Strategies used to measure transaction costs

After estimating the Equation 10, the reduced form of the equation of supply conditioned on the selected strategy can be derived:

$$q = q\,(p\,,z^q\,|\,Prob\,[Net\ Seller\ in\ market\ j]) \tag{11}$$

The estimation of Equation 11 equals an estimation in two stages, where the Mills ratio is introduced (obtained from estimating Equation 10) to take into account the endogenous nature of the decision (sell only at the farmgate or also sell at other locations). To associate transaction costs to the effective price each farmer receives, we chose to estimate a hedonic price equation. The word 'hedonic' is normally used in the economics literature to refer to the underlying profit that is obtained when consuming a good or service. A good that has several characteristics generates a number of hedonic services. Each one of these services could generate its own demand and would be associated with a hedonic price. Rosen (1974) developed the theoretical framework on which hedonic models are based. We interpret the model somewhat differently. The price the farmer receives has a set of 'premia' or 'discounts' for a series of services that have been generated, or perhaps omitted.

Therefore, the average farmgate price can be defined as a function of hedonic prices, which is simply the mathematical relationship between the prices received by this added value (i.e. potato) and the characteristics of the transaction associated with this product. This is:

$$P_j = h\,(z_{1j}, z_{2j}, z_{3j}, ...z_{Kj} \,|\, Prob\,[Net\,Seller\,in\,market\,j]) \qquad (12)$$

where P_j is the average price obtained by j-th farmer for the potato sale; and where (z_{1j}, z_{2j} ...z_{Kj}) represents the vector of characteristics associated with the transactions completed by the farmer. The price function was estimated in accordance with the strategy followed.

It is clear in the literature of hedonic price functions that h(z) does not strictly represent a 'reduced form' of the functions of supply and demand that could be derived from the production or utility functions of the economic agents involved in the transaction (Rosen, 1974; Wallace, 1996). Rather, h(.) should be seen as a restriction in the process of optimization of sellers and buyers. Rosen (1974), and more recently, Wallace (1996) showed that while growing marginal costs exist for some of the characteristics (in this case associated with the generation of information, negotiation and monitoring of the transactions) for farmers and/or sellers, the hedonic function could be non-linear. In this case, the non-linearity would mean that the relative importance of transaction costs is not the same for all farmers.

The estimation of an equation such as the one proposed here permits us to disaggregate the price received by the farmer into a series of components associated with the attributes of the transactions. A complementary way of interpreting this equation is where the constant estimate represents a price indicator that results from following the 'law of one price,' the rest of the equation being the elements that must be discounted from the price due to the differences in the distance of the farmers from the market and other associated transaction costs. Comparing the transaction costs between households with different endowments (private and public assets) will allow us to understand the importance of key assets in reducing transaction costs.

4. Transaction costs in rural Peru

4.1 The study area

For this study, we focused on an area where an important contrast could be found in farmers' way to access markets. To facilitate the analysis and to enable policy decisions to be made, we decided to study farmers living in the same ecological zone who devoted most of their production to a single crop. At the same time we were interested in evaluating the differences that come about when public infrastructure is provided so we focused on farmers with different access to local markets. With these restrictions in mind, we chose as our study area the districts of Pazos and Huaribamba of Tayacaja Province, Huancavelica Department, between 2,500 and 3,500 meters above sea level. This area has 1,400 farmers who grow potato for sale in the local markets of Pichus, Huaribamba and Pazos, the regional market of Huancayo and eventually, Lima. For most of these farmers, the town of Pazos constitutes their main marketing node. However there are two type of road infrastructure that connects rural dwellers to local markets. Part of rural population in this area is connected to Pazos through motorized roads while the other part is connected to the same markets via non-motorized tracks.

Pazos is a Spanish town located in the Mantaro valley, in Peru's central highlands, 70 kilometers south of the City of Huancayo in Junín Department. Only three decades earlier, it was a small village housing small-scale subsistence farmers. Like all Andean towns, residents work mostly in agricultural activities, especially in the production of a variety of potato seeds, due to the favorable conditions of the area. In Pazos, two agro-ecological zones predominate, each with different characteristics of climate, soil and especially, water availability, which permit farmers to obtain yearly potato harvests. Farmers also produce other tubers, grains and cereals.

The area's inhabitants report that since the construction of the Pazos-Pucará highway in the late 1960s, they have been able to reach the central highway that joins the Mantaro Valley (the major production valley of the *Sierra* region) with Lima (the country's largest city). Since that time, important changes have occurred in Pazos. With the highway came electric power and later, potable water service. Then, came people from other regions, interested in marketing potato and other products. New schools and health centers were also built. Dry goods and agricultural supply stores opened up and merchants and drivers permanently settled in the area. All this resulted in an increase in the area's rural-urban population.

By the mid 1970s, Pazos had become a district encompassing 18 villages and small communities. Due to the district's strategic location, it became a center in which the agricultural production of its villages and even those located in the neighboring district of Huaribamba, 22 kilometers from Pazos, converged. Its greater growth and dynamism had considerable effects on nearby communities, especially those connected to Pazos via paved roads. Examples of this include the villages of Chuquitambo, Vista Alegre, Mullaca, Nahuin, etc., in which the construction of the highway connecting them to Pazos resulted in deep changes in the intensity and use of the land. Three major changes took place: a) Seeds of native potato varieties were replaced by improved seed, whose production was destined for the Lima market; b) the potato planted area increased, and c) community pastureland gave way to privately owned land.

However, Pazos district also has villages and communities that are currently connected to the district capital via non-motorized tracks (community roads). The following villages are examples: Pariac, Potacca, Chicchicancha, Yanama, Ñuñunga, etc. These population centers are connected to Pazos via Pichus, a community connected to the district by a recently built highway, where all main non-motorized tracks converge. The farmers of Pazos district and its communities enjoy similar natural conditions. The conditions of altitude, climate, soil, presence of frosts and droughts, availability of irrigation water, etc. are all similar. The main difference is the mode of access to the district capital (paved road/non-motorized track).

4.2 Sample design

As mentioned earlier, the population under study consists of potato farmers living in the districts of Pazos and Huaribamba, Tayacaja Province, Huancavelica Department, at between 2,500 and 3,500 meters above sea level. Using the 1994 Agricultural Census as a reference, 1,396 farmers were identified in the study area.[8] Since we were interested in evaluating the decisions for market integration and transaction costs these farmers face, we decided to use the census question that identified the destination of *the largest percentage of each farm's production* as a key variable to obtain a stratified random sample. In Tayacaja Province 69 percent of the potato planted hectares is sold at market. This indicator was slightly lower in the study area, where owners of the 49.3 percent of potato planted fields reported that their harvest is mainly destined for market.

Taking into account that in the study area there is significant variability partially associated with the size of the agricultural plots or with the characteristics of the main access route to the market, we chose to stratify the population by size and type of access route, as shown in Table 1. 'Small' refers to farmers with potato fields less than one hectare, 'medium' refers to those with plots between one and three hectares and 'large' refers to farmers with more than three hectares. Considering stratification in two domains (access by non-motorized track and access by highway) and the three sizes mentioned, as well as a precision rate equivalent to 21 percent of the mean population by stratum, the optimum sample size is 188 observations, for a confidence interval of 95 percent. Finally, the sample was 'rounded off' to 190 farmers distributed among the strata according to their level of heterogeneity.

---

[8] According to the Peruvian Agricultural Census, there are a total of 2,844 potato producers in the zone; however, of these, 1,448 are outside the study area since they are in different agro-ecological zones.

Table 1. Sample design.

Study domain			Level of articulation with the market				
Size	Population	Extension (Has.)	Mean	Standard deviation	Variability (CV)	Precision[1]	Sample size
Highway							
Small	483	0.6	35.9%	41.1%	114.6%	7.5%	46
Medium	527	1.8	53.8%	37.9%	70.4%	11.3%	46
Large	210	5.8	67.5%	34.6%	51.2%	14.1%	17
Subtotal	1220	2.0	49.1%				109
Non-motorized track							
Small	77	0.6	51.1%	47.0%	92.0%	10.7%	38
Medium	84	1.9	48.9%	41.0%	83.9%	10.2%	37
Large	15	4.6	58.2%	35.9%	61.7%	12.2%	6
Subtotal	176	1.5	50.7%	123.9%	244.6%		81
Total	1396	2.0	49.3%				190

[1] Relative precision is equivalent to 20.95%. Reliability rate is 95%.
Source: survey of Transaction Costs, GRADE.

4.3 Main results

Table 2 shows the mean values of the main variables used in the study, differentiated according to each farmer's principal access route to market. Among the key characteristics evident in this table are the following:

- Farmers living in areas with market access via non-motorized tracks reported more than twice as many bad transactions experiences compared with those connected to the market by highways (4.7 versus 2.3).
- The delay in finding out the price that the transaction resulted in is substantially higher among those who are connected to the market via non-motorized tracks (3.4 days versus 0.7 days).
- The number of merchants visited by farmers before carrying out a commercial operation is much higher among those who are connected to the market via non-motorized tracks (6.5 versus 3.9).
- The level of informality of the transaction is quite higher among farmers who have market access through non-motorized tracks (79 percent versus 55 percent do not exchange any type of documentation).

Table 2. Average and standard deviation of the main variables according to access route.

Variable	Unit	Highway		Non-motorized tracks		Total	
		Average	Stand. dev.	Average	Stand. dev.	Average	Stand. dev.
I. Human capital							
Age of head of household	Years	46.4	9.3	50.0	9.9	48.0	9.7
Educational level[a]		2.4	0.8	2.1	0.8	2.3	0.8
Family size	Number	6.7	1.7	6.5	1.3	6.6	1.6
Gender of head of household	Male=1 Fem=0	0.93	0.26	0.90	0.30	0.92	0.28
II. Organizational capital							
Belongs to an association	Yes=1	61%	49%	36%	48%	50%	50%
Sends or receives money from migrants	Yes=1	55%	50%	52%	50%	54%	50%
III. Physical capital and technology available							
Total land	Has	6.1	3.2	5.4	2.1	5.8	2.8
Value of durable consumer goods	Soles	23,332.1	1,534.9	23,514.2	1,175.7	23,409.8	1,392.7
Uses chemical fertilizer	Yes=1	79%	41%	63%	49%	72%	45%
Uses pesticides or other chemical inputs	Yes=1	70%	46%	59%	49%	65%	48%
Uses improved seed	Yes=1	84%	37%	69%	47%	77%	42%
Uses a tractor	Yes=1	57%	50%	0%	0%	33%	47%
Uses an ox plow	Yes=1	60%	49%	58%	50%	59%	49%
IV. Principle flows							
Total production	Kg	30,499.1	26,147.5	20,067.9	14,738.7	26,052.1	22,569.5
Staple food costs	Soles	163.7	106.5	226.0	138.1	190.2	124.6
V. Transaction costs: information							
Believes it is important to have access to a telephone	Yes=1	62%	49%	7%	26%	38%	49%
Knows the price in Pichus	Yes=1	17.4%	38.1%	100.0%	0.0%	52.6%	50.1%
Knows the price in Huaribamba	Yes=1	11.9%	32.6%	1.2%	11.1%	7.4%	26.2%
Knows the price in Pazos	Yes=1	99.1%	9.6%	100.0%	0.0%	99.5%	7.3%

Table 2. Continued.

Variable	Unit	Highway		Non-motorized tracks		Total	
		Average	Stand. dev.	Average	Stand. dev.	Average	Stand. dev.
Knows the price in Huancayo	Yes=1	100.0%	0.0%	61.7%	48.9%	83.7%	37.0%
Knows the price in Lima	Yes=1	87.2%	33.6%	19.8%	40.1%	58.4%	49.4%
Knows neighbor's price	Yes=1	98.2%	13.5%	100.0%	0.0%	98.9%	10.2%
Calls to learn price	Yes=1	93.0%	26.0%	7.0%	26.0%	56.0%	50.0%
Price is below what farmer knew	Yes=1	27.0%	44.0%	35.0%	48.0%	30.0%	46.0%
No. of merchants who visited the farm	Number	4.6	1.6	0.1	0.6	2.7	2.6
No. of days' delay in knowing price	Days	0.7	1.1	3.4	1.8	1.8	2.0
No. of merchants farmer sold to	Number	2.9	1.4	4.0	1.0	3.3	1.3
Travels to learn price	Yes=1	70%	46%	100%	0%	83%	38%
No. of merchants farmer visited	Number	3.9	1.8	6.5	2.1	5.0	2.3
VI. Transaction costs: monitoring							
No. of times merchant went to pay farmer	Number	1.7	0.8	1.5	0.6	1.6	0.7
Merchant makes payments	Always=1 Never=0	0.80	0.40	0.85	0.36	0.82	0.38
Farmer is discounted extra costs	Yes=1	83%	37%	72%	45%	78%	41%
Farmer can demand that crop quality be recognized	Always=1 Never=0	87%	16%	63%	12%	77%	19%
The price is as agreed upon	Yes=1	66%	48%	58%	50%	63%	49%
No. of times farmer was not paid	Number	2.3	1.8	4.7	2.6	3.32	2.51
No. of sacks not paid for	Number	11.4	13.2	13.6	9.0	12.4	11.7
Merchant does not deliver documents	Yes=1	55%	50%	79%	41%	65%	48%
No. of days merchant delays in making payment	Days	1.9	0.7	2.1	0.4	1.9	0.6
No. of years farmer has known merchant	Years	5.1	3.0	3.6	1.8	4.5	2.6

Table 2. Continued.

Variable	Unit	Highway		Non-motorized tracks		Total	
		Average	Stand. dev.	Average	Stand. dev.	Average	Stand. dev.
VII. Transaction costs: negotiation							
Farmer can sell to another merchant	Yes=1	88%	33%	32%	47%	64%	48%
No. of times farmer went to negotiate price	Number	1.5	0.9	1.1	0.4	1.3	0.7
VIII. Transaction costs: transport							
Distance to Pazos	Kms	24.5	19.3	82.0	11.5	49.0	32.9
Time to Pazos	Min	78.7	82.4	388.2	71.3	210.6	172.0
Merchant provides transportation	Yes=1	32%	47%	35%	48%	33%	47%
Average condition of the road	Bad=0, Good=1	0.6	0.28	0.31	0.26	0.45	0.3
Average distance to the sales point	Kms	3.2	1.51	2.37	1.27	2.82	1.5
Average time to the sales point	Min	40.0	22.7	51.7	23.3	45.0	23.6
IX. Transaction costs: future sales							
Farmer makes future sales	Yes=1	18%	39%	16%	37%	17%	38%
Percentage of future sales	%	4.4%	10.1%	3.8%	9.3%	4.1%	9.8%
No. of years of future sales	Years	0.7	1.81	0.53	1.44	0.63	1.7
X. Other transaction costs							
No. of years farmer has grown potato	Years	18.3	5.0	20.2	4.4	19.1	4.8
Merchant pays farmer on consignment	Yes=1	52%	50%	46%	50%	49%	50%
XI. Other important variables							
Sells at the farmgate	Yes=1	100%	0%	6%	24%	60%	49%
Sells in Huancayo	Yes=1	83%	38%	16%	37%	54%	50%
Sells in Lima	Yes=1	37%	48%	0%	0%	21%	41%
Sells in Pazos	Yes=1	39%	49%	100%	0%	65%	48%
Sells in Pichus	Yes=1	3%	16%	95%	22%	42%	50%
No. of sales destinations	Number	2.6	0.6	2.2	0.4	2.4	0.6
Farmgate price	Soles	0.59	0.06	0.50	.	0.49	0.1

Table 2. Continued.

Variable	Unit	Highway		Non-motorized tracks		Total	
		Average	Stand. dev.	Average	Stand. dev.	Average	Stand. dev.
Price in Huancayo	Soles	0.74	0.04	0.76	0.04	0.74	0.04
Price in Lima	Soles	1.01	0.12	.	.	1.01	0.12
Price in Pazos	Soles	0.58	0.08	0.57	0.04	0.58	0.06
Price in Pichus	Soles	0.50	0.10	0.45	0.06	0.46	0.06
Sales price	Soles	0.46	0.08	0.36	0.05	0.42	0.09
Amount sold at farmgate	Kgs	8,035.9	9,081.5	98.2	485.8	4,651.9	7,919.5
Amount sold in Huancayo	Kgs	5,012.8	6,404.2	607.90	2,437.82	3,134.9	5,542.8
Amount sold in Lima	Kgs	3,313.8	6,889.2	0.00	0.00	1,901.1	5,460.8
Amount sold in Pazos	Kgs	1,534.2	2,495.1	2,862.6	4,402.6	2,100.5	3,492.2
Amount sold in Pichus	Kgs	29.8	236.2	3,101.6	3,275.5	1,339.4	2,625.5
Total sales	Kgs	22,908.3	21,857.5	12,981.5	11,394.2	18,676.3	18,766.5
Total sales value	Soles	12,140.7	14,650.8	3,631.4	4,799.5	8,513.0	12,255.9
Proportion of self-consumption of production	(%)	9%	6%	15%	6%	12%	7%

[a]1: incomplete primary school, 2: complete primary school, 3: incomplete secondary school, 4:complete secondary school.
Source: survey of transaction costs, GRADE.

- While 100 of farmers who have access via non-motorized tracks must travel to learn the product price, 30 percent of those living in areas with highway access do not have to do so.
- While an average of 4.6 merchants visits each producer located in areas with highway access, only 0.12 visits farmers located in the non-motorized track areas.
- None of the farmers who have access via non-motorized tracks report owning a tractor while 56.9 percent of those located in motorized access zones owns or reports using one.
- While only 7 percent of farmers who access the market via non-motorized tracks call to find out about prices, 93 percent of those located in highway access zones do so.
- 87 percent of farmers connected to the market via a motorized road reports being informed on potato prices in Lima, compared to less than 20 percent of those with access via non-motorized tracks.

Finally, while 88 percent of those located in highway access areas reports feeling confident about being able to change merchants, if necessary, only 32 percent of those who access the market via non-motorized tracks believe they have an opportunity to do so.

As Table 3 demonstrates, the type of market integration established and the possibility of obtaining a better selling price seems to depend on the set of assets owned by the farmer, especially human capital assets such as education and family size; organizational assets such as membership in associations, and; physical and technological assets such as plot size and the use of improved seed or chemical fertilizers.

Javier A. Escobal

Table 3. Household assets and market access.

	Production (kgs)	Sale (kgs)	Sales price (soles/kg)	Sales value soles	Sale/prod ratio
Educational level					
Incomplete primary	26,865	18,769	0.37	8,068	0.70
Complete primary	26,687	19,274	0.43	8,997	0.72
Incomplete secondary	24,341	17,455	0.41	7,526	0.72
Complete secondary	25,313	18,000	0.47	9,430	0.71
Gender of head of household					
Female	18,931	12,000	0.40	4,709	0.63
Male	26,707	19,290	0.42	8,920	0.72
Family size					
Fewer than 6	20,059	14,073	0.42	6,277	0.70
Between 6 and 8	28,867	20,684	0.42	9,647	0.72
More than 8	25,461	18,520	0.42	8,327	0.73
Membership in an organization					
Is not a member	29,873	21,658	0.42	10,158	0.73
Is a member	22,232	15,695	0.42	6,974	0.71
Size of farm plot (hectares)					
Less than 1	9,929	5,643	0.38	2,167	0.57
Between 1 and 3	21,337	14,753	0.41	6,233	0.69
More than 3	87,313	69,313	0.53	37,496	0.79
Use of improved seed					
Does not use	17,509	11,477	0.41	4,717	0.66
Uses	28,551	20,782	0.42	9,692	0.73
Use of chemical fertilizer					
Does not use	17,272	11,443	0.40	4,598	0.66
Uses	29,449	21,474	0.43	10,101	0.73

Source: survey of transaction costs, GRADE.

Transaction costs

Transport costs are obviously some of the most important transaction costs. While the households surveyed in areas of highway access require an average of 78 minutes to reach Pazos, those located in areas of non-motorized track access need 388 minutes. Additionally, non-motorized tracks tend to be in worse condition than highways.As Table 4 shows, farmers who live closer to Pazos tend to produce and sell more potatoes at higher prices.

Development economics between markets and institutions

Table 4. Transport costs and market access.

	Production (kgs)	Sales (kgs)	Sales price (soles/kg)	Sales value soles	Sale/prod ratio
Condition of road					
Bad	19,654	13,000	0.36	4,710	0.66
Average	20,958	14,468	0.41	6,102	0.69
Good	39,173	29,700	0.47	15,271	0.76
Distance to Pazos (km)					
Fewer than 15	29,289	21,868	0.49	11,211	0.75
Between 15 and 54.9	31,780	24,218	0.45	11,552	0.76
Between 55 and 74.9	25,615	17,487	0.40	7,729	0.68
75 or more	18,793	12,129	0.36	4,563	0.65
Time to Pazos (min)					
Fewer than 30	31,750	23,933	0.49	12,356	0.75
Between 30 and 180	30,690	23,283	0.46	11,156	0.76
180 or more	21,560	14,335	0.38	5,875	0.66

Source: survey of transaction costs, GRADE.

Moreover, some indicators of information costs incurred, as detailed in Table 5, show that farmers who have more timely access to price information average a higher selling price.

Additionally, farmers who had visited fewer traders before deciding on carrying out the transaction tended to attain higher prices. This is because the sample contains farmers who had previously incurred costs to establish their trade relations and as a result, today they enjoy more stable relationships with merchants in the zone.

Table 6 lists some indicators of negotiation costs and market access. Again we see how farmers who incur higher transaction costs are precisely those who have not been able to establish trusting, stable relationships with potato buyers. These farmers receive a lower price for their crop on average and tend to sell less than those who have managed to establish more stable working relationships and who do not require numerous visits to negotiate their transactions. Interestingly, farmers who go to negotiate a transaction more often believe it is 'risky' to approach other merchants. As a consequence, these farmers believe they are commercially 'tied' to the merchant with whom they negotiate. In effect, as Table 6 shows, farmers who believe they cannot approach other buyers receive a much lower price and tend to produce and sell much smaller quantities than those who feel free to approach other buyers.

Table 5. Information costs and market access.

	Production (kgs)	Sales (kgs)	Sales price (soles/kg)	Sales value soles
Membership in an association				
Is not a member	29,873	21,658	0.42	10,158
Is a member	22,232	15,695	0.42	6,974
Sends or receives cash				
yes	24,919	17,636	0.41	7,725
no	27,029	19,574	0.43	9,291
Price is lower than what farmer knew				
Is not lower	26,616	19,278	0.42	8,833
Is lower	24,737	17,272	0.41	7,941
Travels to inquire for prices				
yes	42,042	32,273	0.48	16,787
no	22,691	15,818	0.41	6,838
No. of days' delay in learning price				
Zero	33,411	25,581	0.48	12,929
One or more days	21,358	14,272	0.39	5,782
No. of traders who visited before selling				
Fewer than 2	33,963	25,500	0.44	12,233
Between 3 and 5	26,813	19,548	0.43	9,244
More than 5	22,149	15,078	0.40	6,405

Source: survey of transaction costs, GRADE.

Table 6. Negotiation costs and market access.

	Production (kgs)	Sales (kgs)	Sales price (soles/kg)	Sales value soles	Sale/prod ratio
No. of times farmer went to negotiate price					
0	52,462	41,077	0.51	21,713	0.78
1	25,417	18,136	0.41	8,178	0.71
2	21,488	14,690	0.42	6,245	0.68
3	20,714	14,500	0.47	6,672	0.70
Possibility of approaching other buyers					
Cannot	21,934	14,787	0.37	5,857	0.67
Can	28,348	20,844	0.45	10,075	0.74

Source: survey of transaction costs, GRADE.

Table 7 lists some indicators associated with the monitoring of contracts. In general, as Table 2 shows, a small percentage (21 percent) of farmers located in areas with non-motorized track access does not establish formal contact with the merchant, while 45 percent of producers located in paved road access areas establish formal contractual relations. In this context, Table 7 shows that farmers who have contractual backing generally obtain higher prices. Additionally, farmers who can demand the merchants to recognize the quality of their crop tend to produce more, to sell more and to receive higher prices. Also noteworthy is that the

Table 7. Monitoring costs and market access.

	Production (kgs)	Sales (kgs)	Sales price (soles/kg)	Sales value soles	Sale/prod ratio
No. of times farmer approached merchant for payment					
1	28,299	20,636	0.43	10,020	0.73
2	24,635	17,169	0.40	7,280	0.70
3	21,889	16,167	0.44	7,211	0.74
4	18,500	12,333	0.41	5,111	0.67
Farmer had problems receiving payments from merchant					
Always	20,279	13,662	0.44	6,253	0.67
Never	27,310	19,769	0.42	9,070	0.72
Farmer can demand that merchant recognize product quality					
Rarely	17,050	10,500	0.34	3,592	0.62
Almost always	21,622	14,626	0.39	5,940	0.68
Always	34,484	26,377	0.48	13,510	0.76
Final price is equal to agreed price					
No	24,359	16,958	0.41	7,283	
Yes	27,062	19,702	0.43	9,331	
Merchant delivers supporting document					
Yes	27,476	19,932	0.44	9,330	0.73
No	25,294	18,008	0.41	8,159	0.71
Days of delay of payment					
1	30,998	23,286	0.46	11,716	
2	24,602	17,250	0.40	7,607	
3	24,833	17,833	0.43	7,927	
No. of years farmer has known merchant					
Fewer than 3	19,351	12,853	0.40	5,297	0.66
Between 4 and 6	24,615	17,615	0.42	7,960	0.72
More than 6	44,721	34,471	0.46	17,456	0.77

Source: survey of transaction costs, GRADE.

longer farmers have known their merchants, the more often contracts are honored (whether formal or informal) and the more farmers produce and sell at a higher average price.

Econometric estimation

Table 8 shows the results of the Two-Limit Tobit Model derived from Equation 10. As mentioned earlier, this estimation will serve as basis for estimating both the supply and price equations. Here we note that the greater the commercial experience (number of years producing potato), the greater the organizational capital of the community where the farmer lives, the greater the social capital (community ties with the outside) and the greater the probability that the farmer will establish more stable trade relations and that the merchant will go the farm rather than the farmer being obligated to go to the local or regional fair to sell his crop.

Tables 9 and 10 show the estimations of the Equations 11 and 12. The supply equation (Table 10) can be interpreted as a reduced form of the model shown in the previous section. The results of the price equation show that the Mills ratio is significant, which means that differences exist in the prices received, depending on the marketing strategy adopted. The price equation shows that the effects of the interaction between transaction costs are key;

Table 8. Determinants of farmgate sales (Probit estimate of farmgate sales).

Explanatory variables	Coefficients		
Constant	-66.177	(34.30)	+
No. of years producing potato	0.406	(0.25)	+
Age of household head	-0.136	(0.08)	+
Family size	0.343	(0.30)	
% of households in the community that belong to associations	34.903	(19.09)	+
Use of chemical fertilizers (1=yes)	-1.672	(1.43)	
Use of pesticides (1=yes)	-3.470	(2.02)	+
% of community households with ties outside the farm	27.686	(16.01)	+
Use of improved seed (1=yes)	1.831	(1.32)	
Number of productive assets	-0.854	(0.57)	
Land size (has.)	0.597	(0.57)	
Average distance to sales point (km)	14.249	(7.15)	~

No. of observations: 190; Pseudo R squared: 0.902

The standard deviations are in parentheses, with p<0.10 = +, p<0.05= ~.
Prepared based on data from the Survey of Transaction Costs.

Table 9. Determinants of Sales Price (OLS Estimation of Sales Price).

Explanatory variables	Coefficients		
Constant	0.545	(0.03)	*
Inverse Mills ratio	-0.011	(0.00)	*
Inverse Mills ratio squared	0.000	(0.00)	*
Frequency of merchant compliance	-0.362	(0.07)	*
Merchant compliance*trust in input supplier	-0.138	(0.07)	~
Possibility of demanding that *merchant recognize quality	0.162	(0.05)	*
Possibility of demanding quality*trust in input supplier	-0.282	(0.10)	*
Possibility of demanding quality*ratio of effectiveness	0.277	(0.11)	*
Mills ratio*delay in learning price	0.002	(0.00)	*
Respect for price agreed upon* trust in input supplier	0.331	(0.07)	*
Respect for price agreed upon *bias of the information[1]	0.055	(0.02)	*
Respect for price agreed upon *type of prices known	-0.109	(0.03)	*
Respect for price agreed upon *ratio of effectiveness[2]	0.076	(0.03)	~
Pays to obtain information*merchant complies	0.229	(0.06)	*
Bias of the information*trust in sellers of inputs	0.200	(0.06)	*
Bias of the information*prices known	-0.136	(0.03)	*
Ratio of effectiveness*merchant complies	0.111	(0.04)	*
Ratio of effectiveness *pays for information	-0.194	(0.08)	~
Ratio of effectiveness *bias of the information	0.094	(0.03)	*
Recognizes product quality*trust in input supplier	0.193	(0.07)	*
Recognizes product quality *respects price agreed upon	-0.139	(0.05)	*
Recognizes product quality *bias of the information	0.120	(0.06)	~
Delay in learning price*ratio of effectiveness	-0.037	(0.01)	*

No. of observations: 190; R squared: 0.613

The standard deviations area in parentheses, with $p<0.05 = $ ~ , $p<0.01 = $ *.
[1] bias of the information: if the effective price is below that known.
[2] ratio of effectiveness: (number of merchants who visit/number of merchants farmer sells to).
Prepared based on data from the survey of transaction costs.

therefore, the direct interpretation of the parameters is not simple. In the case of the sales equation, organizational capital, social capital, technology used, as well as access to public goods and services (highway and paved roads, police post and court of justice) are important determinants of the amount sold at market. We should also consider other transaction costs, such as those associated with information (delay in learning price, level of trust established with the merchant) and with contract monitoring (frequency of merchant compliance, respect for price agreed upon).

Table 10. Determinants of amount sold off the farm. (OLS Estimation of Sales Quantity)

Explanatory variables	Coefficients		
Constant	-0.374	(0.13)	*
No. of years producing potato	0.004	(0.00)	*
Gender of head of household (1=male)	0.060	(0.02)	*
% of community households belonging to associations	0.306	(0.08)	*
% of community households with outside ties	0.281	(0.09)	*
Use of improved seed (1=yes)	0.042	(0.01)	*
Use of ox plow (1=yes)	0.025	(0.01)	~
Size of farm plots (has.)	0.162	(0.01)	*
Existence of a court in the community (1=yes)	-0.082	(0.04)	~
Average distance from sales point (km)	-0.072	(0.03)	~
Inverse Mills ratio	0.006	(0.00)	*
Existence of a health post in the community (1=yes)	-0.023	(0.01)	~
No. of days' delay in learning price	-0.006	(0.00)	
Level of trust in input supplier	-0.218	(0.06)	*
Frequency of merchant compliance	0.027	(0.01)	
Respect for price agreed upon (1=yes)	0.033	(0.01)	~
Existence of a police post in the community (1=yes)	0.052	(0.03)	~
Lives in Chuquitambo (1=yes)	0.243	(0.07)	*
Lives in Collpa (1=yes)	0.097	(0.03)	*
Lives in Mullaca (1=yes)	0.153	(0.04)	*
Lives in Pariac (1=yes)	0.064	(0.02)	*
Lives in Pichus (1=yes)	0.078	(0.04)	~
Lives in Putacca (1=yes)	0.048	(0.02)	~
Lives in San Cristobal de Nahuin (1=yes)	0.150	(0.03)	*
Lives in Santa Cruz de Ila (1=yes)	0.122	(0.04)	*
Lives in Tongos (1=yes)	0.117	(0.03)	*

No. of observations: 190; R squared: 0.856

Standard deviations appear in parentheses, with p<0.05 = ~ , p<0.01=*.
Prepared based on data from the Survey of Transaction Costs.

As described earlier, it is possible to estimate and disaggregate transaction costs using as a base the estimations presented in Tables 9 and 10. While Equation 9 enables us to evaluate to the price increases for potatoes that each household would have received if it had not incurred transaction costs in its relations with merchants, Equation 10 permits us to assess the effect that reducing these costs would have on sales. Table 11 shows the discounts in

Table 11. Discount in sales price by type of transaction cost. (Nuevos Soles per kilogram)

Characteristics	Type of transaction cost			Total	% Price
	Information	Negotiation	Monitoring		
Total	-0.164	0.195	-0.185	-0.154	-36.5
	(0.046)	(0.043)	(0.048)	(0.050)	
Type of access					
Non-motorized track	-0.177	0.212	-0.173	-0.139	-38.4
	(0.062)	(0.046)	(0.047)	(0.057)	
Paved road	-0.154	0.182	-0.193	-0.165	-35.4
	(0.040)	(0.041)	(0.049)	(0.050)	
Type of producer					
Small	-0.165	0.195	-0.190	-0.161	-39.5
	(0.047)	(0.043)	(0.047)	(0.050)	
Medium	-0.161	0.184	-0.174	-0.150	-36.5
	(0.046)	(0.041)	(0.046)	(0.051)	
Large	-0.166	0.231	-0.202	-0.138	-27.6
	(0.044)	(0.053)	(0.055)	(0.049)	

A negative value indicates discounts in the price the farmer receives while a positive value suggests a price increase. Standard deviations appear in parentheses.
Prepared based on data in Table 9.

price perceived by households surveyed due to the transaction costs incurred. The high value obtained is noteworthy. These estimates suggest that prices are 36.5 percent of what they would have been without transaction costs. Standard deviations confirm that the transaction costs estimated here are statistically significant. The table also shows that the most important transaction costs are those associated with monitoring and information costs. Negotiation costs are just the opposite of expected -- as mentioned earlier, the farmers who incur more transaction costs are the same ones who have not been able to establish trusting, stable relationships with potato buyers. Thus, farmers who incur greater monitoring costs obtain lower prices. If this is true, the estimated transaction costs should consider monitoring costs with a negative rather than a positive sign, in which case the total transaction costs would be even higher (equivalent to 82.7 percent of the average price).

Table 12 attempts to measure the impact on sales that a reduction of estimated transaction costs would have. The results are the outcome of a partial equilibrium exercise, for which reason no attempt was made to measure the impact of an increased commercial surplus on

Table 12. Discount in amount sold by type of transaction cost (Kilograms).

Characteristics	Type of transaction cost				Total	% Quantity
	Information	Negotiation	Monitoring	Distance		
Total	-107	-927	425	-1876	-2485	-13.3
	(61)	(235)	(142)	(838)	(948)	
Type of access						
Non-motorized	-200	-909	418	-1523	-2214	-17.1
track	(114)	(231)	(142)	(680)	(817)	
Paved road	-39	-940	430	-2138	-2686	-11.7
	(22)	(239)	(142)	(955)	(1049)	
Type of producer						
Small	-117	-931	416	-1833	-2466	-20.6
	(67)	(236)	(139)	(819)	(933)	
Medium	-107	-956	415	-1874	-2522	-17.5
	(61)	(243)	(138)	(837)	(952)	
Large	-74	-805	495	-2037	-2421	-4.1
	(42)	(204)	(168)	(910)	(989)	

A negative value indicates discounts in the quantity sold while a positive value expresses an increase in the quantity sold. Standard deviations appear in parentheses.
Prepared based on data in Table 10.

the local price. Since the production in the study area only accounts for a small part of the market trading in Pazos, Huaribamba or Huancayo, the proposed exercise is reasonable.

The results of the simulation based on the function of supply show that the quantity sold would have been 13 percent higher if transaction costs had not been incurred. In this case, transport costs (whose proxy is the distance to market) are the most important, followed by negotiation costs. If we combine the effects of price and quantity sold we can obtain a global estimate of what transaction costs represent in the study area. Table 13 shows how much the transaction costs incurred by the study population would have reduced the gross sales value. The estimates suggest that sales were 48.5 percent lower due to transaction costs, with transport costs being the most important, followed by monitoring and information costs. As expected, transaction costs are higher for farmers who are connected to the market via non-motorized tracks and among farmers with lower production levels

Table 13. Discount in the gross value of production by type of transaction cost (Nuevos Soles).

Characteristics	Type of transaction cost				Total	% GVP
	Information	Negotiation	Monitoring	Distance		
Total	-3083	3065	-3347	-789	-4153	-48.5
Type of access						
Non-motorized track	-2334	2226	-2170	-549	-2827	-58.3
Paved road	-3531	3563	-4305	-994	-5267	-46.5
Type of roducer						
Small	-2009	1777	-2195	-745	-3173	-63.2
Medium	-2353	2092	-2408	-773	-3442	-56.6
Large	-9744	12875	-11654	-1020	-9543	-31.3

A negative value indicates discounts in the GVP and a positive value indicates an increase.
Prepared based on data in Tables 11 and 12.

5. Conclusions

Public infrastructure connects to welfare through diverse channels. In this chapter we have evaluated one of those channels: public infrastructure helps to lower transaction costs, that is, the costs to reach markets and establish transaction in those markets. Lowering transaction cost is at the heart of increasing specialization and division of labour and hence is a driving force for improving efficiency and income generating opportunities for the rural poor.

The study used a representative sample of 190 potato farmers living in the districts of Pazos and Huaribamba in Tayacaja Province, Huancavelica Department, at between 2,500 and 3,500 meters above sea level, to attempt to evaluate the importance of transaction costs on market integration decisions. It also made a first estimation of these costs. As the results show, transaction costs in the study area equal almost 50 percent of sales value, being appreciably higher (60 percent) for farmers who have access to the market via non-motorized tracks. Likewise, the results confirm that transaction costs are considerably higher for small-scale farmers than for large-scale ones (67 percent versus 32 percent of sales value). The results show that besides distance and time to the market, key variables for explaining the market integration strategy (i.e. when to sell and to what market) include several indicators associated with how much experience the farmer has with the market in which he operates; how stable his relations are with different agents he trades with, and; how much of an investment he makes to obtain relevant information and monitor compliance with implicit contracts associated with the transactions completed.

Although transaction costs are in absolute value greater the larger the scale of the farm, they represent a larger proportion of the value of output for small farmers thus, policies aimed to improve connections between local and regional markets will have also sizable positive impact for small farmers. The benefits that a small farmer can get from lower transaction costs are multiple. First, they can expect to see more merchants coming to their farmgate asking for their products, increasing their bargaining power. It is very likely that they will learn about the price the same day which in turn, will help them monitoring the compliance of the exchanges they have done. The relationship with those merchants will evolve and will not be as risky as they are, when the information asymmetries are large. They might even decide to reduce the number of merchants they sell to being able to capture a higher expected effective price and at the same time reducing the uncertainties of trade.

In the long term farmers with lower transaction costs will be interested in selling their products not only to local or regional markets, but also to national and, eventually, international markets. In turn, increasing their marketable surplus will allow them to exploit the benefits of specialization. The results showed here are consistent with the idea that larger transaction costs are associated with lower market responsiveness of farmers, especially of small farmers. If public infrastructure reduces transaction costs as has been shown here, it is expected that the farmers will be more able to respond more quickly and effectively to market incentives.

Finally, the literature review carried out suggests that, as far as we know, this is the first study that attempts to estimate directly transaction costs in agricultural markets. However, we believe some pending modifications will permit a better estimation of these costs and the subsequent evaluation of the role that public infrastructure has in lowering those costs. In the first place, we believe that transaction cost should also be analyzed in a dynamic context. If we recognize that contractual arrangement evolve in time, we could have a better understanding of the impact of key elements such as trust in developing contractual arrangements. In addition, the relation between risk bearing behavior and transaction cost minimizing behavior should also be evaluated. Equation (10), which shows the marketing options, can be expanded to consider more than two marketing options and in this way could identify different marketing strategies that can correspond to a risk diversification strategy or to the existence of differential transaction costs for each market.

References

Coase, R.H., 1937. The nature of the firm, Economica, 4, 1-37.

De Janvry, A., E. Saudolet and G. Gordillo, 1995. NAFTA and Mexico´s Maize Producers. World Development Vol. 23 No. 8 pp. 1249-1362.

De Janvry, A., M. Fafchamps and E. Sadoulet, 1991. Peasant Household Behavior with Missing Markets: some paradoxes explained. Department of Agriculture and Rsource Economics. University of California, Berkeley Working Paper 578.

De Janvry, A., E. Sadoulet and M. Fafchamps, 1989. Agrarian Structure, Technological Innovations and the State. In: P. Bardhan (ed.), The Economic Theory of Agrarian Institutions. Oxford University Press, Oxford.

Eggertsson, T., 1990. Economic Behavior and Institutions. Cambridge University Press. New York.

Escobal, J. And V. Agreda, 1998. Análisis de la Comercialización Agrícola en el Perú. Boletín de Opinión No. 33. January. Consorcio de Investigación Económica (CIES). Lima. 1997.

Hobbs, J., 1997. Measuring the importance of transaction costs in cattle marketing. American Journal of Agriculture Economics; Vol. 79, November 1997 pp 1083-1095.

North, D., 1990. Institutions, Institutional Change and Economic Performance. Cambridge University Press. New York.

Rosen, S., 1974. Hedonic Prices and Implicit Markets: Product Differentiation in Pure Competition. Journal of Political Economy. Vol. 82, pp. 34-55.

Udry, C., 1995. Recent Advances in Empirical Microeconomic Research in Poor Countries. Mimeo. February. Department of Economics, Northwestern University.

Wallace, N.E., 1996. Hedonic–Based Price Indexes for Housing: Theory, Estimation, and Index Construction. Federal Reserve Bank of San Francisco Economic Review 1996, Number 3. pp. 34-48.

Williamson, O.E., 1979. Transaction-Cost economics: The Governance of Contractual Relations. Journal of Law and Economics. Vol 22 October 233-261.

Displaced distortions: financial market failures and seemingly inefficient resource allocation in low-income rural communities

Christopher B. Barrett[9]

Abstract

Poor households in rural areas of the developing world commonly lack access to (formal or informal) credit or insurance. These financing constraints naturally spill over into other behaviours and (asset, factor and product) markets as households rationally exploit other market and non-market resource allocation mechanisms to resolve, at least partly, their financing problems. These displaced distortions of financing constraints commonly manifest themselves in allocative inefficiency that may lead researchers and policymakers to mistakenly conclude that poor households routinely make serious allocation errors and to direct policy interventions towards the symptoms manifest in other markets rather than towards the root financial markets failures cause.

1. Introduction

Nature abhors a vacuum, quickly filling it by distributing the pressure over space. Economies work similarly. Where a vacuum exists in rural financial markets in low-income communities, the pressure that results from households' limited ability to smooth consumption across time through insurance or credit, or to finance investment by borrowing against expected future earnings inevitably spreads throughout the rural economy. People lacking access to (formal or informal) credit or insurance markets rationally exploit other markets and non-market resource allocation mechanisms to resolve, at least partly, their financing problems. They may also or instead pass up seemingly remunerative investments or liquidate productive assets that offer high expected future returns. The consequence of impeded access to financial services is thus seemingly significant inefficiencies in resource allocation, but I emphasize these are only 'seemingly' inefficient because they can, in fact, be the rational response of households strapped for finance. These are displaced distortions.

Such distortions are pervasive in low-income rural economies. As this chapter discusses, displaced distortions manifest themselves in factor markets for land, labor and other productive inputs, in product markets, in patterns of natural resources use, and in disinvestment and investment behaviors, including those associated with disadoption

---

[9] This work has been made possible by support from the United States Agency for International Development (USAID), through grant LAG-A-00-96-90016-00 to the BASIS CRSP, the Strategies and Analyses for Growth and Access (SAGA) cooperative agreement, number HFM-A-00-01-00132-00, and the Assets and Market Access CRSP. The views expressed here and any remaining errors are the authors' and do not represent any official agency.

and adoption of improved agricultural production technologies and natural resources management practices. Imperfect or missing markets for financial services lead to substantial and sometime persistent inefficiencies that lower the welfare and impede improvement in the well-being of subpopulations rationed out of formal or informal markets for financial services. Note that 'market failure' in this setting does not require the complete absence of transactions in financial services; rather, following De Janvry *et al.* (1991) and a vast literature on nonseparable household modelling, it implies idiosyncratic, household- or individual-specific rationing or self-selection out of a market in equilibrium. And as with other household-specific market failures, idiosyncratic financial markets failures often lead casual observers to mistakenly conclude that poor households make systematic resource allocation errors (De Janvry *et al.*, 1991, Barrett, 1997). Mistaken inferences too often lead to ill-designed policy, which is the reason this topic of displaced distortions bears reflection and discussion.

2. The gaps in financial networks

Credit offers a means of intertemporal trade while insurance enables simply trade across states of nature. When markets are complete and competitive and such trade is unfettered, there exist unique interest rates and insurance premia in equilibrium and households will, in general, achieve Pareto efficient allocations of resources and risk. When barriers to trade across time or states of nature exist, however, different households or individuals may face different shadow prices of capital and when liquidity constraints bind, this will effect households' current shadow valuation of other factors, goods and services. This naturally leads to errors in assessment of farmer-level efficiency. More importantly, when barriers impede trade across time or states of nature, there are, in general, foregone gains from trade, i.e., potential Pareto improvements may exist.

Casual inspection suggests that formal credit markets are missing or incomplete for many prospective borrowers in low-income economies, so that foregone gains from trade in financial markets is a widespread phenomenon. Financial institutions simply do not exist or the institutions' liquidity is itself constrained by the absence of interbank markets so that they have to ration credit, either by price (e.g., high interest rates) or by quantity (i.e., lending only to a subset of prospective borrowers and/or giving borrowers only a portion of the credit they request). In the case of price rationing, many people will self-select out of the credit market.

More importantly, there will generally be quantity rationing due to adverse selection and moral hazard concerns caused by asymmetric information about borrowers' creditworthiness and their use of funds in the presence of limited liability, as well as the fixed costs of lending (e.g., checking creditworthiness, etc.), which will commonly ration smaller volume (i.e., poorer) borrowers out of lending markets because their average fixed costs are higher for them than for larger borrowers (Stiglitz and Weiss, 1981; Carter, 1988). In the quantity rationing case, there will be unmet excess demand for credit from the formal sector at prevailing credit terms, but the social process of rationing will commonly lead to similar 'rationing from the top', wherein local elites are included and more marginalized subpopulations are commonly excluded.

Formal insurance markets are likewise almost wholly absent in low-income economies, especially among the poorest segments of society. The large-scale, commercial risk pooling among anonymous individuals familiar in wealthy countries – e.g., life, home, or vehicle insurance – rarely exists in low-income economies and when it does, it is usually extremely limited in its reach beyond the cohort of indigenous elite and expatriates. Surely this absence is not because risk does not matter to the poor. If anything, risk matters more to the poor than to the wealthy since empirical tests of risk preferences tend to find support for the decreasing absolute risk aversion hypothesis. Nor is it because low-income economies are relatively riskless environments for which there is little need for insurance. Coefficients of variation of income in developing country agriculture typically vary far more than in the high-income, post-industrial economies. Rather, insurance markets fail idiosyncratically for the same reasons that credit fails: asymmetric information, covariate risk, and high costs of search, transactions, monitoring and enforcement relative to the insured capital.

Informal lending and insurance plug part of the gap left by missing formal markets for financial services in low-income rural economies. By reducing problems of asymmetric information, search, transactions and enforcement costs, etc. social relationships can facilitate lending and mutual insurance arrangements that are commercially unprofitable for formal businesses. Hence the important role such institutions play in rural development.

But financing gaps still commonly remain, especially for the poorest peoples, who too often find themselves excluded from social networks that make informal credit or insurance available. Thus the informal provision of credit or insurance merely shrinks the scope of the problem; informal finance comes nowhere near eliminating the problem of idiosyncratically missing financial markets. Problems such as covariate risk, contract enforcement, identifiable poverty trap thresholds and social invisibility[10] sharply limit many people's access to financial products, especially among poorer and more socioculturally marginalized subpopulations. As a result, there is considerable uninsured risk and widespread unmet demand for credit in rural areas of the low-income world.

The routine absence or limited availability of credit and insurance in low-income communities does not change the underlying fact that individuals often need to borrow or to insure against adverse shocks in the face of uncertain incomes. Without access to formal financial

---

[10] See Santos and Barrett (2006) and Vanderpuye-Orgle and Barrett (2007) on this more recent point about social invisibility, i.e., that within villages not everyone is equally well known or connected to others, therefore access to informal financial services mediated by social networks is likewise highly unequal.

markets, people inevitably find other ways to obtain 'quasi-credit' or 'de facto insurance'. Such behaviors are often mistakenly perceived by outsiders as inefficient, irrational or short-sighted. In my experience, they instead more often reflect the cleverness and industriousness of poor smallholders and the crushing lack of options they too often face.

When poor people's demand for credit or insurance is not met through direct financial services, whether provided through formal financial institutions or by family, friends or neighbors, they resourcefully find other means to resolve their latent demand for financial services. These displaced distortions of financial markets can, however, have a high cost to their or their community's future welfare. For individuals without savings, their choices are often limited to distress sale of the limited non-financial assets they possess or seemingly irrational market participation and investment decisions that effectively provide 'quasi-credit,' by allowing consumption today which has a significant opportunity cost in the future, or 'de facto insurance' by cushioning consumption today by drawing down some (often natural) asset stock.

It is important that researchers and policymakers understand this phenomenon of displaced distortions when they observe seemingly irrational behavior in factor (e.g., labor, land) or product (e.g., food) markets, or in patterns of natural resources exploitation, investment or technology adoption. Appropriate policy responses to such patterns may not involve an intervention directly in the distorted market. The first-best response may instead be in resolving the financial market failures at the root of seemingly irrational individual behavior.

3. Factor market solutions to market imperfections

The development economics literature has long implicitly recognized the efficiency costs and behavioral distortions induced by financial market failures. Stiglitz (1974) famously explained sharecropping contracts as optimal arrangements for balancing the moral hazard of a tenant's labor allocation against the uninsured risk inherent to farming. In that case, uninsured risk causes tenants to surrender part of their residual claim on the fruits of their labor, resulting in Marshallian inefficiency that has spawned a vast literature on the institution of sharecropping and the potential efficiency gains of various sorts of land-to-the-tiller tenurial reforms. Of course, if tenants could freely borrow, such inefficiencies would vanish and debates over land rights and the efficiency effects of sharecropping would naturally disappear. Thus inefficient contracting over land is the consequence of failures in a different, financial market, and the policy debates that emerge around this inefficiency may often be misplaced by the displaced distortion.

Uninsured risk is likewise one of several factors that drives a wedge between the marginal revenue product of labor and the prevailing market wage rate for the same workers' labor, violating the textbook allocative efficiency criterion for utility-maximizing labor allocation under complete markets. For example, Barrett *et al.* (2007) find that labor allocation by rice farmers in Côte d'Ivoire systematically reflects a wedge between on-farm and off-farm productivity, with the difference strongly related to land/labor endowment ratios and access to finance in a way consistent with theories of rational labor allocation in the presence of uninsured risk, credit constraints, or both. But these labor allocation patterns, while rational, reflect real foregone income on the part of poor households.

Poor farmers who need cash commonly work for wealthier farmers during peak planting and harvesting periods. As a result, they choose to mis-time work on their own plots, missing optimal field preparation, planting, weeding and harvesting periods – often by several weeks – because their need for cash necessitates working for others when paying jobs are available. But mistiming on-farm activities on workers' own plots leads to non-trivial productivity losses. Furthermore, cash constrained farmers often have difficulty hiring laborers during periods of peak labor demand and therefore have to leave land idle or work fields at a suboptimally slow rate due to insufficient labor availability. These productivity losses due to labor mistiming and under-hiring are a disguised interest rate on the de facto borrowing these farmers engage in through labor markets. If these farmers could borrow, they would not need to suffer these losses.

Consider, for example, the difference between two neighbors in one village in Madagascar's southern highlands. One is a single mother of four children. She has only two years' education and six ares (600 m²) of rice land, having had to sell off half her land to buy food several years ago. In spite of the limited area she cultivates, she leaves another two ares idle because she cannot afford to keep her eldest two children home to work that land; her son treks eight hours to another village to work for cash for several weeks at a time while her daughter finds unskilled work in the nearest town. And she cannot afford to hire workers or even just to buy the food to feed reciprocal *entraide* laborers.

The school teacher in this woman's village also has only six ares of land, and he has seven children, a wife and an elderly parent to support. But he completed eleven years' education and became a teacher, so he has steady cash and in kind income from his non-farm employment (many families pay his wages in rice). He owns one zebu cattle, uses his salary to buy inorganic fertilizer and to hire workers seasonally to help with his small rice fields, reaping yields more than four times that of his more cash constrained neighbor. These yields, plus the rice he receives for teaching, leave his family food secure and enable him to keep his children in school. Indeed, his oldest is in boarding school now in the provincial capital and hopes to study electronic engineering in college. These two farmers' basic land endowments are identical, but the teacher's regular non-farm income – made possible by his superior education, a past investment financed by missionaries – permit him to manage his land optimally and to accumulate surpluses sufficient to give his children an even better prospect than he enjoys. By contrast, none of the woman's children has finished even four years of school. Lacking land, livestock and education, they almost surely face a lifetime of unskilled labor and grinding poverty.

4. Quasi-finance through commodity markets

Liquidity constrained individuals do not only use factor markets to resolve their financing problems, they use product markets as well. One interesting, common example is the 'sell low, buy high' phenomenon in smallholder grain marketing. The market price for storable staple grains typically exhibits a seasonal cycle, reaching a low during and immediately following crop harvest and typically peaking during the growing season, as the preceding season's accumulated stocks run low. This is natural in even complete and competitive markets, which must account for storage costs and losses. But in many low-income rural areas, seasonal price changes far exceed apparent storage losses or interest rates. For example, in Madagascar we found the mean *quarterly* change in rice prices across the island was 29 percent, at a time when mean *annual* interest rates on lending were only 27 percent (Moser *et al.*, 2005). Plainly, it pays to hold rice stocks in Madagascar.

Yet many Malagasy farmers do not hold rice stocks in anything approaching an optimal quantity. Individuals unable to borrow or to insure themselves against recent losses often seek out 'quasi-credit' by selling at low prices and subsequently buying at far higher prices in commodity markets. People can only optimally time their sales for maximal profitability when they possess sufficient assets to enable them to artbirage the market, i.e., to wait to sell when prices peak during the pre-harvest hungry season and to wait to buy when prices hit their seasonal lows post-harvest. With limited savings or credit access, poor households often cannot afford to wait.

Take for example the case of a smallholder farmer in Iandratsay, a village in one of Madagascar's prime agricultural regions. This gentleman sold paddy at FMG1000/kg to a local collector in the commune who evacuates the paddy by ox cart to an urban wholesaler.[11] Yet he predictably runs out of rice three months before his next harvest. He winds up buying rice back from the same fellow using proceeds from his groundnut and maize crops. Accounting for milling losses, he is paying FMG1850/kg paddy-equivalent. So he effectively buys back in January the rice he sold the preceding June at a premium of 85%. This is the implicit interest rate (including storage losses) he pays on seasonal quasi-credit obtained through the rice market. When the financial markets fail, people find alternative means of engaging in intertemporal arbitrage, even when it proves very costly, in this case, due to storage losses, transport costs and the transactions costs associated with multiple physical exchanges.

Such cases are commonplace and have an effect on broader markets that exposes rural residents to greater price risk than those who live in urban areas with better storage infrastructure and superior access to interseasonal finance (Barrett, 1996). Roughly, one-third of Malagasy rice producers both buy and sell rice in the same year (Moser *et al.*, 2005). This leads to significant seasonal flow reversals, wherein grain flows in the harvest period from rural areas to cities,

---

[11] Exchange rate: 1US$ = 9.05 Malagasy franc, FMG (2006).

where commercial traders store grain interseasonally, shipping food back to the rural food-producing areas in the hungry season, once farmers have depleted their own stocks. In so far as seasonal flow reversals are predictable they reflect significant apparent inefficiency because this round-tripping of staple foods adds transport costs and profit margins for marketing intermediaries. The large scale of seasonal flow reversals in Madagascar reflects major inefficiencies in food marketing due to spatial patterns of interseasonal grain storage – underinvestment in efficient storage capacity in rural areas – and credit availability.

The farmers who routinely sell low and buy high seasonally are most commonly the poorest farmers. Of course, they can never really get ahead when they sell when prices collapse post-harvest and buy when prices peak. This distortion of their rice marketing behavior impedes accumulate of the savings necessary to buy fertilizer or improved seed or livestock and thereby increase their productivity on the farm, or to invest in an ox cart or a small store that they could use to diversify into higher-return non-farm activities. The lack of seasonal credit, even consumption credit in the hungry season, lies at the heart of this problem (Zeller *et al.*, 1997).

Similarly, consider a milk producer in Ambohiambo, who can sell milk to Tiko, the main national dairy processor, for FMG 2000/liter, but there is a two week payment lag on sales to Tiko. Alternatively, she can sell to a local trader for the lesser price of FMG 1750/liter with immediate cash payment. If this were a more conventional loan for FMG 1750, with repayment of FMG 2000 – the opportunity cost of not selling to Tiko – in two weeks, the implicit interest rate of 14.3% for two weeks implies an annualized compound interest rate of more than 3000%! Despite the high rate, she often opts to sell to the local trader, revealing that her immediate need for cash is sometimes worth the extremely high effective interest rate she pays by selling at a low price for cash.

These distortions of produce marketing behavior are by no means exclusive to Madagascar. Stephens and Barrett (2007) develop a simple theoretical model of market participation over multiple seasons in the presence of liquidity constraints and transactions costs to explain the 'sell low, buy high' puzzle. Applying their model to data from western Kenyan maize growers, they find that access to off-farm income and credit indeed seem to influence crop sales and purchase behaviors in a manner consistent with the hypothesized patterns. Financial market failures appear to exert a significant influence of commodity marketing patterns.

5. Liquidity constraints and technology adoption patterns

Financial market failures manifest themselves as displaced distortions of investment and technology adoption behaviors as well. A sizeable literature on agricultural technology adoption routinely points to liquidity constraints and access to finance as key explanatory factors of nonadoption or late adoption of remunerative new production and processing methods.[12] This result is intuitive when the improved technology requires cash outlays, as is common for improved seed, livestock or mineral fertilizers. Smallholders lacking cash savings and access to credit commonly cannot afford higher yielding hybrid seed varieties, mineral fertilizers, productivity-enhancing equipment, etc.

What is less well understood is that non-adoption of improved technologies due to financing constraints occurs even when no cash outlay is necessary. Many improved agricultural production technologies and natural resources management (NRM) practices require only labor inputs initially. But for the same reason that poor farmers commonly mis-time their own on-farm activities, the poor commonly seek out cash wage labor rather than invest time in yield-improving innovations on their own farms.

This has been observed in the non-adoption of the system of rice intensification (SRI), a method developed in rural Madagascar that increases yields by more than 80 percent, on average, with no new seed and no mineral fertilizers, just a change in agronomic practices that increases field preparation, planting and weeding labor demands in the initial few years following SRI adoption (Barrett *et al.*, 2004b). In spite of the considerable expected yield gains, few poorer Malagasy rice farmers have experimented with the method. They cannot afford it, even though it requires no cash outlay. They must instead seek off-farm wage labor in order to get the cash necessary to buy food for their families during the hungry season. Current credit constraints limit their ability to seize on the promise of greatly increased yields several months down the line, while their inability to insure against the added yield risk associated with SRI likewise discourages uptake of this method in spite of the great yield increases it generally offers (Moser and Barrett, 2003, 2006).

Similarly, we observe a strong positive relationship between farmers' wealth and their likelihood of adopting improved NRM practices in land scarce areas of the western Kenyan highlands (Marenya and Barrett, 2007a). Practices such as tilling crop residues into the soil, applying manure to cultivated fields, and terracing require labor but not cash. Yet few of the poorer farmers in Vihiga District adopt such practices. The primary reason is, once again, that they cannot afford to invest the time today in increasing the future productivity of their own farm because they need to find off-farm employment, even at meager prevailing wages (less than US$1/day per adult worker) to meet immediate subsistence needs. The absence of credit for investing in on-farm improvements or consumption credit to meet immediate needs induces underinvestment that results predictably in lower future productivity and persistent poverty.

Such findings extend well beyond rural Africa. Rosenzweig and Wolpin (1993) report that the availability of certain nonagricultural income has a substantial positive effect

---

[12] See Feder *et al.* (1985) and Sunding and Zilberman (2001) for excellent reviews of this literature.

on agricultural output and efficiency, suggesting that residual risk exposure indeed leads to efficiency and output losses due to induced producer response in semi-arid India. In a similar spirit, Rosenzweig and Binswanger (1993) look at the effect of rainfall timing on the composition of productive and nonproductive asset holdings. They find that a one standard deviation increase in weather risk induces Indian households of median wealth to reduce expected farm profits by an estimated 15 percent while households in the bottom quartile reduce expected farm profits by 35 percent. The wealthiest quartile households, on the other hand, have adequate independent risk coping mechanisms, so they adjust input use patterns hardly at all to increased exogenous risk.

Townsend (1995) finds that insufficient insurance in rural Thailand contributes to the nonadoption of improved rice varieties. Failure to adopt improved technologies obviously imposes a welfare cost on nonadopters and on society as a whole, given the general equilibrium effects of reduced output on food prices. Fafchamps and Pender (1997) similarly find that poor Indian households are discouraged from making investments that return 19-22% per year in real terms because of the lumpiness and irreversibilty of the investment in the face of liquidity constraints associated with their poverty. So the consistent finding of most empirical studies of the effect of residual risk attributable to credit and insurance market failures is that it impedes technology adoption and therefore proves costly in terms of foregone output, and diminished productivity and well-being.

6. Asset degradation and ex post response to shocks under liquidity constraints

Not only do financial market failures cause households to forego investments of known high return, they can often force asset depletion, one of the most common and costly methods of dealing with financial market failure, but a quite common one in the wake of an adverse, uninsured shock. A shock hits – drought, flood, a cyclone, pest infestation, wildlife destroys one's fields, or someone falls ill or is injured or killed – and suddenly the family needs cash. In the absence of (formal or informal) insurance, they then commonly have to sell off productive assets in order to meet these immediate needs. But that comes at a high price in terms of foregone future productivity. Because households well understand the future consequences of asset liquidation to cope with uninsured shocks, we observe that households typically destabilize consumption intentionally in order to try to protect their stock of productive assets (Barrett *et al.*, 2006, Hoddinott, 2006). Families commonly reduce the number of meals they consume or cut back on the quantity or quality of meals or other basic expenditures before they resort to sale of assets to smooth consumption. Nonetheless, the financing gap faced by poorer households is often too big to weather through expenditure reduction when current consumption is modest in the best of times. Such households often regretfully make distress sales of key assets, predictably leaving them worse off in the future, simply because they lack access to proper insurance today.

So too do households predictably deplete natural capital – the store of wealth held in forests, soils, water and wildlife – when faced with binding credit constraints that impede their ability to conserve scarce natural capital and thereby invest in their future productivity. This is manifest in deforestation patterns in Madagascar (Barrett, 1999), wildlife harvest in Tanzania (Barrett and Arcese, 1998), and soil nutrient depletion in Kenya. As a result, we see a strong positive relationship between soil quality and household wealth and income measures in Kenya's western highlands (Marenya and Barrett, 2007b). In order to meet immediate needs, farmers sacrifice the quality of the soil on their farms, even past the point where soil rehabilitation is reasonably easy. Of course, this then drives them into a poverty trap wherein they lack incentive to rehabilitate degraded soils or even to apply mineral fertilizers to boost current productivity because the marginal returns to fertilizer application are directly affected by broader soil health (Marenya and Barrett, 2007b).

7. Policy interventions

Poor people face difficult decisions when confronted with financial market failures that leave them searching for alternative, costly means to meet immediate cash needs or that cause them to forego otherwise attractive investment opportunities. We observe that poor people throughout the world commonly choose seemingly inefficient or myopically short-sighted responses that carry a high cost. This high cost is, effectively, the astronomical interest rate or quasi-insurance premium they must pay for credit or insurance not available through more conventional channels. The resulting displaced distortions of production and exchange behaviors impede asset accumulation and help to perpetuate poverty.

This is a development challenge of the first order. Hence the recent Nobel Prize awarded to Muhammad Yunus for his pioneering work in creating the Grameen Bank and effectively launching the microfinance revolution of the past twenty years. He began this effort after he repeatedly failed to persuade Bangladeshi banks to lend money to poor families living near the campus where he taught. Only once he offered to guarantee the loans himself could they obtain credit. And these borrowers proved highly creditworthy in spite of the banks' ex ante assessments, paying back their loans completely and on time (Yunus, 2006).

What can be done to address this problem? Part of the solution lies in activating rural financial markets, to be sure. The standard first reaction is to create micro-finance institutions to try to fill the financial services lacuna that plagues most poor rural communities. But this can be difficult, as reflected in a burgeoning evaluation literature that offers quite mixed evidence on the efficacy of micro-finance interventions (Morduch, 1999, Armendáriz de Aghion and Morduch, 2005).

Farmers' involvement in certain commercial activities may make it easier to tap into financial networks that already exist. In Embu, in the central highlands of Kenya, for example, tea factories have arranged for payments to be delivered to smallholder growers through a formal bank account. This has resulted in much greater participation by farmers in formal financial networks, with significant increases in credit access due to the establishment of a relationship between farmers and the banks via the tea payment scheme.

Given the tendency to use commodity and labor markets to resolve credit constraints, interventions in these markets can also help the rural poor avoid paying extreme implicit rates of interest on quasi-credit. For example, commodity price fluctuations partly reflect poor rural infrastructure and storage capacity. Assisting farmers with the installation of paddy or grain banks, or with better on-farm storage can limit the need to seek credit in the first place by reducing yield depreciation and cutting the costs of distribution.

Well-functioning safety nets – e.g., through public works schemes operationalized through food-for-work projects paying reasonable wages – can also provide a viable means to mop up surplus labor in the face of adverse shocks to crop and livestock production (Barrett *et al.*, 2004a). Market demand for unskilled labor collapses when drought or flooding occurs. Governments and nongovernmental organizations can use pre-planned public works schemes to soak up now-idle labor so as to meet the immediate cash needs that drive people into the unskilled wage labor market, else they will be displaced into distress asset sales, soil mining, deforestation, wildlife poaching, etc.

Another option is one-off subsidies of adoption of improved production technologies so as to obviate credit constraints. For example, food for work schemes for on-farm investment in soil and water conservation structures have been shown to yield increased productivity and complementary private investment in soil and water conservation structures (Holden *et al.*, 2006). Carefully targeted subsidies of this sort can enable poor, liquidity constrained households to get a toehold on the ladder out of poverty by surmounting short-term financing constraints that can otherwise trap them indefinitely in low levels of productivity.

When appropriate, affordable financing is unavailable, people find alternative ways to address current consumption needs. Credit and insurance market imperfections thereby lead to displaced distortions of other markets that negatively impact the productivity and welfare of low-income rural communities. If people don't have access to financial services, they finance necessary expenditures through other markets, notably asset, factor and product markets and by drawing down non-marketed assets, including natural resources in which distant, far wealthier populations take a keen interest. This can have undesirable long-term consequences as it reduces productivity and accumulation, helping to trap people in chronic poverty. It may also have undesirable externalities, both pecuniary externalities such as higher food prices that predictably result from non-adoption of improved agricultural production technologies, and real externalities such as those associated with natural resource depletion associated with deforestation and wildlife poaching. The good news is that ways exist to help people break out of the poverty traps associated with these displaced distortions.

Christopher B. Barrett

References

Armendáriz de Aghion, B. and J. Morduch, 2005. The economics of microfinance. MIT Press, Cambridge, MA.

Barrett, C.B., 1996. Urban bias in price risk: The geography of food price distributions in low-income economies. Journal of Development Studies 32 (6): 830-849.

Barrett, C.B., 1997. How credible are estimates of peasant allocative, scale or scope efficiency? A commentary. Journal of International Development, 9 (2): 221-229.

Barrett, C.B., 1999. Stochastic food prices and slash-and-burn agriculture. Environment and Development Economics 4(2): 161-176.

Barrett, C.B. and P. Arcese, 1998. Wildlife harvest in integrated conservation and development projects: Linking harvest to household demand, agricultural production and environmental shocks in the Serengeti. Land Economics 74(4): 449-465.

Barrett, C.B., S. Holden, and D.C. Clay, 2004a. Can food-for-work programs reduce vulnerability? In: S. Dercon (ed.) Insurance against poverty. Oxford University Press, Oxford, UK.

Barrett, C.B., C.M. Moser, O.V. McHugh and J. Barison, 2004b. Better technology, better plots or better farmers? Identifying changes in productivity and risk among Malagasy rice farmers. American Journal of Agricultural Economics 86 (4): 869-888.

Barrett, C.B., P.P. Marenya, J.G. McPeak, B. Minten, F.M. Murithi, W.Oluoch-Kosura, F. Place, J.C. Randrianarisoa, J. Rasambainarivo and J. Wangila, 2006. Welfare dynamics in rural Kenya and Madagascar. Journal of Development Studies 42 (2): 248-277.

Barrett, C.B., S.M. Sherlund and A.A. Adesina, 2007. Shadow wages, allocative inefficiency, and labor supply in smallholder agriculture. Agricultural Economics, in press.

Carter, M.R., 1988. Equilibrium credit rationing of small farm agriculture. Journal of Development Economics 28(1): 83-103.

De Janvry, A., M. Fafchamps and E. Sadoulet, 1991. Peasant household behavior with missing markets: Some paradoxes explained. Economic Journal 101(409): 1400-1417.

Fafchamps, M. and J. Pender, 1997. Precautionary saving, credit constraints, and irreversible investment: Theory and evidence from semiarid India. Journal of Business & Economic Statistics 15(2): 180-194.

Feder, G., R.E. Just and D. Zilberman, 1985. Adoption of agricultural innovations in developing countries: A survey. Economic Development and Cultural Change 33(2): 255-298.

Hoddinott, J., 2006. Shocks and their consequences across and within households in rural Zimbabwe. Journal of Development Studies 42(2): 301–321.

Holden, S., C.B. Barrett and F. Hagos, 2006. Food-for-work for poverty reduction and the promotion of sustainable land use: Can it work? Environment and Development Economics 11 (1): 15-38.

Marenya, P.P. and C.B. Barrett, 2007a. Household-level determinants of adoption of improved natural resources management practices among smallholder farmers in western Kenya. Food Policy 32 (4): 515-536.

Marenya, P.P. and C.B. Barrett, 2007b. State-conditional fertilizer yield response on western Kenyan farms. Cornell University working paper.

Morduch, J., 1999. The microfinance promise. Journal of Economic Literature 37(4): 1569-1614.

Moser, C.M. and C.B. Barrett, 2003. The disappointing adoption dynamics of a yield-increasing, low external input technology: The case of SRI in Madagascar. Agricultural Systems 76(3): 1085-1100.

Moser, C.M. and C.B. Barrett, 2006. The complex dynamics of smallholder technology adoption: The case of SRI in Madagascar, Agricultural Economics 35(3): 373-388.

Moser, C.M., C.B. Barrett and B. Minten 2005. Missed opportunities and missing markets: Spatio-temporal arbitrage of rice in Madagascar. Cornell University working paper.

Rosenzweig, M.R. and H.P. Binswanger, 1993. Wealth, weather risk and the composition and profitability of agricultural investments. Economic Journal 103(416): 56-78.

Rosenzweig, M.R. and K.I. Wolpin, 1993. Credit market constraints, consumption smoothing, and the accumulation of durable production assets in low-income countries: Investments in bullocks in India. Journal of Political Economy 101(2): 223-244.

Santos, P. and C.B. Barrett. 2006. Informal insurance in the presence of poverty traps: Evidence from southern Ethiopia. Cornell University working paper.

Stephens, E.C. and C.B. Barrett, 2007. Incomplete credit markets and commodity marketing behavior. Cornell University working paper.

Stiglitz, J.E., 1974. Incentives and risk sharing in sharecropping. Review of Economic Studies 41(2): 219-155.

Stiglitz, J.E. and A. Weiss, 1981. Credit rationing in markets with imperfect information. American Economic Review 71(3): 393-410.

Sunding, D. and D. Zilberman, 2001. The agricultural innovation process: Research and technology adoption in a changing agricultural sector. In: B.L. Gardner and G.C. Rausser (eds.) Handbook of Agricultural Economics, volume 1A. Elsevier, Amsterdam, the Netherlands.

Townsend, R.M., 1995. Financial systems in northern Thai villages. Quarterly Journal of Economics 110(4): 1011-1046.

Vanderpuye-Orgle, J. and C.B. Barrett. 2007. Risk Management and Social Visibility in Ghana. Cornell University working paper.

Yunus, M., 2006. Nobel Lecture. Accessible online at http://nobelprize.org/nobel_prizes/peace/laureates/2006/yunus-lecture-en.html.

Zeller, M., G. Schrieder, J. von Braun and F. Heidhues, 1997. Rural finance for food security for the poor: Implications for research and policy. IFPRI, Washington D.C., USA.

Sustainable development and peace

Jan Pronk

Abstract

Searching for a new paradigm in international policy making, the mutuality of peace and development is increasingly recognised. However, in many developing countries globalisation and exclusion are leading to persistent feelings of injustice, neglect and alienation. These may become a source of increased instability, insecurity, conflict and violence. Globalisation has also resulted into ecological distortions, sharpening of inequalities, a greater conflict potential and a weakening of the capacity of the polity to deal with these concerns. It is, therefore, imperative to strive for sustainability as an inclusive concept that reinforces mutual thrust that justice will be maintained and secured for all people, without any discrimination, thus providing ultimate guarantees of mutual security. Shortly after the fall of the Berlin wall the then UN Secretary General, Boutros Boutros Galil, drew the attention to the need for a new paradigm in international policy making: there is no peace without development and there is no development without peace. It may sound as a truism. However, it had been ignored during the Cold War between East and West as well as during many years of lukewarm peace between North and South in the aftermath of decolonisation.

1. Mutual interests

The neglect of this truism had already become clear during the discussions about the conclusions presented by the Brandt Commission in the two reports *North-South: A Programme for Survival* (1980) and *Common Crisis: Cooperation for World Recovery* (1983). These reports dealt with issues concerning development, poverty and new relations between North and South, a new international economic order. They reflected a new philosophy: interdependence between nations, mutual dependencies resulting in a common interest of all nations, not only in order to establish recovery, but also to manage development and even to guarantee survival. The reports claimed that all nations shared a global responsibility for world social and economic development. A world public sector was advocated, parallel to orderly international market operations, in particular in energy, food, trade in general as well as in finance. This was complemented by the statement that there was a need for world institutional reforms. Would all this be possible? Yes, as Willy Brandt said in the preface to the report: 'One should not give up the hope that problems created by men can also be solved by men'.

This programme has not been implemented. Why not? Maybe the minds were not yet ripe. The confrontation between East West during the Cold War did not create a climate in favour of global cooperation. There was not yet a feeling of global communality.

There was a second reason as well. The world economic recession of the second half of the seventies and the eighties was not conducive to new approaches. Instead countries followed a pattern of adjustment to what was felt as the economic reality. This adjustment took place through expenditure cuts, rather than investments resulting in growth and development. This led to an economic philosophy consisting of elements which were not in accordance with those proposed by the Brandt Commission: market liberalisation, deregulation, a smaller public sector and more reliance on unbridled market mechanisms.

In the second half of the eighties and in the nineties the world has changed drastically. There was the end of the Cold War, followed by a new phase of globalisation. The new chances for world peace after the Cold War were a big boost to world economic growth, benefiting both the US, the countries of the former Soviet Union as well the countries in Eastern and Western Europe. The economies of these countries were benefiting from a fast and intense globalisation following the opening of national borders. There was a new mutuality of peace and economic progress. Could it be extended towards developing countries as well?

2. Conflict and disaster

The conflict between the East and the West had geopolitical consequences. Both the East and the West had wanted to contain a possible extension of each other's sphere of influence in the Southern part of the world. Both had been in favour of keeping the status quo and had tried to prevent change within countries that might result in alliance hopping. Both had given political, economic and military or intelligence support to friendly regimes, irrespective whether these regimes were representing the interests of their own population. Some of these regimes were oppressing their citizens, violating human rights and reaping the fruits of progress for their own benefit. The end of the Cold War put an end to this. Change within countries was no longer prevented or obstructed by powers from outside. However, such change very often took place in a violent manner. Conflicts which already existed in many societies of the so-called then Third World, but which had been contained with the help of outside powers, re-emerged. The new civil wars were a serious blow to the chances for economic development. However, many of these conflicts were not purely political. They had social and economic roots as well. For this reason Boutros Boutros Galil rightly pleaded linking peace and development not only internationally, but also within countries.

So, instead of international conflicts in particular in the developing world there were more domestic conflicts, within countries. They were quite complex, due to the fact that in many nations people were attaching greater importance to the cultural (religious or ethnic) dimension of their identity, next to a social and economic (class) dimension. These tendencies led to new poverties: more exclusion than exploitation, less reversible. Poverty was no longer a form of collateral damage on the way to general economic progress, but calculated neglect, a built-in default in the global system.

Since then the mutuality of peace and development has been confirmed in numerous resolutions of the General Assembly of the United Nations and in other international declarations. However, in the same period we can count more violent conflicts within countries than before. Many developing countries have gone through a period of higher economic growth than in the two decades directly following decolonisation, but this economic growth has not resulted in less poverty. Still two billion people live below the poverty line, with no more than two dollars a day.

A second important new paradigm endorsed in the ninety nineties is the precautionary principle. Environmental pollution, the imminent depletion of natural resources, the loss of biodiversity and, last but not least, climate change gave rise to an awareness that we had to change our attitude. Policies aiming at high economic growth could endanger the lifeline between the earth and the people living on it. There was no room anymore for careless frivolity. Maintaining the carrying capacity of the earth required precaution.

Politicians, signing the UN Convention on Climate change, promised their citizens: *'we will take precautionary measures to anticipate, prevent or minimize the causes of climate change and mitigate its adverse effects. Where there are threats of serious and irreversible damage, lack of full scientific insights should not be used as a reason for postponing such measures'.* This was wise and forward looking indeed. Politicians made the promise: even if and when there is no full scientific proof, not hundred percent certainty, even if some scientific doubt is still legitimate, we as politicians commit ourselves and promise our citizens that we will not delay, that we will act, that we will not seek a pretext for postponement and non-action.

The promise has not been kept. Despite the fact that scientific insights have improved, despite an increased likelihood about ongoing global warming, melting snow and ice, rising sea levels, temperature extremes, changing rainfall patterns, severe droughts and a greater intensity of cyclones, economic activities have gone on without restrictions. Governments committed themselves to reduce Greenhouse gas emissions, but the emissions have increased since then.

Why have we not been more seriously tried to live up to these commitments? Why didn't we really address questions concerning the destiny of the earth, the needs of poor people and the risks confronting future generations? Was it a lack of imagination? Was it due to old-fashioned impotent machinery and procedures within the UN system? Was it scepticism, a lack of political will, a lack of insight into changed world conditions or a lack of capacity to translate new insights into a new approach? Have we fooled ourselves by believing that both political conflicts as well as natural disasters are exceptions to the rule? It seems as if policy-makers still base their policies on the assumption that conflicts and disasters are not structural, that neither of these two phenomena is man-made, that both are exogenous to human behaviour. This assumption is nonsensical, of course. However, our policies have yet to be guided by the opposite assumption: that conflict is not an exception, but that it is inherent to development and that the depletion of the earth's capacity is the result of human activities. Or do we know all this, but have we in the meantime become obsessed by other dangers: threats to national security and the war against terrorism?

3. A paradigm dispute

Maybe all these reasons and motives have played a role. But I would like to offer another explanation. Maybe there is still lack of clarity about the paradigms themselves. There always has been disagreement about paradigms. Throughout history dominant paradigms have been contested. It helped sharpening convictions beyond a justification of interests. The paradigms of those in power are always different from the paradigms of the non-elite. When paradigm disputes turn into an ideological confrontation, conditions in society may become stifled, because groups draw back into their bulwarks. However, a straight and genuine paradigm dispute can help putting a clash between interest groups at a higher political level. It can help disarming powerful elites, undermining their self-justification, unravelling the case in favour of the status quo, by focusing on the longer turn interest of society as a whole.

That is true for paradigm disputes both within nations and world wide. In international relations such a major dispute had taken place earlier, after the decolonisation in the sixties. There was a risk that the newly won independence of the young nation states would not be followed by a reasonable degree of political and economic autonomy. At the time the answer was a threefold new paradigm: self-reliance plus the fulfilment of basic human needs plus a new international economic order. Neither of the three became reality. Instead the world went through a period of neo-colonialism, widening gaps between rich and poor and a return to the old order. In the eighties this led to complete stagnation. The South was told to adjust to new realities set by the North. There was no international cooperation to address world problems such as mounting debt burdens, a deteriorating environment and increasing world poverty. All possible efforts were paralysed by the last convulsions of the Cold War between East and West. However, after the fall of the Wall another new paradigm for development cooperation emerged, again defined with the help of three concepts: democracy, eradication of poverty and sustainable development.

The end of the Cold War had meant that there was room for change everywhere. Change to the good: freedom, democracy, human rights and disarmament. These perspectives, together with the themes of precaution and peace and development, were brought together under one roof: sustainability. Sustainability meant: progress for the present generation in all respects and everywhere, without discrimination, but on the understanding that any next generation would be entitled to at least the same opportunities. Any living generation was obliged to use the resources at its disposal in such a way that these would be fully sustained or renewed for the benefit of the next generation. Optimism prevailed, a belief that the future could be shaped within the framework of a new and more just order and that choices could be made in an atmosphere of harmony.

The optimism did not last long. The world lacked the capacity to translate the new dream into reality. Unbridled globalisation of markets, together with an economic spirit characterised by greed and by the-sky-is-the-limit notions resulting from rapid technological innovation, blurred the commitment to sustainability. Conflicts within nations seemed to be unmanageable. And the erosion of the international public system, in particular after the US invasion in Iraq, following its self declared right to strike pre-emptively, weakened the capacity to bring about democracy, poverty eradication and sustainable development in conflict ridden countries.

4. Brandt revisited

The philosophy of the Brandt Report is still relevant. The need to base policies on a sense of global communality is even greater than before. However, the programme proposed in the report is partly overtaken by events, partly not sufficient to deal with the new challenges.

The analysis of the Brandt report is still adequate. A quote from the introduction by Willy Brandt may serve to illustrate this: 'We are confronted, whether we like it or not, with more and more problems which affect mankind as a whole, so that solutions to these problems are inevitably internationalized. The globalization of dangers and challenges – war, chaos, and self-destruction – calls for a domestic policy which goes much beyond parochial or even national items. Yet this is happening at a snail's pace. A rather defensive pragmatism still prevails, when what we need are new perspectives and bold leadership for the real interests of people and mankind. The 'international community' is still too cut off from the experience of ordinary people, and vice versa'. It could have been written today.

Brandt presented the second report of his commission under the title *Common Crisis*. The character of that crisis has changed. Unlike at he time of the publication of the Brandt Commission reports there is presently no more general world economic stagnation or a deficient world economic growth, no more the traditional North South divide that followed the decolonisation, no more a major ideological confrontation between a communist East and a capitalist West, leading to an arms race between the two blocks, blocking each form of global cooperation. Each of these three major issues could have been addressed and attenuated through international cooperation. Each of them impeded such cooperation. As a matter of fact, the three crises no longer persist. There are others, that block international cooperation and that themselves should be on the agenda of international institutions.

The present crisis is not due to stagnating world economic growth, but the result of unbridled growth leading to ever greater imbalances and inequalities. Structural world economic and social inequality has increased and this has social, cultural and political ramifications. Many developing countries have become emerging economies or even major players. China is a major player in world finance, Brazil in trade, West Asia and some new oil producers in energy, India in finance, trade and investment. All these countries have become major importers of consumption goods, because of an increased purchasing power of their expanding middle classes.

The international market mechanism is thriving. International trade is blossoming. International energy markets presently seem to function quite efficiently. International financial markets are functioning globally, with less and less restrictions. This results in increased and quick access of more and more companies, everywhere, to short and long term capital, which is originating from more and more sources, anywhere. The conglomerates and the multinational banks merge and take over economic activities everywhere where they consider this profitable, sometimes in the form of aggressive buy outs. They find new ways to operate, as hedge funds and private equity funds. They cross borders as if these do not exist. As a matter of fact, they don't anymore, at least in economic terms. In economic terms borders do indeed not exist anymore. It means that the capacity to control and to institutionalize checks and balances has been eroded tremendously.

The new challenges thus can be described not only as new phenomena, for instance in the field of information technology, creating vast new opportunities, but also as greater risks and uncertainties. These economic uncertainties are compounded by others. As shown by Stern, the costs of inaction and postponement of action to address climate change will be unexpectedly high. Aids and other poverty diseases lead to huge social deficits in many countries, which presently still seem to be underestimated, with irreversible consequences. In many societies globalisation and exclusion are leading to persistent feelings of injustice, inequality, neglect and alienation. These feelings are a source of increased instability, insecurity, conflict and violence. Such violence may lead to terrorism, a phenomenon which nowadays does not stop at borders, but has been globalised as well. And, finally, as recently was spelled out by Richard Holbrooke, there is the risk of blundering into a new World War. That risk has always existed, but it has been reduced by the existence of the United Nations with strong values, widely shared, strong institutions and well accepted procedures to manage international conflicts. Both the values, and the procedures and institutions have been eroded since '9/11' and since the intervention in Iraq at the beginning of the new millennium. International cooperation in the spirit of mutuality and communality, as advocated by the Brandt Commission, has become weaker rather than stronger. However, in the light of the new and greater risks and uncertainties such global cooperation is even more needed than at the time of publication of the Brandt report.

Some reform of international institutions has taken place. However, it has been uneven, more related to facilitate and further economic globalisation, rather than checking and containing the negative fall out. In so far reform has taken place, it is stagnant now. Within the institutions the political will to use proper governance instruments is fading away. The

IMF is not being granted an adequate institutional capacity to deal with global financial instabilities, for instance those related to the US deficit. The WTO Doha Trade Round is stagnating for years already. The World Bank cannot deal adequately with one of the most importing financing needs: resources for reconstruction after conflict – despite the fact that the Bank's original function, as International Bank for Reconstruction and Development, was exactly this. It is still focused on national development finance, for which many private commercial financial resources can be attracted at capital markets, and fails to address adequately priorities such as poverty eradication, the development of the global commons or adjustment to climate change. The European Union has been enlarged, but an agreement to reform its institutional structure in order to render the extended Union more effective has failed and a new Constitution has been rejected. The United Nations have endorsed the Millennium Development Goals but it is clear already that these laudable goals will again fail to be met. The talks about UN reform that were meant to give substance to its 50th anniversary failed as well. The Security Council is not up to its task, the reform of the Human Rights Council is only a token reform, the new Peace Building Commission has no powers and the Criminal Court is not effective. United Nations peace keeping has been stretched to the limits of its capacity and seems to be less effective the greater the need to keep the peace and to protect the victims of human rights violations in case of violent conflict. The UN system itself is still a hotchpotch of agencies, without a unified approach. On the contrary, increasingly centrifugal tendencies can be noted. As far as climate change is concerned there is still no full endorsement of the Kyoto Protocol. The political will to adhere to what has been agreed seems to get weaker while an agreement about a follow up to the Protocol that would better reflect the need mitigate the emission of greenhouse gases is still out of reach. Finally, international policy making in other environmental domains, such as biodiversity, fisheries, pollution and the preservation of water systems is stagnant.

5. Four crises

In my view the paradigmatic crisis is fourfold. First, there is a crisis in policy making. In international policy making conclusions could not be reached because international interest diverged too much and international institutions were too weak to counter the imbalances. And in those cases where agreement was reached policies were not adequately put into practice or did not work out, because the implementation process was constrained by unequal power relations.

Second, there was a crisis in the world system itself. The deficiencies mentioned above in the international institutions render it alarmingly inadequate to address the global challenges of today. After World War II the new system of the United Nations had been created in order to contain the use and abuse of power by individual countries by establishing a power sharing system on the basis of consensus, meant to address common insecurities and other challenges to the world as a whole. That system has been eroded. Unchecked power inequalities are back on the international stage.

Third, underneath the crisis in global policy making and in the world system there is a crisis ridden process, characterized by more poverty, not as an unintended corollary of progress but as a result of intended exclusion, economically as well as politically. The global economy becomes more and more dualistic.

All this reflects a basic crisis in ideas and values, in political theory about the relation between man and society, in the economic theory about what constitutes welfare, in the thinking of people all around the world about the relation between man and nature and the earth's resources, in ideas about the legitimacy of violence in order to reach one's objectives. Throughout history such basic questions have been answered differently in different countries. Such differences have existed throughout history, but mostly in different societies. Presently paradigmatic differences have globalised themselves, due to world wide migration, more intense communication, mass information and fast economic as well as technological modernisation. This is bound to lead to cultural and political conflicts, everywhere. This is the fourth crisis.

The basic questions have become quite pertinent: will growth foster development or inherently jeopardize the resource basis for future sustainable human development? Is it ever possible that those who are not in power will benefit from progress, or are they bound to remain victims?

If with regard to both questions the latter answer is the most probable, will there be any chance that conflicts can be managed, let alone prevented?

6. Conflict and globalization

As argued above, since the end of the Cold War conflicts rose mostly within nations, not between them. Some of those conflicts were not new at all. They did not emerge, but re-emerge, often after decades of silence. Most of these conflicts had both economic and cultural dimensions. Economic conflicts can be managed within a reasonable period of time, by a good combination of economic growth and (re)distribution of assets and income, creating a perspective of progress for everybody, both future and present generations. On the other hand, cultural, social, ethnic, religious or sub-national domestic conflicts last long. They are rooted deeply in society. Cultural conflicts, whether or not accompanied or sharpened by economic inequalities, outlive generations. They are less manageable than economic conflicts, because there is no way out by means of sharing or redistribution. In an economic conflict there is always a win-win solution feasible: the right path of investment, growth and distribution can make all parties gain. An economic increase of one party does not necessary have to result in a welfare loss for others. Cultural identity conflicts are different. People and their identity groups - be it a tribe, an ethnic group, a religious denomination, a social class, a sex, a tongue, a colour, a caste, an elite, a nationalistic clan, or any group defining its identity in other than purely economic terms - are inclined to define their identity not as a share of total potential welfare in a society, but as absolute positions, demonstrably different from other groups. From this perspective a stronger position of one group in a society always means that another group will lose. However, when individual people and the groups to which they belong consider their identity not threatened – and thus potentially diminished – but enriched through communication with other identity groups, conflicts can be avoided.

Welfare is a relative concept. It can be increased, also through intelligent distribution. Power is an absolute concept. Total power cannot be increased by means of redistributing it. Power is a zero sum game. Only when cultural identity conflicts are not seen as power conflicts but as welfare conflicts, a solution is possible. Mutually enriching cultural communication can transform an identity, once a straightjacket, into a stepping stone towards value added, resulting from a process in which values are shared rather than used as a defence mechanism. This requires cultural confrontation to be transformed into mutual cultural exchange. Without an exchange on the basis mutual curiosity rather than a confrontation based on aversion and fear, cultural conflicts are longer lasting, less manageable and more violent than either economic conflicts or international disputes. That is what happened in the ninety-nineties, after the euphoria of the end of the Cold War had evaporated. Old conflicts re-emerged, weapons were wetted and violence struck many countries from within.

Violence was not contained to the original location of the conflict. It was brought to other countries by the same forces which brought about globalization. That was the second major new phenomenon in the ninety nineties. Of course, there had been internationalization throughout: intercontinental transport, foreign investment and trade, international finance, imperialism and colonisation, world wars, efforts to build international alliances, a League of Nations, the UN itself. Globalisation was not a new process, we had seen it for centuries, and had witnessed a stronger pace in the four decades since World War Two. But in the final decade of the last century it got a new shape. Internationalisation had been an economic and a political process, steered and fostered by means of concrete decisions of policymakers and entrepreneurs. It was man-made. But somewhere in the nineties internationalisation turned into globalisation. It got a momentum of its own, became less a consequence of demonstrable human decisions, more self-contained and self-supporting. The driving force was twofold. First: technological advance, enabling full and fast information and communication everywhere, physically and virtually. Second: economic, the global market, linking production, investment, transportation, trade, advertisement and consumption anywhere in the world to any other place. The result was a disregard for national frontiers, a strengthening of global corporations and an erosion of nation states.

Globalisation became a cultural affair as well. A reality in the mind of the people: time differences and long distances are no longer barriers for communication. Technology has solved this. What used to be far away has come close, what lays at walking distance is not being noticed or is even shut out of people's consciousness. The factual distance and the actual time difference are no longer relevant, only the distance within the human mind counts. When we travel, most of us feel through our air ticket, cell-phone, e-mail, credit card and CNN much closer connected with people in comparable conditions in metropolis abroad than with poor people around the corner, landless people in the countryside or jobless people in the shantytowns nearby. And at home most of us will relate in our minds more with surfers on Internet on the other side of the globe than with poor people around the corner. For everybody in the world everything happening anywhere else is happening at the same moment and may affect everybody, wherever. It is a real time world with real time connections and we feel part of it.

7. Global apartheid

That is to say: provided that we have access to modernity. Provided that we are not excluded. Many people are excluded. Globalisation was neither coherent nor complete. It was a globalisation of markets and of greed. In the ninety-nineties economic growth was higher than ever since World War Two. It got a boost from new technologies, rising expectations and a mounting demand at the global market. This unprecedented growth could have helped enlarging the capacity of the international community to address poverty and sustainability questions. It did not. Instead, globalisation led to an even more unbalanced development: not more, but less sustainable, at least in social and ecological terms. Globalisation also made international cooperation lopsided by directing political attention mainly towards facilitating the workings of the world market and neglecting other concerns. The intentions at the global market are a mix of a belief in the blessings of modern technology and a selfish, materialistic and commercial approach to notions of welfare and progress.

What did this mean for the poor? During long periods of capitalist expansion poor people were exploited. But they had an opportunity to fight back, because the system needed them: their labour and their purchasing power, the power to buy the goods produced by the system and thereby sustain the very system that exploited them. This common strength of the poor helped to modify exploitation. Development became potentially also beneficial to the poor. They got a perspective: to have a more liveable life than their parents and to give an even better life to their children. That is development: change for the better, even if small, but in the perspective that improvements will last. Everybody within the system was entitled to such a perspective. Everybody had the right to hope. That hope is no longer justified. Globalisation has changed the character of capitalism. There are more people excluded from the system than exploited in the system. Those who are excluded are being considered dispensable. Neither their labour nor their potential buying power seems to be needed. That is the reason why they cannot fight back anymore. They lost a perspective. If you know that your life is worse than that of your parents and if there is no hope that your children will do better, but instead will be even worse of than yourself, than there is no perspective whatsoever.

For many people this is today's reality. Better than in the past they know how life could be. Modern communication tells them that. But deeper than before they realize that such a life is not within reach, because they have lost solid ground. They have no land to work on, no job, no credit, no education, no basic services, no security of income, no food security, but ever more squalor, an ever greater chance to be affected by HIV/AIDS, a house without electricity, water and sanitation. Despite unprecedented world economic progress during the last decade, for about two billion people there is only the experience of sinking further and further into quicksand. In the World Summit on Sustainable Development, in Johannesburg in 2002, President M'beki called this Global Apartheid. The gap between rich and poor in the world can no longer be explained in terms of a strikingly unequal distribution of income and wealth which could be modified through world economic growth and a better distribution of the fruits of growth. The gap appears to have become permanent. Rich and poor stand apart, separated from each other. Under the Apartheid regime people were either white or black. So, they were part of the system, or they were not. Today people belong to modernity, or not. The world of modernity is western of origin, but stretches towards islands and pockets of modernity in the east and in the south. The worlds of modernity are linked with each other by means of modern communication, physical or virtual communication. Through the culture of modern communication people feel that they belong to modernity, that they are part of it, part of the globalized uniform western neo-liberal culture of mass-consumption, materialism, greed, images and virtual reality. That modern world is separated from the world next door, physically sometimes just around the corner, but far away in terms of time, mentality, experience and consciousness: poverty, hunger, unemployment, lack of basic amenities in the shantytowns of a metropolis, at the countryside and in the periphery, where pollution is permanent, where the soil is no longer productive, water scarce, life unhealthy. Poor people have to live in the worst places of the earth. 'A world society based upon poverty for many and richness for some, characterised by islands of welfare surrounded by a sea of poverty is not sustainable', President M'beki said. Indeed, that is Apartheid. On the one side security and luxury, on the other deprivation, hardship and suffering. At the beginning of the new Millennium for many people life has never been so good. At the same time for many people in our direct global neighbourhood life is not liveable.

Like in the past, under the South African Apartheid regime, security and luxury on the one side of the fence is being sustained and protected by continuing the squalor, suffering and poverty elsewhere. Not by exploiting the poor. There still is exploitation - low commodity prices for instance and indecent wages for migrant labour - but the globalised Western culture has so much capital and purchasing power, that it can sustain itself without exploitation However, poor people, instead of being exploited, are excluded. The Western world is afraid that they will cost more than they can contribute. They do not fit into western cost benefit calculations. People living in the slums of Calcutta, Nairobi and Rio de Janeiro, AIDS affected in Africa, landless people in Bangladesh, subsistence farmers in the Sahel, illegal migrants crossing the Mediterranean, all of them lack the capacities needed to contribute to the modern Western economy and the buying power for its products. That is why these people are considered dispensable. Well-to-do people are not interested in the ideas of the poor, let alone their feelings or their fate. The poor are a burden and should not try to come close. They are being kept away by connecting the islands of wealth with each other, using the means provided through globalisation. In doing so we deprive them of space and soil, in particular good soil: productive, fertile or commercially attractively located. We deprive them of water, forest and natural resources. We burden them with sky-high debts. We deny their enterprises fair access to the market by favouring foreign companies, providing them with more licenses, higher credits and tax holidays. We deny them basic provisions for survival, such as affordable medication against AIDS. President M'beki was right: globalisation is Apartheid. Globalisation takes away living space. Globalisation is appropriation. Globalisation means fencing off. Globalisation is occupation. Occupation of space - living space -, expropriation of resources, sealing off societies, subjecting cultures. The poor are told to stay in their homelands, in occupied territories, separated from each other by boundaries drawn by those who do have access to the resources, the capital and the technology which lay at the basis of the modern Western culture.

The revolution of globalisation has made winners and losers. Real losers and those who see themselves as losers. Globalisation is shaking established structures and cultures. Some have the skills to gain access to the modern world market and play up fully. Others adjust themselves. For again others, it is either sink or swim. Many of them, economic asylum seekers for instance, are struggling with the waves of modernity and sink into the undercurrent of the new dynamics. For other people, single females with children in Africa for instance, modernisation means entire uprooting. Their existence was fragile and gets shattered. Many of them are dragged away and go down.

8. Resistance

Others resist. Such a resistance can take different forms: protest, economic action, migration, forming alliances, a political counteroffensive at high level. It can also imply the strengthening of a vulnerable culture or an effort to tie religion with politics. It can result in violence, first against those within that culture who choose in favour of modernity and assimilate themselves with the foreign, western culture. Later on violence may be directed at the foreign culture itself. That is a final stage. The more the centre of globalisation disregards the periphery, not only the economic and social needs of the periphery, but also its traditions, culture, religion and aspirations, the harsher the resistance. A Western attitude of self-sufficiency and self complacency is seen as arrogant, as an insult, a slap in the face. That breeds resentment. The excluded feel not only poor and dispossessed, but also defeated and humiliated.

In the 18th century such a haughty attitude of the elite brought about a revolution. Today revolt is in the air. 'If you don't visit your neighbourhood, it will visit you', Thomas Friedman wrote. That visit can take different forms. One is migration to the towns. Another one is crime and violence in any metropolis with a dazzling city-centre next to favelas and shantytowns with breathtaking poverty. A third reaction can be terrorism. Migration does not lead to crime, and crime does not result in terrorism. But all of them are consequences of uprooting. Even when there is no direct link between poverty and violence, systematic neglect of aspirations and feelings of injustice creates conditions within which violence can flourish. People may acquiesce to violence when they feel humiliated, personally and as a group, once they feel not to be taken seriously, not respected or recognized as a culture or as a society, once they feel excluded by the new world system orchestrated by the West. Then they may give a willing ear to violence. Some approve silently, others give support or shelter. Others show themselves receptive to a message of violent action. Why not, they may think, if the world does not leave us an alternative?

Those who feel that the system does not care about them may try to seek access to the system, try to clear themselves a way into the system. That was the aspiration of migrants and of emancipation movements. Often they were successful. But if you experience that the system does not only ignore you, but brushes you aside, doesn't want you, cuts you off, excludes you, then you may become inclined to consider it your turn: to turn away from the system. 'If the system doesn't want me, then I do not want the system' is a form of logic. People who come as far as this do not even seek access to the system any more. They turn their back upon the system, denounce that system. One step further is to resist and oppose it, to want it being undermined, to attack and undermine it themselves.

Poverty does not lead straight on to violence. Poverty without any perspective whatsoever, plus the experience of exclusion and neglect, the perception to be seen as lesser people with an inferior culture, to be treated as dispensable by those who do have access to modernity, to the market, to wealth and power, all that together will lead to aversion, resistance, hate, violence and terrorism. Resistance against globalisation which is perceived as perverse, as a curtailment of living space, as occupation. Aversion against Western dominant values, which steered that process of globalisation in the direction of Global Apartheid. Hate against leaders of that process and against those who hold power within the system. Violence against its symbols. Deadly violence against innocent people within that system. Unscrupulous violence, unsparing nothing and nobody, uncompromising also towards oneself, fanatically believing: 'this is the only way'.

Is it fully incomprehensible that people, who consider themselves desperate, without any perspective, become receptive for the idea that they have been made a desperado by a system beyond reach? One step further and they become receptive for the whisperings of fanatics that they have nothing to lose in a battle against a system that is blocking their future. One more step and they believe that they will gain by sacrificing themselves in that battle. It is hideous, beyond justification, but the notion exists. It should and can be fought, but the most effective way to do so is not by resorting to counter violence alone, but by taking away the motives and reasons people may have when surrendering to the temptations of fanatics.

Most people, however poor and desperate, dislike violence. They are disillusioned, but in doubt. Many people in the world have developed a hate/love relation towards the West and its culture. They do not want to make a choice for or against the West. Unless they are forced to do so, for instance by the West itself. Then resentment overtakes doubt.

After September 11, 2001, world leadership has the task to disarm the fanatics without alienating those who doubt. That requires, as was pleaded by Secretary General Kofi Annan when he received the Nobel Peace Prize, building a sustainable, democratic and peaceful world society, within which humanity is seen as indivisible. That concept of sustainability, Kofi Annan added, ought to be based upon the dignity and inviolability of all human life, irrespective of origin, race or creed.

9. Lifelines

That is what is really at stake in the political dispute about the sustainability paradigm: indiscriminate access for all people to basic conditions of life itself. These are health, water, biodiversity, agriculture and energy. Health means survival, crossway between life and death. Water provides people with a lifeline with the present environment. It is the lifeline between people, nature and resources. Biodiversity provides men, women and children with a lifeline with the past as well as the future. It is the ultimate guarantee of the continuity of life. Agriculture stands for life itself. Agriculture provides people with food, work, an income and a home. Energy is the lifeline with progress: a more efficient use of resources, more food, more work, a better home, a higher income and the preservation of life and the postponement of death. Water, energy, health, agriculture and biodiversity, they form a string of lifelines. Together they give human survival a meaning and life a sense of direction, by freeing people from the fight for mere survival, to overcome the constraints set by space and time, to enable them to prevent and conquer misery and to develop instead, to reflect on the sense and meaning of human existence, to divide labour and exchange the fruits of labour, to philosophize, write poetry, make love, create images, tell stories, collect knowledge, play games. A sustainable world society means that people are free to do all this together with other people, within the family and with partners in society, coming from different backgrounds, with different cultures: different experiences, different insights, different languages, different poems, stories, images and games, and to share all that with each other.

Water, energy, health, agriculture and biodiversity. Together they are the lifeline between people and the planet. They shape the essential conditions for a sustainable development of human life, provided of course that they themselves are being preserved, sustained and developed in equilibrium with each other. That is crucial if we are to cut poverty in half or eradicate it all together. But those conditions for sustainability can only be met if there is a common determination towards the goal, common values and a shared belief concerning the system within which the endeavour should take place, full agreement about rights and duties, a willingness to comply with the norms and to live up to the principles, being confident that all this will be adhered to by everybody else, whether rich and powerful or not. Not only the task itself is complex, to achieve sustainable development including poverty eradication, but so is the setting within which the task has to be accomplished. That can not be imposed by a government or a bureaucracy. It cannot be ordered from above. It can only be achieved together with all other partners in society, bottom up, in a participatory approach, so that each and every individual and all peoples groups can trust that all will benefit more or less equally from a common endeavour to make life worth to be lived, now and in the future.

10. Inconvenient truths

In the last two decades we have seen globalisation resulting into ever-greater ecological distortions, a sharpening of inequalities, a greater conflict potential and a weakening of the capacity of the polity to deal with these concerns, rather than a strengthening of that capacity. One might call that stumbling into disaster. Economic conflicts have been complicated by religious and cultural conflicts, violence could not be contained but has spread around the globe and all this became further complicated by the violence of terrorism.

Al Gore, in his book *An Inconvenient Truth* has depicted the state of mind resulting from denial, doubt and disinformation, deliberate disinformation about climate change. These three D's have produced a fourth: Delay, the delay of action, precautionary as well as remedial action. Indeed, what should be done hasn't been done. It was this delay, this total lack of action, which brought people into a state of despair: they see that there is a lot of talking going on, many conferences, many UN resolutions, but no action. In turn this leads to a sixth D: Distrust.

The same six D's dominate the discussion about poverty. Denial that poverty exists; Doubt whether something can be done about it; Disinformation about the nature and causes of poverty; Delay of action; Despair amongst the victims; total Distrust in leaders.

The result is deeper conflict and less peace. In many countries the result is also political alienation: people who do not trust their leaders, leaders disregarding people's needs, people suspicious about values, models and doctrines propagated by leaders, leaders manipulating views and opinions of citizens, people's groups sharpening their identity, fencing themselves in, keeping others out, thereby creating a general climate of distrust. Uncertainty, disbelief, distrust, alienation and fear, which goes far beyond disagreement. People don't trust their leaders anymore when they speak about sustainability, precaution, peace, security, democracy and human rights.

Since the 11th of September 2001 the world stands at the crossroads. The choice is between two paradigms: security or sustainability. Security is exclusive: 'our' security, which we presume to be threatened by others - outsiders, foreigners, potential enemies - and which we try to protect through exclusion. The other paradigm, sustainability, is inclusive: a safe and secure place for all human beings, a safe habitat, a safe job, secure access to food, water and health care, secure entitlements to resources which are essential for a decent and meaningful life, worthy of human beings. Sustainability as an inclusive concept implies the mutual thrust that justice will be maintained and secured for all people, without any discrimination, the ultimate guarantee of mutual security.

In international policy security is all the go now. This implies a predominance of inward attitudes, more exclusion, pre-emptive strikes, retaliation, more violence, more terrorism, war. Going for absolute security kills. Embracing sustainability means sowing the seeds of life.

Sustainable management of natural resources

Policy making and land use changes: facing new and complex realities[13]

Teunis van Rheenen and Tolulope Olofinbiyi

Abstract

Land use policies are increasingly challenging. Recent global developments such as climate change, urbanisation, and the strong policy focus on the multifunctional role of agriculture further highlighted the need to revisit the tools and methods of analysis used so far to support land use policies. This chapter presents a number of global drivers that have far reaching impacts on the way that land is used, particularly in the developing world. The primary role of agriculture in the developing world is to feed the poor. Food security and poverty alleviation in the developing world remain key issues that continue to need global attention. Current tools and methods of analysis for land policies need to be improved to enhance the way that the poor can adapt to and mitigate global developments. Global developments have further enhanced the role of agriculture as a producer of food and in doing so further diminished the other roles of agriculture. In our conclusions we warn for land policies that pursue short-term benefits without giving consideration to long-term social costs are dangerous to the poor.

1. Introduction

Developing appropriate land use policies has always been a highly challenging endeavour. There are many reasons why this is the case including the fact that so many people depend directly on land for their livelihoods, complicated land use rights in developing economies, and increased competition for land. Tools used to support land use policies typically include linear programming and econometric models. The extent to which these models have actually been useful has been subject to debate for a long time. This debate has become even more relevant considering a number of developments at the global level. There has been an increasing disconnect between models that have been developed at the local level and those that have addressed issues at the regional or global level.

Recent developments that will impact land use include globalisation, climate change, negotiations at global podia such as the World Trade Organisation (WTO) Doha which address trade liberalisation and environmental issues in tandem. Urbanisation, the emerging bio-fuel economy, as well as the greater prominence given to the multifunctional roles of agriculture are all having an impact on the way that land is used.

---

[13] This chapter draws heavily on: Van Rheenen T and T Mengistu 2007. Rural areas in transition: a developing world perspective. Paper presented at the International Workshop on 'Understanding relations in nature and economy: an application to the rural countryside', May 31 – June 2, Wageningen, The Netherlands.

Globalisation is significantly transforming the world agri-food system – including agricultural production – having direct effect on land use patterns. Changes linked to globalisation include trade liberalisation, changing consumer preferences, and the supermarket revolution. Global developments, such as climate change, cause land use change which will have effects on the condition, quality, and functions of agriculture. For instance, intensification can cause rapid changes in landscape dynamics and hydrologic processes that affect the human well-being. An understanding of these interconnected processes is needed to guide policy actions that contribute to the deterrence of negative effects (Kok *et al.*, 2007).

Globalization and climate change have direct impacts on agricultural production and livelihoods, as well as the viability of rural economies. Both processes occur simultaneously and work together to create dynamic conditions that influence vulnerability and resilience. As a result, globalisation and climate change have important implications for economic and environmental sustainability (O'Brien and Leichenko, 2005), as well as future land use.

The chapter starts by presenting some mega drivers of change that have enormous direct and indirect impacts on land use. This is followed by a discussion where land use change is placed in a broader political debate. In this section we show how, particularly in the developed world the role of agriculture in rural development has been re-evaluated. In the developed OECD countries the concept of multifunctionality of agriculture is emphasised, while in developing countries the concept of 'roles of agriculture' has been emphasised. At the end of the section, we draw some conclusions regarding the policy challenges in developing countries.

2. Globalisation of the agri-food chain

The world has become an increasingly small place where the forces of globalisation more than ever before are having an impact on land use strategies, particularly of smallholders in developing countries. Indeed, the term globalisation has many connotations and the changes that are linked to land use include amongst many others trade liberalisation, changing consumer preferences, the expansionary role of multinationals and the supermarket revolution, as well as the agenda for the reduction of domestic agricultural subsidies and supports. While the majority of developing countries have become involved in the wider process of globalisation, the impact of this process has been varied across regions experiencing changes in different ways (Mertz *et al.*, 2005).

2.1. Trade liberalisation

Trade liberalisation is considered one of the important drivers of globalisation. While proponents of trade liberalisation argue that the reduction of agricultural supports can yield significant economic gains, critics argue that it can undermine small farmer production in developing countries. They suggest that small farmers may be forced to abandon agriculture, thereby creating negative effects on their livelihoods and security (O'Brien and Leichenko,

2005). According to Van Meijl *et al.* (2006), recent studies suggest that trade liberalisation can play a positive role in poverty reduction. While this may be the case in the long run, policy makers are faced with the challenge to facilitate a transition that will not have a negative impact on the poor.

A number of studies have attempted to show how further liberalization of trade will have an impact on land use. For example, in a study done by Grepperud *et al.* (1999), the authors looked at the effects of maize trade liberalisation and fertilizer subsidies in Tanzania using a computable general equilibrium model (CGE). Results show that trade liberalisation would accelerate the rate at which new lands would be put into use for crop production. The results also show that by the year 2010 in comparison to a baseline scenario, trade liberalisation increases agricultural land use by roughly 2 percent.[14]

In another study by Sun and Liqiao (2006), the impact of liberalisation on land use and payments for watershed services in China was examined. The authors argue that the WTO entry and outcome of the Doha Round will strongly affect land use and management in rural China. They suggest that increased trade - agricultural imports in particular - as a result of the WTO accession will involve shocks to the Chinese agricultural sector, which in turn will bring about structural transformations such as changes in agricultural land use. The authors expect farmers to shift production from cereals to high value crops such as fruits and vegetables. Consequently, the switch to high value agriculture will lead to an increase in land value, reduction in transaction costs of negotiating land use rights, and clear-cut definition of property rights.

Environmental concerns cannot be isolated from trade liberalisation and land use issues. According to a study by Van Meijl *et al.* (2006), the environmental implications of agricultural trade liberalisation for global land use outcome remains uncertain. The authors suggest that land-use modelling can address this gap because land can move in or out of agriculture for many reasons. The study was conducted in an attempt to 'quantify the economic impact of different agricultural trade liberalisation regimes until 2030.' Their results show that 'developing regions such as Africa, Asia, South and Central America obtain the highest growth in total agricultural land use.' Because idle agricultural lands are available in these developing regions, an increase in demand for land results in modest increases in rental rates. Therefore, the expansion of agricultural lands does not generate high costs.

Further, the Africa region is expected to see the most dramatic changes in land use. Projections suggest that agricultural land use will increase up to 70 percent by 2030, and crop production growth will increase by roughly 3 percent annually. The authors emphasise that the realisation of these expansions is heavily dependent on the sustainability of agricultural land. In contrast, in developed regions like Europe where much of the agricultural land is

---

[14] See Grepperud *et al.*, 1999 for model specification and scenarios.

already in use, an increase in demand for land results in high increases in rental rates and thus, reduction in land use. However, the results show that instead of abandonment, it leads to extensification[15] (Van Meijl *et al.,* 2006).

2.2. The upcoming strength of the retail sector (supermarket revolution)

Many countries in the developing world are confronted by a so-called 'supermarket revolution', which has taken off in Asia and Latin America, and also to some extent in Africa. In India – for example – the growth in food retail outlets increased from 2,769 thousand in 1996 to over 3,680 thousand in 2001 (see Table 1). This revolution is having a large impact on the way that land has been used and on the livelihood strategies of the poor. The changes in the food chain system present opportunities and challenges. Some will benefit, many will not. As a consequence of the supermarket revolution, the supermarkets increasingly control the segments of food retail markets. More intervention is needed to connect farmers to markets, so that they do not fall further into poverty. With the diffusion of supermarkets and rapid consolidation of their procurement systems, small farmers need to expand their supply capacity (Weatherspoon and Reardon, 2003), thereby, changing agricultural land use.

As the retail sector in the developing world has grown dramatically, this has gone hand in hand with far reaching effects and welfare implications on smallholders. As supermarkets become a greater force to be reckoned with in many developing countries, it is important to realise that this means more and more farmers will be supplying their produce to retailers and will have to meet the quality requirements of these supermarkets. Many who will not be able to do so will become the losers of this development and will be driven further into poverty.

Weatherspoon and Reardon (2003) point out that supplying to supermarkets is very different from the business as usual for many small farmers. While many changes provide great opportunities for many farmers, studies have shown (e.g. Reardon and Berdegue, 2002) that adapting to these changes has proven to be very challenging, especially for the smaller and poorer farmers. Policy makers have a role here to facilitate transition processes in such a way that the poor can also benefit from the upcoming strength of the retail sector. In Central and Eastern Europe (CEE), Dries *et al.* (2004) hypothesize that the supermarket revolution will drive a larger proportion of small farmers out of business because they are not able to meet up with the standards' requirements set by the retail sector.

---

[15] 'Extensification is the process of introducing production into land areas that were previously unused or used for less intensive purposes' (TAMU, 2000).

Table 1. Growth of retail outlets (in Thousands) in India during 1996-2001.

	No. of retail outlets					
	1996	1997	1998	1999	2000	2001
Food	2,769.0	2,943.9	3,123.4	3,300.2	3,480.0	3,682.9
Non-food	5,773.6	6,040.0	6,332.2	6,666.3	7,055.5	7,482.1
Total	8,542.6	8,983.0	9,455.6	9,966.5	10,535.5	11,165.0

Source: Guruswamy, *et al.,* 1999.

3. Climate change

For a long time little credibility was given to the link between climate change and land use change. Recently, within and outside of the scientific community, this link has been acknowledged, and we are realising how big the challenges will be in the coming decades. As land use changes, the physical and biological properties of the land surface change, thereby resulting in changes in the climate system (Houghton *et al.,* 2001). In a reverse manner, climate change also impacts land use patterns, particularly in agriculture. As a result, the agricultural sector takes on a dual role – one, as a source of climate change and two, as a recipient of the impacts of climate change (FAO, 2002).

Climate change impacts are dependent on the systems in which they occur (Van Aalst, 2006). As an example, the author suggests that global warming may boost agricultural productivity at the outset. But, with more climate change, the impacts may become more detrimental and irreversible. Further, the author calls attention to the fact that developed countries compared to developing countries will have greater capacities to adapt to climate change because of differences in skill, technology, institutional capacity, and wealth.

Easterling and Apps (2005) also assert that the success of adaptation to climate change will vary on a regional basis. A study by (Winters *et al.,* 1999; cited in McCarthy *et al.,* 2001) which examined the impacts of climate change on Africa, Asia, and Latin America with the use of a general equilibrium model revealed that income impacts on the most vulnerable groups (poor farmers and urban poor consumers) with adaptation would more likely be negative and in the range of 0 to -10 percent. Those who are least equipped to cope with the challenges of climate change are hit the hardest.

Furthermore, recent results from the Intergovernmental Panel on Climate Change (IPCC, 2007) raise a number of concerns which may have large effects on land use patterns. Crop productivity will decrease in seasonally dry and tropical regions, thus increasing the risk of hunger. Droughts and floods are also expected to increase. This will harm local crop production. Figure 1 shows projections of the global incidence of droughts for IPCC's A1B scenario (see Houghton *et al.*, 2001 for description of scenarios).

Developing regions which are in the worst condition to respond to climate change will be hit hardest. For the purposes of this chapter, we will present a summary of impacts for the areas of the world where most of the hungry and poor reside in:

In Africa, by 2020, between 75 and 250 million people are projected to be exposed to an increase of water stress due to climate change. If coupled by increased demand, this will adversely affect livelihoods – which are mainly in agriculture – and exacerbate water-related problems. New studies confirm that Africa is one of the most vulnerable continents to climate variability and change because of multiple stresses and low adaptive capacity. Some adaptation to current climate variability is taking place; however, this may be insufficient for future changes in climate.

In Asia, freshwater availability, particularly in Central, South, East, and Southeast Asia large river basins, is projected to decrease due to climate change which, along with population growth and increasing demand arising from higher standards of living, could adversely affect more than a billion people by the 2050s. Climate change is projected to impinge on sustainable development of most developing countries of Asia, as it compounds the

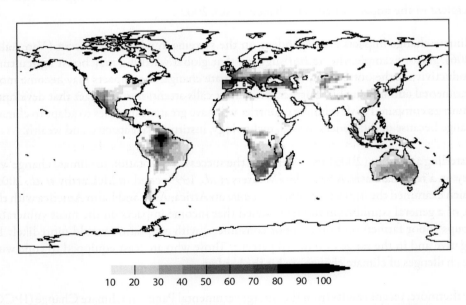

10 20 30 40 50 60 70 80 90 100

Figure 1. Drought projections for IPCC's A1B scenario (adapted from Holdren, 2007).
Note: percentage change in average duration of longest dry period, 30-year average for 2071-2100 compared to that for 1961-1990.

pressures on natural resources and the environment associated with rapid urbanisation, industrialisation, and economic development.

In Latin America, by mid-century, increases in temperature and associated decreases in solid water are projected to lead to gradual replacement of tropical forest by savannah in eastern Amazonia. Semi-Arid vegetation will tend to be replaced by the arid land vegetation. There is a risk of significant biodiversity loss through species extinction in many areas of tropical Latin America. In dryer areas, climate change is expected to lead to salinisation and desertification of agricultural land. Productivity of some important crops is projected to decrease and livestock productivity is expected to decline, with adverse consequences for food security.

What do these mean for future land use and agriculture in developing countries? Drawing from modelling studies conducted by Easterling and Apps, 2005, 'any warming above current temperatures will diminish crop yields in the tropics, while up to 2-3° C of warming in the mid-latitudes may be tolerated by crops, especially if accompanied by increasing precipitation.' The authors suggest that without significant price increases, a rise in global temperature greater than 2.5 °C may surpass the capacity of global food production system capacity to adapt to climate change. Thus, the threat posed by climate change for food production may have significant implications for agricultural land use.

4. Urbanisation

The world's urban population is growing far more rapidly compared to the rural population (see Figure 2). In the next three decades, the world's urban population is projected to increase by nearly 2 billion. In contrast, the world's rural population is projected to see a small decrease from 3.3 billion in 2003 to 3.2 billion in 2030 (Cohen, 2006). This trend has important implications for the use of land.

In the developing world, the rapid urban transformation presents difficult challenges. The increasing numbers of the poor is of major concern (Cohen, 2006). However, Ravallion *et al.*, 2007 point out that 'the urbanisation process has played a quantitatively important, positive, role in overall poverty,' with the exception of Sub-Saharan Africa where the role of urbanisation is less obvious (see Table 2).

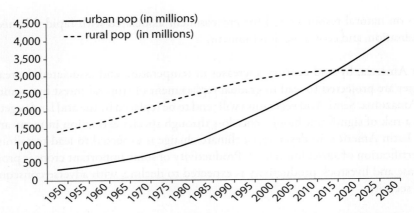

Figure 2. Estimated and projected size of the world's urban and rural populations, 1950-2030 (Cohen, 2006).

Table 2. Number of urban population living under $1.08 a day (PPP).

Region	1993	2002
East Asia & Pacific	436	234
China	342	179
Latin America & the Caribbean	55	65
South Asia	499	542
Sub-Saharan Africa	273	328

Source: Ravallion *et al.*, 2007.

5. The emerging bio-fuel economy

With energy prices at an all time high, a great deal of attention has been given recently to the increasing popularity of bio-fuels. As more agricultural land will be used for bio-fuel production, the question is whether this will have a positive or a negative impact on small scale farmers and people living in the rural areas. Von Braun and Pachauri (2006) argue that while there are several challenges that will need to be met, a greater production of bio-fuels will not necessarily be harmful for the poor, they can become more food secure with the adoption of proper production technologies. They conclude their essay with the following: 'As countries move to strengthen their energy security by increasing their use of bio-fuels, they should also work to ensure poor people and small farmers participate in the creation of a more sustainable bio-energy system. With sound technology and trade policies, win-win solutions – that is, provide outcomes for the poor as well as for energy efficiency – are possible with bio-fuels in developing countries.'

Three things still remain to be seen. First, the extent to which farmers will actually be able to have access to the technologies needed to participate in the production of bio-fuels. Second, with the breakdown of the WTO negotiations at this moment, one wonders how much trust one should be put in the creation of sound trade policies in the near future. Thirdly, high oil prices have encouraged large oil companies to accelerate investments in explorations and there are already indications of new reserves, which - if true - could lead to lower prices making it unattractive for farmers to switch to bio-fuels. Inevitably, there is a great risk for farmers to make the switch to bio-fuel production.

6. Agriculture's role in society: an unfinished debate

At different stages of development, agriculture will play different roles and this hypothesis is presented in Figure 3. In the initial stages of economic development, agriculture will play an important role in poverty reduction.

The fact that agriculture is multifunctional or has several roles to play is in itself not new. However, more than ever before in the developed world is the focus shifting from agriculture as a producer of food to agriculture as a provider of several other functions (Maier and Shobayashi, 2001; Shobayashi, 2003). DeVries (2000) mentions that it emerged on the international stage as early as 1992, at the Rio Earth Summit, which concluded:

'...multifunctional aspects of agriculture, particularly with regard to food security and sustainable development' (Agenda 21, Chapter 14)

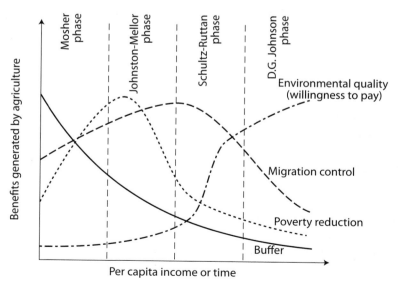

Figure 3. Roles of agriculture in development (Sakuyama, 2007).

The author points out that the benefits of agriculture can cover a very broad spectrum, but in common usage generally include:

- viable rural communities;
- environmental benefits;
- food security;
- landscape values;
- food quality and safety;
- animal welfare.

The competition for land and the notion that agriculture is multifunctional has implications for global land use policy. For instance, rural areas in developed countries have rapidly been changing and there are numerous factors that have contributed towards this, including increasing population densities, agricultural and rural area policies, and higher levels of wealth leading to different demands on rural areas for recreation.

Policy makers continue to have an enormous impact on land use. It goes beyond the scope of this chapter to examine in great detail, the different ways that policy makers have had an impact on land use. The Common Agricultural Policies of the EU have to a large extent shaped many rural areas in the EU. Agriculture continues to be the largest user of rural land in the union and the role of the CAP has become even stronger when then EU expanded to 27 countries.

There is a general acceptance in the union that the two pillars of the CAP are essential to keep rural areas socially, economically and environmentally viable. To give a flavour of recent policy making regarding policies for the rural areas of the union we briefly mention here the recent plans as they were published in the Official Journal of the European Union (2006/144/EC), which states that:

> The guiding principles for the CAP, market and rural development policies, were set out by the European Council of Gotenburg (15 and 16 June 2001). According to its conclusions, strong economic performance must go hand in hand with sustainable use of natural resources and levels of waste, maintaining biodiversity, preserving ecosystems and avoiding desertification. To meet these challenges, the CAP and its future development should, among its objectives, contribute to sustainable development by increasing its emphasis on encouraging healthy, high quality products, environmentally sustainable production methods, including organic production, renewable raw materials and the protection of biodiversity.

Clearly policy makers in the EU are committing themselves strongly in the next 7 years to facilitate rural development in a way that they consider desirable, rather than follow a *laissez-faire* model. To achieve this they have set themselves a number of broad objectives, including:

- improving the competitiveness of the agricultural and forestry sector;
- improving the environment in the countryside, including the preservation of landscapes and their aesthetic value;
- improving the quality of life in rural areas and encouraging diversification of the rural economy, which includes building local capacity for employment and diversification;
- insuring biodiversity and conserving genetic resources.

The gradual shift from focusing on agriculture primarily as a provider of food to its multifunctional role in the rural areas is to some extent controversial. Agriculture has repositioned itself (Van Huylenbroeck and Durand, 2003) and the question has been raised if the multifunctionality of agriculture can actually justify the continuation of domestic subsidies to farmers, while these subsidies may cause trade distorting effects. Potter and Burney (2002) conclude that the multifunctionality argument indeed is genuine and in certain ways a unique feature of European agriculture. The concept is particularly relevant to environmental issues. At the same time, the concept has been criticized for being a smoke screen for the continuation of protectionist agricultural policies (Potter and Burney, 2002).

The concept of multifunctionality of agriculture is generally acknowledged in developing countries. For example, at the conference on Non-Trade Concerns (NTCs) in Agriculture held in Ullensvang, Norway, in July 2000, the paper submitted by Mauritius defined multifunctionality as follows:

> 'The role of agriculture in all countries is not limited to the production of food and fibre. In many cases, it underpins the socio-economic fabric of rural areas and often, that of countries themselves. In a number of developing countries, provides an instrumental link to the development of ecotourism, production of energy, avoidance of the use of fossil fuels, the provision of social amenities and the fostering of research and technology development. Moreover, it has an important role in the protection and preservation of the environment and biodiversity.' (PROSI Magazine, 2000, quoted from DeVries, 2000).

Apart from the OECD concept of multifunctionality, there are other concepts that have been introduced which acknowledge the fact that agriculture has several functions in society. The ROA project, for example, describes the concept of 'Role of Agriculture' which is defined as:

> 'The function that agriculture has or is expected to have in society'

Table 3 compares certain aspects of the Role of Agriculture concept with that of multifunctionality.

Table 3. Comparison between FAO's role of agriculture and OECD's multifunctionality.

	Role of agriculture	Multifunctionality
Scope	Developing countries	Developed countries
Objective	Providing policy guidance to make best advantage of the roles of agriculture in development strategies	Establishing good policy principles to harmonize multifunctionality objectives with trade liberalization
Definition	Indirect functions with externality characteristics that agriculture has or is expected to have in society	Non-commodity outputs with externalities and public goods characteristics that are jointly produced with commodities
Key concepts	Indirect linkages Externalities	Joint production Externalities Public goods

Source: Sukuyama, 2007.

6.1. Food security: the primus inter pares[16]

Despite the debate about the multifunctionality of agriculture, the major developmental concern in many developing countries is achieving food security and reducing poverty. According to the Food and Agriculture Organization (FAO), in 2004 approximately 830 million people across the developing world were undernourished, with the majority located in South Asia (300 million) and Sub-Saharan Africa (213 million) (FAO, 2006). Additionally, despite the Millennium Development Goals, since 1996, the number of undernourished in developing countries is increasing (Figure 4).

In terms of poverty, the developing world – without China – has not seen significant reductions in absolute poverty. In Sub-Saharan Africa, poverty has actually increased (Chen and Ravallion, 2007). Table 3 shows the number of people living on less than $1 a day in the developing world and China. Between 1981 and 2004, the number of the poor in Sub-Saharan Africa increased by 131 million people.

---

[16] 'First among equals.'

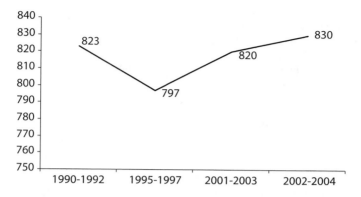

Figure 4. Undernourishment in developing countries in millions (FAO, 2006).

Table 3. Number of people living under $1 a day in millions.

Region	1981	1990	1999	2004
East Asia & Pacific	796	476	277	169
China	634	374	223	128
Latin America & the Caribbean	39	45	49	47
South Asia	455	479	463	446
Sub-Saharan Africa	168	240	296	298

Source: adapted from Chen and Ravallion, 2007.

7. Conclusions

The majority of the poor still live in the rural areas of the world and their livelihoods are expected to change at an ever increasing speed. These changes are being caused by mega drivers of change and we have briefly discussed a selection of them in this chapter. In the developed world, agriculture has re-positioned itself from being solely a provider of food to playing a broader role in the rural areas. However, it clearly would be incorrect to assume that each functional role carries an equal weight. Different functions carry different weights in the development phase of a country and the extent to which this is the case needs to be well understood, lest short term benefits present long term social costs.

Challenges such as climate change, shifting rural-urban populations, the uncertainty associated with rising energy prices and persistent levels of poverty all contribute to the challenges that policy makers face in developing countries. In the context of developing countries, the three simultaneous goals of (a) agriculture production, (b) poverty alleviation, and (c) environmental sustainability have to be achieved in order promote sustainable growth and development. As developed by Vosti and Reardon (1997), these three policy issues present a 'critical triangle,' which entail trade-offs among the three goals. Policymakers therefore have to implement the right mix of policies and find the right institutions and technologies to make the three goals more compatible.

For example, new farming arrangements as well as an information-driven management of the food chain (from production to distribution) have emerged. One negative externality that has resulted from the intensification of agricultural production is environmental degradation, including water and soil contamination by agrochemical residues and agro-industrial wastes. Many developing countries are not well-equipped to deal with these issues. Broader-scale strategies, including technologies and policies are needed for these countries that experience fast global integration (Viglizzo, 2001).

Mega-drivers of change, as described in this chapter, have in the developed world led to an increased recognition of the multifunctional role of agriculture in rural areas, and policies have played an important role in supporting and enhancing different functions. At the same time, in the developing world, the mega-drivers of change have given the food security function greater prominence and made it the *'first among equals.'* As long as the survival of human beings is at stake, the food security function will continue to gain in prominence. It will be necessary to strive for a proper balance of policies, so that the other functions of agriculture are also recognized and supported. If we pursue one goal of the 'critical triangle,' while neglecting the other goals, then we will be putting the livelihoods of future generations at stake. Many rural communities in developing countries have internalized the concept of multifunctionality in their livelihood strategies and it will be important for policy makers to support and harness this approach multifunctionality.

References

Chen, S. and M. Ravallion, 2007. Absolute poverty measures for the developing world, 1981-2004. Development Research Group, World Bank.

Cohen, B., 2006. Urbanization in developing countries: current trends, future projections, and key challenges for sustainability. Technology in Society 28 (2006) 63-80.

DeVries, B., 2000. Multifunctional Agriculture in the International Context: A Review. The Land Stewardship Project.

Dries, L., T. Reardon and J. Swinnen, 2004. The rapid rise of supermarkets in central and eastern Europe: Implications for the agri-food sector and rural development. Development Policy Review 22 (5): 525-556.

Easterling,W. and Apps, M., 2005. Assessing the consequences of climate change for food and forest resources: A view from the IPCC. Climatic Change 70: 165-189

FAO (Food and Agriculture Organization), 2006. State of Food Insecurity in the World. Rome.

Food and Agriculture Organisation (FAO), 2002. The state of food and agriculture. Food and Agriculture Organization, Italy, Rome, 227p.

Grepperud, S., H. Wiig and F. Aune, 1999. Maize trade liberalization vs. fertilizer subsidies in Tanzania: A CGE model analysis with endogenous soil fertility. Statistics Norway discussion paper no. 249, Oslo, Norway.

Guruswamy, M., K. Sharma, J. Mohanty and T. Korah 1999. FDI in India's retail sector more bad than good? Centre for Policy Alternatives, New Delhi, India.

Holdren, J., 2007. Science and technology for sustainable well-being. Presidential Lecture at the Annual Meeting of the American Association for the Advancement of Science, San Francisco, February 15, 2007.

Houghton, J.T., Y. Ding, D.J. Griggs, M. Noguer, P.J. van der Linden, X. Dai, K. Maskell and C.A. Johnson (eds.), 2001. Climate change 2001: The scientific basis. Cambridge University Press, Cambridge, United Kingdom.

IPCC (Intergovernmental Panel on Climate Change), 2007. Climate Change 2007: Impacts, Adaptation and Vulnaribility. Working Group II Contribution to the Intergovermental Panel on Climate Change Fourth Assessment Report. Summary for Policy Makers.

Kok, K., P. Verburg and T. Veldkamp, 2007. Integrated assessment of the land system: The future of land use (Editorial). Land Use Policy 22: 517-520.

Maier, L. and M. Shobayashi, 2001. Multifunctionality: Towards an analytical framework. OECD, Paris, France, 158p.

McCarthy J., O. Canziani, N. Leary, D. Dokken and K. White, 2001. Climate Change 2001: Impacts, adaptation, and vulnerability. Cambridge University Press, Cambridge, United Kingdom.

Mertz, O., R.L. Wadley and A.E. Christensen, 2005. Local Land Use Strategies in a Globalizing World: Subsistence Farming, Cash Crops and Income Diversification. Agricultural Systems 85 (3): 209-215.

O'Brien, K. and R. Leichenko, 2005. Global Environmental Change, Globalization, and Food Systems. In International Human Dimensions Programme (IHDP) on Global Environmental Change Update. IHDP Newsletter, Bonn, Germany.

Potter, C. and J. Burney, 2002. Agricultural multifunctionality in the WTO-legitimate non-trade concern or disguised protectionism?

Ravallion, M., S. Chen and P. Sangraula, 2007. New Evidence on the Urbanization of Global Poverty (April 1, 2007). World Bank Policy Research Working Paper No. 4199 Available at SSRN: http://ssrn.com/abstract=980817 <http://ssrn.com/abstract=980817>.

Reardon, T. and J. Berdegue, 2002. The rapid rise of supermarkets in Latin America: Challenges and opportunities for development. Development Policy Review 20 (4): 371-388.

Sakuyama, T., 2007. The roles of agriculture in development: Findings, lessons and policy implications from a FAO project. Roles of Agriculture Project. Project Brief No. 2. Rome: Food and Agriculture Organization of the United Nations.

Shobayashi, M., 2003. Multifunctionality: the policy implications. OECD, Paris, France, 109p.

Sun, C. and C. Liqiao, 2006. A Review of Watershed Environmental Services in China. IIED Watershed Services Diagnostic Study, desk study report.

Texas A&M University (TAMU), 2000. Impact methods to predict and assess contributions of technology (IMPACT). Texas Agricultural Experiment Station, TAMU, College Station, Texas.

Van Aalst, M., 2006. The impacts of climate change on the risk of natural disasters. Disasters 30 (1): 5-18.

Van Huylenbroeck G. and G. Durand (eds.), 2003. Multifunctional Agriculture: A new paradigm for European agriculture and rural development. Ashgate publishing, Aldershot, United Kingdom, 239p.

Van Meijl, H., T. van Rheenen, A. Tabeau and B. Eickhout, 2006. The impact of different policy environments on agricultural land use in Europe. Agriculture, Ecosystems and Environment 114: 21-38.

Viglizzo, E., 2001. The impact of global changes on the rural environment in eco-regions of the southern cone of South America. In: O.T. Solbrig, R. Paarlberg and F. di Castri (eds.) Globalization and the rural environment. Harvard University Press, Cambridge, Massachusetts.

Von Braun, J., 2004. Small-scale farmers in a liberalized trade environment. Prepared for a seminar on agriculture producers in a liberalized trade environment, Helsinki, Finland.

Vosti, S. and T. Reardon, 1997. Sustainability, growth, and poverty alleviation: a policy and agro-ecological perspective. Johns Hopkins University Press, Baltimore, Maryland, 432p.

Weatherspoon, D. and T. Reardon, 2003. The rise of supermarkets in Africa: Implications for Agri-food Systems and the Rural Poor. Development Policy Review 21 (3): 333-355.

Development strategies for less-favoured areas

Peter Hazell, Ruerd Ruben, Arie Kuyvenhoven and Hans Jansen

Abstract

This chapter explores development strategies for so-called less-favoured areas, or areas 'neglected by man and nature' in the sense that both agro-ecological conditions for agriculture as well as the degree of market development are modest. Several potential avenues for future development are discussed, ranging from technological innovations to institutional change. The diversity of less-favoured areas, however, precludes the formulation of simple one-size-fits-all policies.

1. Introduction

Less-favoured areas (LFAs) account for significant shares of the developing world's total poor and agricultural output, and encompass significant areas of environmentally fragile or valued natural resources. LFAs have typically gained little from past agricultural successes and most are characterized by low levels of agricultural productivity, widespread poverty and environmental degradation. It follows that strategies for managing LFAs areas have an important bearing on possibilities for developing countries to achieve their economic, social and environmental goals, including Millennium Development Goals. But what strategies to adopt, including what role for agriculture, and how much to invest in LFAs compared to more-favoured areas are important policy questions that have generated a lot of debate. This chapter considers these issues in some detail and argues that because of the diversity of situations found in LFAs, strategies need to be tailored to local conditions. It also argues that there are increasing opportunities for successful investment in many LFAs because of recent developments in technology and natural resource management practices, accumulating experience with the kinds of institutional and governance approaches needed for community led development, and a greater willingness on the part of society to pay for some of the environmental services provided by LFAs.

2. What are less-favored areas?

Following Pender and Hazell (2000) less-favoured areas (LFAs) are broadly defined in this chapter to include lands that have been neglected by man as well as nature. They include marginal lands that are of low agricultural potential due to low and uncertain rainfall, poor soils, steep slopes or other biophysical constraints; as well as areas that may have higher development potential but which are presently under-exploited due to poor infrastructure and market access, low population density or other socio-economic constraints. Conceptually they include all the shaded areas in Table 1.

Peter Hazell, Ruerd Ruben, Arie Kuyvenhoven and Hans Jansen

Table 1. Classification of favoured and less-favoured areas.

Access to markets and infrastructure	Agricultural potential	
	High	Low (biophysical constraints)
High	Favoured areas	Marginal areas (LFA)
Low	Remote areas (LFA)	Marginal and remote less-favoured areas

Translating this concept into a definition that can facilitate the identification and characterization of LFAs in practice is necessarily subjective and we adopt a farming systems approach. Table 2 shows six major farming systems that encompass most of the LFAs in developing countries. These systems are defined according to resource use regimes (CGIAR/TAC, 2000) and an assessment of poverty incidence in tropical farming systems by FAO-World Bank (2001). These farming systems cluster into dryland and upland farming systems, and together account for about 40 percent of the agricultural land area and 42 percent of the rural population in the developing world.

Based on associated demographic and ecosystems data (Wood *et al.*, 1999), we estimate that LFAs account for about 1.2 billion (42%) of the total 3 billion people living in rural areas of the developing world, and 360 million (40%) of the 900 million rural poor. They are predominantly concentrated in Asia and Africa.

Our farming systems approach does not explicitly capture the role of poor infrastructure and market access in defining LFAs – these will only be captured to the extent they affect the type of farming system practiced. This may lead to an underestimate of the importance of LFAs because many regions with good agricultural potential but poor infrastructure and market access may not have been included. Recent developments in GIS and spatially referenced data sets enable the overlaying of data on agro-ecological conditions with data on market access, and hence provide a basis for a more complete approach to defining LFAs. Table 3 summarizes recent results obtained along these lines (Sebastian, 2007). Market access is defined in terms of the average time taken to reach the nearest market town, while the agricultural systems are defined primarily on the basis of irrigation, moisture and temperature regimes.

If we take LFAs to include all low potential rainfed areas plus irrigated and good rainfed areas with low market access, then as with the farming systems approach about 40% of the rural population is found to live in LFAs. But the GIS approach also results in about 70% of the total agricultural area falling within LFAs, which is much higher than the 40% obtained

Development economics between markets and institutions

Table 2. Predominant farming systems in less-favoured areas.

Agro-ecological zone	Production system	Share of developing countries'		Main locations
		rural population (%)	agricultural land (%)	
Highlands / upland areas	Perennial / tree crops	3	2	East African highlands, Central America, Andean hillsides, Asian uplands
	Shifting cultivation	2	5	East and Central Africa, Southeast Asia
	Mixed upland systems	24	9	Semi-humid highlands of Southern Africa, Southeast Asia and Central America
Drylands / arid areas	Migratory herders	6	12	Arid areas of Sub-Saharan Africa, Middle East and North Africa, Southeast Asia
	Agro-pastoral	4	8	Semi-arid areas of Sub-Saharan Africa, Middle East and North Africa, Southeast Asia
	Mixed rainfed	3	4	Central and Southern Africa, South Asia, coastal North Africa, Northeast Brazil and Yucatan peninsula in Mexico
Total		42	40	

Source: based on FAO-World Bank (2001).
Note: estimates based on FAO expert judgments (Delphi method).

with the farming systems approach. There are of course important definitional differences between the two approaches but it seems likely that the GIS approach is capturing large areas of grazing and forest areas that are not captured in the farming systems approach.

Table 3. Distribution of rural population and agricultural area in the developing world by type of agricultural system and market access.

	Market accessibility							
	High access		Medium access		Low access		Total	
Rural population								
Type agricultural system	Million persons	%	Million persons	%	Million persons	%	Million persons	%
Irrigated	632	24	491	18	49	2	1,172	44
Good rainfed	197	7	297	11	174	7	668	25
Low potential rainfed	227	8	388	15	217	8	831	31
Developing world total	1,056	40	1,176	44	439	16	2,671	100
Agricultural area								
Type agricultural system	10^3 km^2	%	10^3 km^2	%	10^3 km^2	%	10^3 km^2	%
Irrigated	2,726	6	2,222	5	584	1	5,573	12
Good rainfed	3,501	7	5,986	13	6,489	14	15,976	34
Low-potential rainfed	3,456	7	7,789	17	14,344	30	25,589	54
Developing world total	9,724	21	15,996	34	21,417	45	47,137	100

Source: Sebastian, 2007.

3. Characteristics of less-favoured areas

LFAs encompass a rich array of low-input farming systems, including migratory herding in arid areas; agro-pastoral systems in dryland areas; integrated crop, tree and livestock production in hillside areas; and slash and burn cultivation in forest margins. They cover vast areas of environmentally fragile lands in terms of their soils, vegetation and landscapes. Some, especially upland and forest areas, also protect watersheds, regulate water flows in major river basin systems, sequester large amounts of carbon above and below ground, and are host to a rich array of biodiversity. Most of these environmental benefits are not valued in the market place leading to powerful externality problems.

Population is growing in many LFAs and this is placing considerable pressure on the natural resource base. Until a few decades ago, natural resources were commonly abundant in these areas, and once they were used, farmers could allow resources time to recover through fallows and shifting cultivation. Many of the more fragile lands were not farmed at all or were extensively grazed by nomadic herders. Forests were sparsely settled and provided hunter/ gathering livelihoods for indigenous tribal peoples. Today, many of these lands support moderate to high population densities, providing food, fuelwood, water, and housing. Without adequate increases in land or animal productivity to secure their livelihoods, farmers expand their crop area by reducing needed fallows and clearing new land -- much of which is environmentally fragile and easily degraded -- and add livestock to already over-stocked pastoral areas.

Already many LFAs exhibit some serious environmental problems. Land degradation, which affects some 20 percent of the total land area in developing countries, is heavily concentrated in LFAs. Land degradation is particularly troublesome in hillside and mountain areas because of its impact on downstream watersheds. Soil fertility mining is endemic in many LFAs, especially in areas with poor infrastructure and marketing institutions where use of inorganic fertilizers is uneconomic. In much of Africa, for example, average use of inorganic fertilizers is only 8 kg/ha (compared to 107 kg in all developing countries), and many areas are losing 40-60 kg/ha/year of NPK nutrients (Wood *et al.*, 1999). Water scarcity in many dryland areas is leading to the unsustainable mining of ground water and aquifers. More than 60 percent of deforestation is related to subsistence farming in hillside areas, where declining yields force poor people to rely on shifting cultivation. (FAO, 1997, 2005). Finally, poor people in LFAs are custodians of important agricultural biodiversity, both rare crops and traditional landraces. Yet at same time by expanding their activities (cropping, livestock grazing, wood collection, etc.) into forests, woodlands, wetlands, parks and other environmentally valued sites, they contribute to a more general loss of biodiversity.

Many disadvantaged social groups are concentrated in LFAs (Bird *et al.*, 2001; Cleaver and Schreiber, 1994). These include women and female-headed households facing unequal opportunities for access to land, education, employment and asset ownership, aggravated by male migration leading to the further feminization of agriculture and increases in women's work burden. They also include indigenous peoples whose traditional land rights are threatened by the encroachment of migrants (e.g. forest areas and traditional grazing lands); and forest dwellers in mountain areas that have become deprived from rights of collection.

4. What drives resource degradation in LFAs?

Some resource degradation in rural areas has little if anything to do with agriculture, such as logging, mining, agro-industry and tourism. But farmers are major contributors. And any policy and technology interventions to overcome environmental problems in agriculture need to be based on a proper understanding of why individual resource users manage resources the way they do.

Many factors impinge on private incentives for managing natural resources, including prices, subsidies, interest rates, market access, property rights, technology, and the difficulties of sustaining effective collective action. These are all reasonably well understood drivers, though globalization is changing some of them in new ways. Market forces, for example, are more powerful today, and farmers are under enormous pressure to cut production costs, even if this means using resources in unsustainable ways. But new niche markets for green or organic labelled products are providing new win-win opportunities (good for both agriculture and environment) for farmers who can sell in such markets. Most of these factors can be modified through policy reform and public investment.

Three more difficult issues are poverty, population, and externalities. Poverty is more likely to be an important driver of resource degradation in LFAs with poor quality and fragile soils and high population densities. In most other contexts, poor people control only small shares of the total resources available and hence are relatively minor contributors to their degradation. An important implication is that reducing poverty will often not be enough to reverse resource degradation. However, the poor are typically most affected by resource degradation wherever it occurs since they depend most on common property resources that are widely degraded and because they have the fewest assets and options for coping with degradation.

Population pressure has mixed impacts on resource degradation, depending mainly on the technology options available. In the absence of suitable technology options for intensification, then as Malthus observed in eighteenth century England, population pressure leads to agricultural encroachment into ever-more marginal areas, reducing average yields, degrading resources and worsening poverty. But if suitable technologies exist, population growth can lead to their adoption and sustained improvements in resource conditions and yields. Since many NRM technologies are labour intensive (e.g. terracing or contouring land and building irrigation structures), population growth can actually assist their uptake because it lowers labour costs (Boserup, 1965; Tiffen *et al.*, 1994). When population pressure is combined with high initial levels of poverty and few technology options for boosting productivity, downward spirals of degradation and poverty can arise (Cleaver and Schrieber, 1994). This seems to be happening in some high population density areas of Africa (such as Ethiopia) where farms are now too small to support a family and there are too few non-farm job opportunities.

Externalities crop up when the costs of resource degradation are not fully borne by the individuals causing the problem. For example, removing trees that protect watersheds may be privately beneficial to the individual farmers who do it but can also lead to silting and flooding downstream that are costly to society. The difficulty with externalities is that their avoidance requires cooperation among those who cause the damage and those who are affected, and this may need to involve quite large numbers of people separated in space, time and interests. Intergenerational externalities are even more intractable, arising when farmers use resources today with insufficient regard for the resource heritage they are leaving future generations.

Local institutions set up to manage externalities on a participatory basis often fail because of free rider problems.[17] And while government regulation is an obvious alternative, it has proved difficult in many developing countries. Externalities in agriculture are important in explaining water pollution, groundwater mining, deforestation, over-grazing rangelands, and the loss of biodiversity.

5. Development options

Public interventions to promote more sustainable development pathways are warranted in many LFAs on both poverty and environmental grounds. But what form should these interventions take? Much depends on the type of LFA being targeted and their national economic context, and there is considerable diversity on both counts. Strategic options vary from encouraging additional out migration, promoting income diversification into non-farm activities, increasing recurrent expenditure on safety nets programs, supporting more intensive pathways of agricultural development, and introducing payment schemes for environmental services. Non-agricultural options are generally more viable in transforming and industrializing countries with dynamic non-agricultural sectors, and less viable in poor agrarian countries with stagnant economies (Haggblade *et al.*, 2002).

Given the diversity of conditions found in LFAs and the need to tailor development strategies to local conditions, GIS techniques and spatially referenced data sets are proving useful in defining and mapping different types of LFAs at detailed scales in terms of the basic livelihood options available at community and household levels (Ruben and Pender, 2005; Omamo *et al.*, 2006; Jansen *et al.*, 2006).

6. Agricultural development strategies

Agricultural development in LFAs is constrained to varying degrees by fragile and often sloped and already degraded soils, erratic and often low rainfall, poor market access, and high transport costs. Despite this, what is typically most required is a shift towards more intensive agricultural production systems that can raise land and labour productivity and reduce or reverse the need for further crop area expansion. The challenge is to achieve this while ensuring the sustainable use of resources at local levels and avoiding negative environmental externalities at higher scales.

---

[17] Also known as the prisoner's dilemma problem, the difficulty is that when a group cooperates to regulate an externality, each member has incentive to break the rules as long as the others do not.

Agricultural development strategies for LFAs need to be built on four pillars, each of which presents its own problems and challenges. The first is technology. To a large degree, farmers and local communities have already exploited the available productivity enhancing technologies and NRM practices, often with considerable indigenous adaptation and improvement over the years. But the low productivity levels of most LFAs require major new technology breakthroughs if resource degradation is to be reversed and livelihoods significantly improved. Unlike the Green Revolution, these breakthroughs have to build on improved NRM practices and this is a complex and site specific undertaking.

Second, local communities need to be in the driving seat. This is in part because of the enormous agro-ecological diversity within LFAs and the need to select and adapt technology and NRM investments to fit with local needs and conditions. It is also because most NRM investments require secure long term property rights and collective action for their adoption, both of which depend on strong community based organizations.

Third, the public sector has to provide a supportive policy environment. This must include investments in agricultural R&D and improvements in rural infrastructure and market access so that farmers in LFAs can compete in markets, access key farm inputs and diversify into higher value products. LFAs also face significant weather risk like drought which cannot be adequately managed at local levels.

Fourth, because of the very significant environmental externalities that arise from farming in many LFAs, mechanisms need to be found to overcome these market failures. Recent developments in Payment for Environmental Schemes (PES) look promising. We discuss each of these pillars.

6.1. Improved technologies and NRM practices

After years of relative neglect, LFAs have recently attracted more agricultural R&D attention by a variety of public, NGO and private agencies. Given existing soil and/or water constraints and the high cost of transporting modern inputs into many LFAs, efforts have initially been targeted at improving NRM practices that conserve and efficiently use scarce water, control erosion, and restore soil fertility while using few external inputs like fertilizer. Plant breeding has focused on varieties that are more tolerant of drought and poor soil conditions and that have greater pest and disease resistance rather than on yield responsiveness to high input use. These kinds of technology improvements can lead to significant gains in productivity as well as reverse some types of resource degradation. They can also help create more favourable crop growing conditions in which it becomes profitable to use higher yielding crop varieties and inorganic fertilizers, leading to even more significant long term gains in land and labour productivity.

One promising technology that has already had far ranging impacts in many hillside and agro-pastoral areas in Africa is agroforestry (Leaky and Tchoundjeu, 2001; Otsuka and Place, 2001). Agroforestry can provide multiple benefits to farmers, including useful and marketable tree products like fruits, poles, fuelwood and fodder at relatively low cost and risk, improved soil fertility and greater soil water storage capacity (Wiersum, 1997; Leaky and Tchoundjeu, 2001). In Western Kenya and Eastern Zambia, for example, incorporation of fast growing tree species (*tephrosia, sesbania*) into fallow systems for maize and vegetables restored soil fertility while saving weeding labour (Swinkels and Franzel, 1997). Other examples include the benefits of non-timber forest products (NTFP) in Cameroon where such products almost double farmers' incomes (Leaky & Tchoundjeu, 2001)); drought-resistant trees (pistachio, almond) used for soil moisture upgrading and fodder provision in agro-pastoral regions in Tunisia; living fences used in marginal dryland regions of the Sahel and Central America; and contour farming on acid soils in upland areas in the Philippines (Garrity, 2002).

High nutrient deficits are common in subsistence-oriented rain-fed cropping systems that use almost no fertilizers and are located in more remote areas. Sustained treatments with only inputs of organic matter (green or animal manure, crop residues) are usually not sufficient to halt the decline in yields. Strategies for Integrated Nutrient Management (INM) rely on the combination of appropriate organic and inorganic fertilizer applications, soil and water management practices, and agronomic and soil conservation measures to increase yields and maintain the ecosystem stability of the environment (Vanlauwe *et al.*, 2002).

The impact of soil and water conservation measures on farmers' income varies between different settings. In the Sahel countries, simple and low-cost technologies (earth bunds, vegetation strips, windshields) that retain soil nutrients and reduce erosion can lead to higher (and more stable) yields and income (Reij and Steeds, 2003; De Graaff, 1996). In Northern Africa and in the East African Highlands regions, land management is more focused on soil and water conservation, and agronomic measures (intercropping, agroforestry) have been able to provide win-win solutions with reduced erosion and increased productivity (Shiferaw and Holden, 1999).

In the steep hillsides of the Chiapas region in Mexico, the combination of conservation tillage and crop mulching has increased net returns to land and labour by 13 and 28% respectively (Erenstein, 1999). In the Central American hillsides, average smallholder maize yields were 3-9 times higher after a period of 10-22 years relying on cover crops and green manure (velvet beans) as a method for improving soil fertility (Bunch, 2002). Intensive training and support for these programs has been provided by local NGOs and diffusion took place through a 'farmer to farmer' methodology. Recent studies point, however, to some dis-adoption of green manuring due to plant diseases and difficulties in adapting the approach to changing agro-ecological conditions (Neill and Lee, 2001)

Watershed development can make significant contributions to agricultural productivity and poverty alleviation in LFAs (Kerr *et al.*, 2001). Water harvesting techniques like small dunes and planting pits (*tassa*) have been widely adopted by smallholders in Niger (Hassane *et al.*, 2000). Surface drainage with broad-beds and furrows are widely used in the central highlands of Ethiopia to protect crops from water-logging (Deckers, 2002).

Small-scale farmer-controlled irrigation programs that use simple and low-cost technologies like river diversion, lifting with small (hand or rope) pumps from shallow groundwater or rivers, or seasonal flooding enjoy localized successes in Africa, Central America and the Andes region. In Zimbabwe, low-cost indigenous water management systems for *dambo* gardens managed by individual farmers on land allocated by local communities allow flexible water management, based on shallow wells and water channels between beds. Initial investments ($ 500/ha) are 4 to 20 times lower than conventional gravity irrigation, while returns are twice as high because high-value horticulture crops are grown instead of grains. Dambo gardens are about 10 times more productive than dryland farming, prevent erosion and protect downstream water flows, and can relieve pressure on upland resources.

The intensification of traditional livestock systems is most commonly constrained by the availability of feed (McIntire *et al.*, 1992). Alternatives have been developed to reduce feed and water constraints, based on improved pasture management (e.g. area rotation, silvo-pastoral systems), production of leguminous fodder crops, and the use of crop residues and industrial sub-products (e.g., feed blocks in Northern Africa, cottonseed in West Africa).

Despite the availability of many improved technologies and NRM practices, the experience with their uptake has been mixed.[18] Some improved NRM practices simply do not offer sufficient gains in land and labour productivity to make the investment worthwhile. Many are labour intensive and are incompatible with seasonal labour scarcities, ageing populations and the increasing role of women farmers. Fallow and green manures also keep land out of crop production and composting and manuring compete for scarce organic matter with household energy use, both of which are difficult for many small farms. NRM practices are also knowledge intensive and farmers, especially women farmers, may not have access to appropriate agricultural extension or training. Learning from neighbours turns out not to be very effective for complex NRM practices (Tripp, 2006).

Additionally, unlike technologies based on use of single season inputs like fertilizer and improved seed, investments in improved NRM practices are long term. This requires secure long term property rights over resources. Farmers will be reluctant to plant trees, for example, if they are uncertain of being able to retain possession and reap the eventual rewards. And communities are more likely to make investments in improving common properties like

---

[18] See Tripp (2006) and Ruben and Pender (2004) for useful reviews.

grazing areas and woodlots if they have secure use rights over those resources and can exclude or control outside users (Knox *et al.*, 2002; Kerr *et al.*, 2001).

Many NRM investments also need to be undertaken at landscape levels and this requires cooperation by groups of farmers or even entire communities (Kerr *et al.*, 2001). For example, contouring hillsides to control soil erosion and capture water requires a coordinated investment by all the farmers on the same hillside. Watershed development requires cooperation amongst all the key stakeholders in a watershed, and this may involve one or more entire communities. Collective action is also needed for the successful management of common properties.

Land rights are rarely legally defined and registered in lagging regions. More often they are based on customary tenure systems that are recognized and enforced at community and tribal levels (Migot-Adholla *et al.*, 1991). In much of SSA, for example, land is traditionally owned by the village and the village elders allocate land to farmers and uphold and enforce their use and transfer rights and resolve disputes. The provision of secure property rights thus depends on effective community leadership. Similarly, the ability to organize and sustain collective action is also a key function of local communities. These are important reasons why local community based organizations must lie at the heart of agricultural development in lagging regions.

6.2. Community based approaches to NRM

Community-based approaches are important for facilitating the secure property rights and collective action needed for improving NRM. They can also serve as an important vehicle for managing local externalities and as an intermediary between local people and the project activities of governments, donors and NGOs, helping to inform and adapt investments and policies to local needs and conditions and representing local interests. There has been a veritable explosion in the formation of community based organizations for NRM in recent years, driven largely by NGOs that have become active in many lagging regions. They have also been encouraged by some donors (such as IFAD) as a way to empower the poor, particularly poor women, and to ensure that they participate in new growth opportunities. Some governments have also turned to local communities to take over NRM roles formerly fulfilled – usually very inadequately – by the state, such as the management of forests in India and rangelands in the Middle East and North African (MENA) region.

To be effective for managing NRM, community orgainzations need to include all the key stakeholders involved (Uphoff, 2001). In some cases this may need only involve a group of farmers within the community, as for example in contouring part of the landscape. In other cases it may need to involve the whole village, as in watershed development projects. In some cases they need to embrace other villages too, as in the management of open rangelands which may be shared with other local communities or even distant tribes.

Community approaches require effective local organizations to plan and implement agreed interventions. To avoid elite capture and to be able to resolve local disputes, broad representation is needed in their governance. Some successful community organizations are led by poorer members of the community, often women. The start up and early development of local organizations is typically difficult because of a lack of leadership and technical and administrative skills in the community. Technical training and leadership support from outside agencies (NGOs) have often proved crucial in the early stages of organizational development.

The growing role of community based organizations is proving a problem for some government ministries responsible for agriculture and natural resources. These agencies often do not have the organizational culture and human resources appropriate or necessary to institutionalize and support participative approaches. New specialist structures may need to be created, cutting across disciplines and relevant ministries. Or organizations could be contracted from the private sector and civil society to link central policy and procedures with practice on the ground.

Despite their promise, community approaches on their own are not a panacea. There are often situations where acute resource loss, irreconcilable social conflict, the lack of capacity, or simply the absence of a valid community, require more centralized interventions or at least support from outside agencies. For example, resolution of the conflicting interests between pastoralists and expansionist crop farmers in many dryland areas, or management and control of water resources beyond the immediate watershed, are problems that may demand more than community approaches can deliver. Much remains to be learned about the conditions in which the approach can succeed and be scaled up.

6.3. A supporting policy environment

Since the development potential of most lagging regions is constrained by poor infrastructure and access to markets, the right kinds of public investments can unleash considerable growth. There are many examples of how a new road connection has transformed a lagging village or area by opening up new opportunities for diversifying into higher value agricultural products and non-farm activities. Public interventions are also needed to ensure that farmers have access to long term credit for NRM investments, to reinforce indigenous property rights systems where necessary and to ensure that communities have secure ownership rights over all communally owned resources. Supportive risk management aids are also needed in many drought prone areas, such as simple forms of regional weather insurance or safety net programs (Skees *et al.*, 1999).

The impact of market reform policies on LFAs has been mixed and often detrimental to the poor (Kuyvenhoven, 2004; Chaherli, 1999). Where agricultural prices have gone up as a result of the policy reforms, farmers with net surpluses to sell have gained, but often at the expense of the poor who are net buyers of food. But farmers in LFAs have benefited relatively less from more favourable prices because they have poorer access to markets and must pay substantial transport and marketing costs when purchasing inputs and selling products. Physical access is also a problem, with traders proving reluctant to collect and ship the relatively small quantities of goods produced in remote LFAs or to deliver inputs to their thin regional markets. These problems have been aggravated for LFAs by the withdrawal of input subsidies and public delivery systems (e.g. for fertilizer and credit) as part of the policy reforms. As a result, farmers in LFAs are often at a considerable disadvantage when competing in national or export markets, and may not have the same opportunities as farmers in more accessible areas. LFAs are also more environmentally fragile, and are more subject to degradation arising from land use patterns driven by market prices that do not take adequate account of environmental costs and benefits. The reduced use of fertilizers on smallholder farms and in less-favoured areas as a result of the policy reforms is also contributing to greater soil nutrient mining and threatening future yields.

While countries should adopt market and trade policies that make most sense for the country and agricultural sector as a whole, there needs to be greater awareness of the adverse impact these policies can have in LFAs and on their poor, and complementary policies and investments need to be put in place to buffer or offset potentially adverse outcomes. Key complementary interventions include public investments in infrastructure, human capital and technology to enhance the competitiveness of LFAs, and policies to improve their market access. Reforms might also be better sequenced and phased to give people in LFAs more time to adjust, such as the phased withdrawal of input subsidies and other price supports, and implementation of effective safety net programs to cushion temporary and permanent loss of livelihood options.

Many such interventions have been neglected in the past because of the perception that rates of return to public investments are better in high potential areas, as was indeed true during the early phases of the Green Revolution in Asia and may be true in Africa today. But some public investments (e.g. roads, education, irrigation and some types of R&D) can give competitive rates of return (Fan and Hazell, 2001), indicating possibilities for win-win outcomes in terms of growth and poverty and environmental goals. The public agenda today also places higher priority on poverty and environmental goals (e.g. the MDGs) even where tradeoffs do arise

IFPRI work shows high marginal returns to roads, R&D and education, even in many LFAs. But Fan *et al.* (2000) and Fan and Chan-Kang (2004) show how in India and China the marginal impact of specific measures differs among different regions as well as in time. The right mix of investments can therefore not be determined without reference to the regional context, the extent of past investment and the expected returns at the micro level.

A challenge for policy makers is that many interventions are not win-win for growth and the environment. For example, government attempts to help herders manage droughts and grazing areas in the agro-pastoral systems of the MENA region ended up worsening land and rangeland degradation (Hazell *et al.*, 2001). Moreover, given the large externalities that arise in lagging regions, policies that promote sustainable farming and reduce poverty do not necessarily ensure that environmental degradation does not continue. In the forest margins of the Brazilian Amazon, interventions to help smallholder settlers by improving the profitability and sustainability of their farming system and increasing their access to markets contribute to additional settlement and deforestation (Vosti *et al.*, 2002). There are few technology fixes for these kinds of tradeoffs and solution requires much more effective mechanisms for managing environmental externalities.

6.4. Managing externalities

Many LFAs have landscapes that produce a wide range of valuable environmental services: sequestering carbon, harbouring biodiversity, protecting watersheds, and providing clean water downstream. But in the absence of markets that can compensate farmers for these wider benefits there is an inevitable underinvestment in NRM practices that can generate additional environmental services. Many traditional approaches to increasing environmental services are based on the good will and livelihood interests of farmers and local communities, or seek to regulate what farmers can and cannot do. Neither approach has worked well or can be sustained over time. Occasionally win-win technologies can help fix the problem, but these are few and far between.[19] The bottom line is that farmers need some form of environmental payment if they are to make the NRM investments that society at large desires.

This has been attempted at small scales in the past, for example by providing concessionary loans for NRM investments, using food for work programs for NRM constructions, and supplying key inputs like trees without charge. But these efforts are far too small to address the really big externality problems, especially those with international or global dimensions. Emerging markets or payment schemes for environmental services are an exciting new development that aims to fill this void by compensating farmers and communities for part of the externality benefits of environmental services they provide. Most of the payment schemes attempted so far in developing countries have focused on retaining forest, but interest is growing in agricultural areas. In China, the government launched the Sloping Lands Conversion Program (SLCP) in 1999 which is now paying farmers to plant and conserve trees in watershed protection areas. The program has a target of about 15 million hectares by 2010 and a total budget of about $40 billion (Bennett and Xu, 2006).

---

[19] As shown in a recent CGIAR/SPIA study, more powerful win-win options are elusive (CGIAR, 2006).

If payment schemes are to be used more widely, they will have to overcome several important challenges. One is ensuring that the funding base is sustainable for the long term, linking service users and providers directly. This is easier when there are just one or two large service users with fairly clear actual or potential environmental threats and when the causes and effects between farmers' activities and environmental outcomes are relatively well understood. Small watersheds with a downstream hydropower plant (usually most vulnerable to sedimentation) or a domestic water supplier (affected by contamination and sedimentation) are good candidates. Large basins with multiple users, where downstream impacts are the cumulative impact of myriad upstream uses are poor candidates. Biodiversity conservation is also difficult because of the lack of a stakeholder with strong financial interests.

7. Conclusions

The scale and importance of the poverty and environmental problems of LFAs requires that they receive greater attention in national and global development agenda. Agricultural development will not always be the appropriate strategy, but where it is then new technologies and natural resource management practices are needed that can significantly shift the productivity envelope. Some new and promising technologies have emerged, but widespread adoption of better natural resource management is complicated by their long term nature and the need for collective action. There are also difficult institutional constraints to overcome. One of the more promising recent developments has been local organizations for community-based natural resource management. Learning how to scale them up should be a priority. Also required is a supportive policy environment, especially strengthening property rights, providing long-term credit for natural resource management, policy instruments to help manage climate risks, and additional investments in rural infrastructure.

Many LFAs have landscapes that produce a wide range of valuable environmental services that can be of considerable benefit to society, yet farmers and communities are not rewarded for their efforts in the market, so they do not provide enough of those services. In the absence of more powerful win-win technologies, they need some form of financial compensation. The emergence of new markets and programs for payments for environmental services is a promising new development that should be urgently pursued by the international community.

References

Bennett, M., and J. Xu, 2006. China's Sloping Land Conversion Program: Institutional Innovation or Business as Usual. Paper presented at the Workshop on Payments for Environmental Services-Methods and Design in Developing and Developing Countries. November 2. Titisee, Germany.

Bird, K., D. Hulme, K. Moore and A. Sheperd, 2001. Chronic poverty and remote rural areas. CPRC Working paper no. 13. Manchester: IDPM.

Boserup, E, 1965. The conditions of agricultural growth. New York: Aldine Publishing Company.

Bunch, R, 2002. Increasing productivity through agroecological approaches in Central America: experiences from hillside agriculture. In Agroecological innovations: Increasing food production with participatory development, ed. N. Uphoff. London, Earthscan.

CGIAR/TAC, 2000. CGIAR research priorities for marginal lands. Consultative Group on International Agricultural Research/Technical Advisory Committee. TAC Secretariat, Food and Agriculture Organization of the United Nations, Rome.

Chaherli, N., 1999. Impact of Market Reforms on the Low Rainfall Areas in West Asia and North Africa. In: N. Chaherly, P. Hazell, T. Ngaido, T. Nordblom and P. Oram (eds.), Agricultural Growth, Sustainable Resource Management, and Poverty Alleviation in the Low Rainfall Areas of West Asia and North Africa. Feldafing, Germany: German Foundation for International Development.

Cleaver, K. and G. Schrieber, 1994. Reversing the spiral: The population agriculture, and environment nexus in sub-Saharan Africa. Washington, DC: World Bank.

Deckers, J., 2002. A systems approach to target balanced nutrient management in soilscapes of Sub-Saharan Africa. In: B. Vanlauwe, J. Diels, N. Sanminga and R. Merckx (eds.) Integrated plant nutrient management in Sub-Saharan Africa: From concept to practice. Wallingford: CAB International.

De Graaff, J., 1996. The price of soil erosion. An economic evaluation of soil conservation and watershed development. Mansholt Studies No. 3. Wageningen: Wageningen Agricultural University.

Erenstein, O.C.A., 1999. The economics of soil conservation in developing countries: The case study of crop residue mulching. PhD Thesis. Wageningen, the Netherlands: Wageningen University.

Fan, S. and P. Hazell, 2001. Returns to Public Investments in the Less-favored Areas of India and China. American Journal of Agricultural Economics, 83(5): 1217-1222.

Fan, S. and C. Chan-Kang, 2004. Returns to investment in less-favored areas in developing countries: A synthesis of evidence and implications for Africa. Food Policy (29) 4: 431-444.

Fan, S., P. Hazell and T. Haque, 2000. Targeting public investments by agro-ecological zone to achieve growth and poverty alleviation goals in rural India. Food Policy (25): 411-428.

FAO, 1997. State of the World's Forests 1997. Rome: Food and Agriculture Organization of the United Nations.

FAO, 2005. Global Forest Resource Assessment. Rome, Italy: Food and Agriculture Organization of the United Nations (FAO). Available on line at http://www.fao.org/forestry/site/fra/en/.

FAO/World Bank, 2001. Farming systems and poverty: Improving farmers' livelihoods in a changing world. Washington, D.C.: Food and Agriculture Organization of the United Nations (FAO) and World Bank.

Garrity, D., 2002. Increasing the scope for food production on sloping lands in Asia: Contour farming with natural vegetation strips in the Philippines. In Agroecological innovations: Increasing food production with participatory development, ed. N. Uphoff. London, Earthscan.

Haggblade, S., P. Hazell and T. Reardon, 2002. Strategies for stimulating poverty-alleviating growth in the rural nonfarm land economy in developing countries. Environment and Production Technology, Discussion Paper No. 92. Washington, D.C.: International Food Policy Research Institute.

Hassane, A., P. Martin and C. Reij, 2000. Water harvesting, land rehabilitation amd household food security in Niger. Amsterdam/Rome: VU/IFAD.

Hazell, P., P. Oram and N. Chaherli, 2001. Managing Livestock in Drought-Prone Areas of the Middle East and North Africa: Policy Issues. In: H. Löfgren, (eds.), Food and Agriculture in the Middle East: Research in Middle East Economics, vol. 5. New York: Elsevier Science.

Jansen, H., J. Pender, A. Damon and R. Schipper, 2006. Rural development policies and sustainable and sustainable land use in the hillside areas of Honduras. Research Report 147. IFPRI, Washington DC.

Kerr, J., G. Pangare and V.L. Pangare, 2001. The role of watershed projects in developing rainfed Agriculture in India. Research Report. Washington D.C.: International Food Policy Research Institute.

Knox, A., R. Meinzen-Dick and P. Hazell, 2002. Property Rights, Collective Action, and Technologies for Natural Resource Management: A Conceptual Framework. In: A. Knox, R. Meinzen-Dick and P. Hazell (Eds). Innovation in Natural Resource Management; The Role of Property Rights and Collective Action in Developing Countries.. Johns Hopkins University Press, Baltimore.

Kuyvenhoven, A., 2004. Creating an Enabling Environment: Policy Conditions for Less-Favored Areas. Food Policy 29(4):407–30.

Leaky, R.R.B. and Z. Tchoundjeu, 2001. Diversification of tree crops: Domestication of companion crops for poverty reduction and environmental services. Experimental. Agriculture 37: 270-296.

McIntire, J., D. Bourzat and P. Pingali, 1992. Crop-livestock interaction in Sub-Saharan Africa. Washington, D.C.: World Bank.

Migot-Adholla, S., P.B.R. Hazell, B. Blarel and F. Place, 1991. Indigenous Land Rights Systems in Sub-Saharan Africa: A Constraint on Productivity? World Bank Economic Review, 5(1):155-75.

Neill, S.P. and D.R. Lee, 2001. Explaining the adoption and disadoption of sustainable agriculture: the case of cover crops in Northern Honduras. Economic Development and Cultural Change 49 (4): 793-820.

Omamo, W, X. Diao, S. Wood, J. Chamberlin, L. You, S. Benin, U. Wood-Sichra and A. Tatwangire, 2006. Strategic priorities for agricultural development in eastern and central Africa. Research Report 150. IFPRI, Washington DC.

Otsuka, K. and F. Place, 2001. Land tenure and natural resource management: a comparative study of agrarian communities in Asia and Africa. Baltimore and Washington: The Johns Hopkins University Press and IFPRI.

Pender, J. and P.Hazell (eds.), 2000. Promoting sustainable development in less-favored areas. Focus No. 3 Policy briefs. Washington D.C.: IFPRI.

Reij, C. and D. Steeds, 2003. Success Stories In Africa's Drylands: Supporting Advocates And Answering Skeptics. Amsterdam: Centre for International Cooperation, Amsterdam. Available on line at http://www.drylands-group.org/Articles/803.html.

Ruben, R. and J. Pender, 2005. Rural Diversity and Heterogeneity in Less-favored Areas: The Quest for Policy Targeting. Food Policy 29(4):303–20.

Sebastian, K., 2007. GIS/Spatial Analysis Contribution to 2008 WDR: Technical Notes on Data and Methodologies. Background paper for WDR 2008. World Bank, Washington DC.

Shiferaw, B. and S. Holden, 1999. Soil erosion and smallholders' conservation decisions in the highlands of Ethiopia. World Development (27) 4: 739-752.

Skees, J., P. Hazell and M. Miranda, 1999. New approac hes to crop yield insurance in Developing Countries. EPTD Discussion Paper No. 55. Washington D.C.: International Food Policy Research Institute.

Swinkels, R., and S. Franzel, 1997. Adoption potential of hedgerow intercropping systems in the highlands of western Kenya. Part II. Economic and farmers' evaluation. Experimental Agriculture 33: 211-223.

Tiffen, M., M. Mortimore and F. Gichuki, 1994. More people, less erosion: Environmental recovery in Kenya. Wiley, Chichester, UK.

Tripp, R., 2006. Self-Sufficient Agriculture: Labour and Knowledge in Small-Scale Farming. Earthscan, London.

Uphoff, N., 2001. Balancing Development and Environmental Goals through Community-based Natural Resource Management. In: D. Lee and C. Barrett, (eds.), Tradeoffs or Synergies? Agricultural Intensification, Economic Development and the Environment. Wallingford, U.K.: CAB International.

Vanlauwe, B., J. Diels, N. Sanginga and R. Merckx, 2002. Integrated plant nutrient management in Sub-Saharan Africa: From concept to practice. Wallingford: CAB International with IITA.

Vosti, S., J. Witcover and C. Carpentier, 2002. Agricultural Intensification by Smallholders in the Western Brazilian Amazon; from Deforestation to Sustainable Land Use. Washington, D.C.: International Food Policy Research Institute (IFPRI), Research Report 130. Available on line at http://www.ifpri.org/pubs/abstract/130/rr130.pdf.

Wiersum, K.F, 1997. From natural forest to tree crops, co-domestication of forests and tree species. Netherlands Journal of Agricultural Research 45: 425-438.

Wood, S., K. Sebastian, F. Nachtergaele, D. Nielsen and A. Dai, 1999. Spatial aspects of the design and targeting of agricultural development strategies. Environment and Production Technology, Discussion Paper No. 44. Washington, D.C.: International Food Policy Research Institute.

Sustainable technology adoption on Central American hillsides: the case of minimum tillage in Honduras

Peter Arellanes and David R. Lee

Abstract

Recent years have seen a growth of interest in research on the adoption and diffusion of low-input sustainable agricultural technologies among smallholder agriculturalists in developing countries. This chapter examines the adoption of one such technology, minimum tillage (*labranza minima*), among resource-poor agricultural households in villages in central Honduras. We analyze the determinants of adoption of minimum tillage among a sample of 256 agricultural households. The results show that plots with irrigation, plots farmed by their owners and plots with steeper slopes were more likely candidates for minimum tillage adoption. Farmer household characteristics are not generally found to represent significant influences on adoption. Importantly, household income does not appear to be a determinant of adoption, suggesting that minimum tillage is an appropriate low-input technology for resource-poor households. The results also indicate that previous use of leguminous cover crops and soil amendments (including chemical fertilizers) are also associated with minimum tillage adoption. Results from studies like this are useful in targeting low-input technologies and programs promoting them among the farm household population.

1. Introduction

Sustainable agricultural systems have been characterized as those that can 'indefinitely meet demands for food and fibre at socially acceptable economic and environmental costs' (Crosson, 1992). Producers, researchers and non-governmental organizations (NGOs) have increasingly sought to identify, validate and implement practical farming technologies and methods which meet 'sustainability' criteria, although the challenges of doing so have been great. This has been especially true in developing countries where chronic rural poverty is often closely linked to a rapidly degrading resource base.

In Honduras, a small, mountainous country in Central America which is ranked among the Western Hemisphere's poorest, hillside farmers face many of the same problems found elsewhere in the developing world. Degraded soils and increasing land scarcity have made the traditional slash-and-burn farming system increasingly inefficient. Production increases for the two main staple foods, maize and beans, have historically been below that of population growth, threatening household food security. Modern technologies such as improved varieties and chemical inputs have helped spur yields among some farmers, but these do not prevent erosion nor do many farmers possess the financial resources to use them. With increasing population and decreasing availability of new land to exploit, maintaining adequate fallows has become increasingly difficult and continuous cropping has become commonplace. This has resulted in a 'vicious cycle' of soil degradation, crop yield declines, further pressure on available lands to generate required food supplies, and often, migration out of agriculture entirely.

To address the many constraints faced by resource-poor hillside farmers, development NGOs and other organizations have increasingly promoted limited external input or 'sustainable' agriculture technologies such as conservation tillage and the use of leguminous cover crops. It is widely believed that these low-cost innovations, typically not requiring large capital investments and relatively easy to implement, can help poor farm households become more productive by improving fallow management and increasing yields. The interest in these low-cost innovations among NGOs and development organizations has been significant, and, over the past decade or so, increased attention has been devoted by researchers to formal analysis of the adoption and diffusion of these technologies and systems (Lee, 2005).

This chapter addresses the assessment of the determinants of adoption of two conservation agriculture practices – minimum tillage and green manures/cover crops among a sample of hillside farm households in central Honduras. In addition to providing insights regarding agricultural sustainability on Central American hillsides, the results have implications for more effectively evaluating, targeting and disseminating these technologies in the future.

2. Adoption of sustainable agriculture technologies

Economists have historically devoted great attention to the technology adoption process at both individual farmer and aggregate levels (Feder *et al.*, 1982). At the individual farmer level, considerable work has focused on identifying biophysical, human capital and economic determinants of adoption of modern agricultural innovations such as high-yielding 'Green Revolution' varieties and complementary inputs such as irrigation, fertilizers and pesticides (Gollin *et al.*, 2005). However, while the adoption of low external input technologies has received considerable attention in developed countries (see, for example: Rahm and Huffman, 1984; Smit and Smithers, 1992; and Weersink, *et al.*, 1992), applications of this research in the developing world have, until the past decade, been considerably fewer, especially compared to the importance of these systems among farmers globally.

In recent years, there has been an increase in the analysis of adoption of low-external input and other 'sustainable' technologies and systems in developing countries. As far back as 1990, Anderson and Thampapillai found that a wide variety of factors including land tenure arrangements, access to credit and farmers' risk attitudes influence soil erosion and the rationality of adopting soil conservation practices among farmers. Hwang, *et al.* (1994) also suggested that poor access to credit and lack of secure tenure, as well as low output prices, were limiting the adoption of soil conservation practices among farmers in the Dominican Republic. In the case of the Philippines, farmers recognized the soil-regenerating and erosion-limiting properties of cover crops or 'green manures,' but declined to plant them because of additional labor expense (Fujisaka, 1993). Other studies have, for farmers in various locations, confirmed the influences on technology adoption of factors related to underlying farm characteristics (Polson and Spencer, 1991; Nkonya, *et al.*, 1997; Clay, *et al.*, 1998), economic and labor market factors (Feder, *et al.*, 1992; Caviglia and Kahn, 2001; Neill and Lee, 2001), demographic and human capital variables (Sureshwaran, *et al.*, 1996; Shively, 1997), and social and institutional variables (Foster and Rosenzweig, 1995; Caviglia and Kahn, 2001). Kuyvenhoven and Ruben (2002) provide a thorough review of the basic agro-ecological and management principles guiding farm household decision-making regarding yield-increasing technologies and practices. Lee (2005) provides a comprehensive review of recent research on the adoption of sustainable agriculture technologies and systems.

In the hillside agriculture systems of Central America, the focus of this study, recent research has partially addressed these and similar issues. One particular system -- the maize-*mucuna* (velvetbean) system of the North Coast region of Honduras – has received the attention of several researchers including Ruben *et al.* (1997), Buckles *et al.* (1998), and Neill and Lee (2001). However, the conditions influencing this 'success story' among sustainable agriculture systems are fairly unique and it is difficult to generalize this well-known example to elsewhere in the region. Other research also been devoted to analysis of adoption of specific agricultural technologies and broad-based farming systems and development pathways of hillside agriculturalists (Pender *et al.*, 2001; Jansen *et al.*, 2006)

3. Sustainable agriculture adoption in the Honduran hillsides

In the research reported here, we examine the determinants of farmer household adoption of two specific sustainable agriculture technologies widely used by central Honduran farmers: minimum tillage (*labranza minima*), and cover crops and green manures. A total of 256 farm households (both adopters and non-adopters) farming 388 plots distributed over 16 communities in Central Honduras were interviewed with the goal of identifying the key factors influencing their adoption of several sustainable agriculture practices. These farmers are almost universally smallholders, poor, and have been the target of extension and development efforts by various NGOs in recent years.

Table 1, drawn from the survey, summarizes the various 'sustainable' practices which were identified as practiced by substantial numbers of farmers. These range from the practice of simply avoiding burning their fields (95%) to the construction of stone conservation barriers (10%). Minimum tillage, or *labranza minima,* as interpreted in this region of Honduras includes a set of practices including contour planting, incorporation in the soil of manure and other organic matter, and greater density of maize seed planting. Nearly 30% of the sample households engaged in this set of practices, which has been widely promoted by NGOs. Another practice of particular interest is the planting of leguminous cover crops or green manures to improve fallow management, soil quality and crop yields. This was used

Table 1. Adoption of sustainable technologies and practices in central Honduras.

Technology or practice	% of sample plots where practice was adopted (n=388)
No longer burning	95
Labranza minima *(minimum tillage)*	29
Applying animal manure to maize plots	10
Planting green manure in maize plot	11
Live soil conservation barriers	17
Stone soil conservation barriers	10
Drainage ditches for soil conservation	17

Table 1 does not reflect adoption rates of all farmers in the Cantarranas area since *labranza mínima* adopters were selected using a choice-based sampling strategy.

by only 11% of our sample of households in Central Honduras, but is much more widely used on the North Coast and in other regions of Honduras.

Table 2 shows the variation in maize yields from the sample survey. While yields are quite low overall, the use of *labranza mínima* practices resulted in a roughly 40% increase in yields above those obtained from conventional tillage practices. The use of green manures and cover crops, although confined to a smaller set of adopters, resulted in even higher yields compared to conventional practices. In trying to understand the factors at work in influencing these differential outcomes.

Table 3 shows a comparison of key summary statistics from plots (not households) using traditional maize and bean cultivation and those employing *labranza mínima* practices. It is clear that the latter were generally smaller in size, more commonly irrigated, more likely

Table 2. Yields for Cantarranas maize (quintales/hectare).

Plot type	Mean	Std. dev.	Median
All	16.61	13.40	14.29
Conventional	14.84	11.79	12.87
Labranza minima (minimum tillage)	20.85	15.90	17.14
Green manure	22.80	13.78	22.86

One quintal = 100 lbs.

Table 3. Plot characteristics.

Variable		Conventional plots (n = 275)	*Labranza mínima* plots (n = 113)
Size[c]		0.79 ha	0.42 ha
Irrigated (yes = 1)[b]		12%	35%
Tenure status:	Owned (yes = 1)[b]	65%	89%
	Rented (yes = 1)[b]	27%	11%
	Borrowed (yes = 1)[b]	8%	0%
Land quality:	Good (yes = 1)	39%	32%
	Fair (yes = 1)	55%	61%
	Poor (yes = 1)	6%	6%
Slope:	Flat (<10% slope, yes = 1)[b]	34%	20%
	Medium sloped (10-40%, yes = 1)[a]	52%	63%
	Steeply sloped (>40%, yes = 1)[b]	13%	17%

[a] Difference significant for z-test of binomial proportions at $\alpha = 0.05$.
[b] Difference significant for z-test of binomial proportions at $\alpha = 0.01$.
[c] Difference significant for two-sample t-test at $\alpha = 0.01$.

to be owned, and more typically characterized by medium or steep slopes. Of the 256 households in the survey, 105 practiced *labranza mínima* on some or all of their plots, while 151 households did not. Table 4 shows selected summary statistics for households either doing conventional farming (n = 151) or those employing *labranza minima* practices (n = 105). Heads of households which employed *labranza mínima* technologies were more likely to be literate (77% vs. 68%), had received slightly more schooling (2.8 vs. 2.1 years), were less likely to have been recent in-migrants to the area (11% vs. 20%). The importance of information-related factors is especially evident. Those households adopting *labranza mínima* were far more likely to have been visited by an extensionist (72% vs. 22%) and to have participated in an agricultural field day (15% vs. 6%) in the year preceding the survey. Further details are provided elsewhere (Arellanes and Lee, 2003).

Table 4. Household characteristics.

Variable	Conventional households (n = 151)	*Labranza mínima* households (n = 105)
Age of farmer	45.2 years	44.4 years
Farming experience	24.1 years	24.6 years
Family size	6.6 persons	6.5 persons
Men working in maize production	1.8 persons	1.9 persons
Women working in maize production	0.1 persons	0.1 persons
Reads (yes = 1)	68%	77%
Years of schooling[b]	2.1 years	2.8 years
Migrant (yes = 1)[a]	20%	11%
Participated in agricultural field day in prior year (yes = 1)	6%	15%
Visiting by extensionist in prior year (yes = 1)	22%	72%
Days of agricultural wage labor (prior year; per household)	116.9	90.7
Days of non-agricultural employment (prior year; per household)	29.6	46.8
Used credit for maize production (prior year) (yes = 1)	0%	0%

[a] Difference significant for z-test of binomial proportions at $\alpha = 0.05$.
[b] Difference significant for z-test of binomial proportions at $\alpha = 0.01$.
[c] Difference significant for two-sample t-test at $\alpha = 0.01$.

4. Empirical results

The determinants of adoption of (1) *labranza minima* practices and (2) the use of cover crop/green manures were analyzed econometrically using the cumulative logistic probability function or logit model. This approach assumes that the dichotomous choice of whether or not to adopt the technology on each plot (yes = 1; no = 0) can be represented by a logistic regression (logit) model which explains the probability of adoption (Greene, 1996) as:

$$\text{Probability of adoption} = P(y = 1) = \frac{e\beta_0 + \beta_1 x_i}{1 + e\beta_0 + \beta_1 x_i}$$

The logit transformation of the probability of adoption, $P(y = 1)$ can be represented as

$$\log\left[P(y=1)\,/\,(1-P(y=1))\right]=\beta_0+\beta_1 x_i$$

which gives the logarithm of the 'odds' of technology adoption conditional on the various explanatory variables influencing adoption. The variables which were here hypothesized to influence adoption are summarized in Table 5, as are the expected signs in the adoption equations. Most of the expected signs follow from the previous literature and the authors' familiarity with hillside agriculture in Central Honduras; space constraints do not permit elaboration here.

One aspect of this analysis needing further explanation is that in order to estimate the conditional probability of adopting a subsequent technology, an estimate of the probability of adopting *labranza mínima* (variable Y1PRED) was used as a regressor in the equation for cover crop adoption. Because decisions to adopt several innovations may be simultaneous, using actual observations of adoption or non-adoption of *labranza mínima* may lead to correlation with the error term in subsequently estimated models. To avoid this problem and the inefficient coefficient estimates which result, an instrumental variable approach was followed, where the predicted value from the *labranza mínima* adoption was included as a regressor in the subsequent adoption equation for cover crops/green manures. The underlying hypothesis (confirmed by field observations and discussions with NGO representatives) is that farmers view *labranza mínima* as a 'first step' to the adoption of a wider set of sustainable agriculture practices, and they may be more amenable to recognizing the benefits of and then adopting other technologies.

Table 5. Definition of variables for adoption study.

Variable	Description	Expected sign
Dependent variable:	Sustainable agriculture technology adopted on plot: 1 = yes; 0 = no	N/A
AGE	Age of farm household head (years)	-
EXPER	Farming experience (years)	+
EDUC	Formal education (years)	+
TOTALINC	Annual household income (Lempiras))	?
Y1PRED	Probability of adopting *labranza mínima* (est.)	+
IRRIGATE	Plot has irrigation: 1 = yes; 0 = no	+
OWN	Producer owns plot land: 1 = yes; 0 = no	+
MEDSLOPE	Plot slope 10 – 40%: 1 = yes; 0 = no	+
MUYSLOPE	Plot slope > 40%: 1 = yes; 0 = no	+
QUALITY	Farmer considers land 'fair' or 'poor' for crops: 1 = yes; 0 = no	+

Table 6 shows maximum likelihood estimates of the logistic regression models, estimated odds ratios, measures of goodness-of-fit and changes in probabilities associated with each coefficient. All of the nine estimated coefficients in the *labranza mínima* adoption model exhibit the expected signs and six are significant at the 10% level or better. The coefficient of age of head of household (variable AGE) is negatively associated with adoption, indicating some lack of receptivity of older farmers toward newly introduced technologies. The coefficient measuring the availability of irrigation (IRRIGATE) is positively signed and significant, with adoption (as measured by the odds ratio) more than four times as likely with irrigation available than without. (In fact, lack of water and irrigation potential is widely considered as the key constraint to sustainable agriculture adoption in hillside agriculture in Honduras). Plots which are owned (variable OWN) are also more than four times as likely to employ *labranza mínima* techniques, most likely because the security of land access is necessary to induce farmers to make the necessary investments in their land. Increased adoption is also estimated to be positively associated with plot steepness (variables MEDSLOPE and MUYSLOPE) and negatively with land quality (QUALITY), indicating that this technology is indeed 'appropriate' for hillside farmers, as promoted by NGOs and other development organizations. Farmers are able to reduce risk exposure by trying new techniques on their more marginal lands, typically sloped, relatively less productive parcels (at least initially) adjacent to their residences. Farm household income is not a significant

Table 6. Maximum likelihood estimates and goodness-of fit measures for the Labranza mínima *model.*

Variable	Coefficient	Odds ratio	t-ratio	Change in probability
Constant[a]	-2.7980		-3.993	
AGE[c]	-0.0329	0.72	-1.677	-0.0154
EXPER	0.0277	1.32	1.455	0.0170
EDUC	0.0786	1.08	1.204	0.0004
IRRIGATE[a]	1.5416	4.67	4.944	0.1642
OWN[a]	1.4916	4.44	4.148	0.1557
MEDSLOPE[b]	0.7567	2.13	2.489	0.0575
MUYSLOPE[a]	1.1316	3.10	2.691	0.1015
QUALITY[c]	0.4771	1.61	1.732	0.0320
TOTALINC	0.1404 ($\times 10^{-4}$)	1.01	0.510	0.0007

Hosmer-Lemeshow Test: 3.19 (p = 0.922)[*].
Likelihood Ratio Test: 66.19[a].
(%) Correct Predictions: 283 (75%).
Explanation of significance levels (a)-(c) follows at end of Table 7.

determinant of adoption, contrary to the results of most studies of 'Green Revolution' and related technologies. This appears to confirm that *labranza mínima* techniques are indeed accessible to resource-poor farmers regardless of differences in income levels.

The estimation equation for the cover crops adoption equation (Table 7) shows, as expected, that the coefficient of the instrumental variable representing *labranza mínima* adoption is positive and significant; the presence of *labranza minima* increases the odds of adopting green manures over six times. In fact, both are promoted by NGOs to hillside farmers as low-cost technologies providing multiple benefits. Plot ownership yields a positive and significant coefficient, consistent with the fact that farmers typically view green manuring as a long-term investment in soil improvement, the benefits of which are more likely to be realized with land ownership. Neither farmers' age, experience or income levels are estimated to have a uniquely distinguishable effect on green manure adoption (as distinct from their effects already incorporated in the instrumental variable). Again, in the case of the income variable, this seems to confirm the underlying 'appropriateness' of the technology among a wide variation of resource-poor farmers.

Table 7. Maximum likelihood estimates and goodness-of fit measures for cover crop/green manure adoption.

Variable	Coefficient	Odds ratio	t-ratio	Change in probability
Constant[a]	-3.9949		-3.851	
Y1PRED[c]	1.8603	6.43	1.797	0.0877
AGE	-0.0240	0.98	-0.764	-0.0004
EXPER	0.0178	1.02	0.604	0.0003
OWN[c]	1.2954	3.65	1.872	0.0450
QUALITY[b]	1.0735	2.93	2.381	0.0330
TOTALINC	0.7236 ($\times 10^{-5}$)	1.00	0.233	0.0000

Hosmer-Lemeshow Test: 0.24 (p = 0.889).
Likelihood Ratio Test: 28.07 [a].
(%) Correct Predictions: 335 (89%).
[a] Indicates significance at $\alpha = 0.1$.
[b] Indicates significance at $\alpha = 0.5$.
[c] Indicates significance at $\alpha = 0.10$.

5. Conclusions

Previous research has demonstrated that the determinants of sustainable agriculture adoption in Central America are many fold, including plot, household, economic and policy-based factors (Neill and Lee, 2001). The results of this analysis show that, in this particular case of Central Honduran hillside farm households, adoption of conservation tillage and cover crops practices are influenced primarily by maize plot characteristics, including irrigation, plot ownership, plot slope and perceptions of soil quality. Assured land access in the form of plot ownership, as in prior studies, is shown to be highly important in influencing adoption. The use of leguminous cover crops or green manures was found to depend on farmer adoption of minimum tillage and the factors underlying its adoption. Interestingly, and unlike many studies of adoption of Green Revolution technologies, human capital variables and farm household incomes appear to play a limited role in influencing adoption. In part this is likely due to the simple to understand, low-cost nature of these technologies. Minimum tillage and the use of cover crops and green manures appears to be relatively insensitive to distinctions in farm household heads' age, degree of farming experience, educational levels, and household income.

With the introduction of structural adjustment programs and associated policy and institutional reforms – in the case of Honduras, the introduction of the Agricultural Modernization Law in 1991 – formal agricultural extension programs have been widely reduced, eliminated, or privatized. In this changing policy environment, low-cost technologies and farming practices, appropriate for small-scale, labor-intensive systems, have been aggressively promoted by local non-governmental organizations and international development organizations. Innovative farmer-led extension methods have also been widely employed in Honduras and other Central American countries. It appears that these practices are being adopted among the resource-poor, small-scale, hillside farm households for whom they were intended. The devotion of greater resources to technology dissemination efforts by government and non-governmental organizations would be expected to result in an even greater level of their adoption.

References

Arellanes, P. and D.R. Lee, 2003. The Determinants of Adoption of Sustainable Agriculture Technologies: Evidence from the Hillsides of Honduras. Paper presented at XXV Conference of International Association of Agricultural Economists, Durban, South Africa, August.
Buckles, D., B. Triomphe and G. Saín, 1998. Cover Crops in Hillside Agriculture: Farmer Innovation with Mucuna. IDRC and CIMMYT, Mexico City, Mexico.
Caviglia, J.L. and J.R. Kahn, 2001. Diffusion of Sustainable Agriculture in the Brazilian Rain Forest: A Discrete Choice Analysis. Economic Development and Cultural Change 49: 311-333.

Clay, D., T. Reardon and J. Kangasniemi, 1998. Sustainable Intensification in the Highland Tropics: Rwandan Farmers' Investments in Land Conservation and Soil Fertility. Economic Development and Cultural Change 46: 351-377.

Crosson, P., 1992. Sustainable Food and Fiber Production. Presented at the Annual Meeting of the American Association for the Advancement of Science, Chicago, February 9, 1992.

Feder, G., R.E. Just and D. Zilberman, 1982. Adoption of Agricultural Innovation in Developing Countries: A Survey. World Bank Staff Working Paper No. 542. The World Bank, Washington, D.C.

Feder, G., L. Lau, J. Lin and X. Luo, 1992. The Determinants of Farm Investment and Residential Construction in Post-reform China. Economic Development and Cultural Change 41: 1-26.

Foster, A. and M. Rosenzweig, 1995. Learning By Doing and Learning from Others: Human Capital and Technical Change in Agriculture. Journal of Political Economy 103: 1176-1209.

Fujisaka, S., 1993. A Case of Farmer Adaptation and Adoption of Contour Hedgerows for Soil Conservation. Experimental Agriculture 29: 97-105.

Gollin, D., M. Morris and D. Byerlee, 2005. Technology Adoption in Intensive Post-Green Revolution Systems. American Journal of Agricultural Economics 87: 1310-1316.

Hwang, S., J. Alwang and G.W. Norton, 1994. Soil Conservation Practices and Farm Income in the Dominican Republic. Agricultural Systems 46: 59-77.

Jansen, H.G. P., J. Pender, A. Damon, W. Wielemaker and R. Schipper, 2006. Policies for sustainable development in the hillside areas of Honduras: a quantitative livelihoods approach. Agricultural Economics 34.

Kuyvenhoven, A. and R. Ruben, 2002. Economic Conditions for Sustainable Agricultural Intensification. Ch. 5 in N.T. Uphoff, ed. Agroecological Innovations: Increasing Food Production with Participatory Development. London: Earthscan Publications, Ltd., pp. 58-70.

Lee, D.R., 2005. Agricultural Sustainability and Technology Adoption: Issues and Policies for Developing Countries. American Journal of Agricultural Economics 87: 1325-1334.

Neill, S. and D.R. Lee, 2001. Explaining the Adoption and Disadoption of Sustainable Agriculture: The Case of Cover Crops in Northern Honduras. Economic Development and Cultural Change 49: 793-820.

Nkonya, E., T. Schroeder and D. Norman, 1997. Factors Affecting Adoption of Improved Maize Seed and Fertilizer in Northern Tanzania. Journal of Agricultural Economics 48: 1-12.

Pender, J., S.J. Scherr and G. Durón, 2001. Pathways of Development in the Hillside Areas of Honduras: Causes and Implications for Agricultural Production, Poverty and Sustainable Resource Use. In: D.R. Lee and C.B. Barrett (eds.), Chapter 10: Tradeoffs or Synergies? Agricultural Intensification, Economic Development and the Environment. Wallingford, U.K.: CABI Publishing, pp. 171-195.

Polson, R.A. and D.S.C. Spencer, 1991. The Technology Adoption Process in Subsistence Agriculture: The Case of Cassava in Southwestern Nigeria. Agricultural Systems 36: 65-78.

Rahm, M.R. and W.E. Huffman, 1984. The Adoption of Reduced Tillage: The Role of Human Capital and Other Variables. American Journal of Agricultural Economics 66: 405-413.

Ruben, R., P. van den Berg, M. Siebe van Wijk and N. Heerink, 1997. Evaluación Economica de Sistemas de Producción con Alto y Bajo Uso de Recursos Externos: el Uso de Frijol de Abono en la Agricultura de Ladera. Paper presented at Kellogg Center Workshop, Zamorano, Honduras, February 5-7.

Smit, B. and J. Smithers, 1992. Adoption of Soil Conservation Practices: An Empirical Analysis in Ontario, Canada. Land Degradation and Rehabilitation 3: 1-14.

Sureshwaran, S., S.R. Londhe and P. Frazier, 1996. A Logit Model for Evaluating Farmer Participation in Soil Conservation Programs: Sloping Agricultural Land Technology on Upland Farms in the Philippines. Journal of Sustainable Agriculture 7: 57-69.

Weersink, A., M. Walker, C. Swanton and J. Shaw, 1992. Economic Comparison of Alternative Tillage Systems Under Risk. Canadian Journal of Agricultural Economics 40: 199-217.

Farm household behaviour and agricultural resource degradation: using econometric and simulation approaches for policy analysis

Gideon Kruseman

Abstract

In this chapter we discuss how the relationship between household behaviour and agricultural resource degradation in developing countries can be analysed for adequately addressing policy questions in the field of sustainable agricultural resource management. Policy advice requires a consistent and theoretically sound analytical framework that captures both behavioural and material system relations. Bio-economic programming models and econometric analysis represent two alternative approaches useful for analyzing these aspects. We discuss an econometric analysis of resource degradation in Kenya and a bio-economic model for Ethiopia to demonstrate the possibilities and constraints of both approaches. We argue that bio-economic models are especially useful in determining the effects of different policy instruments for environmental externalities, and econometric analysis provides better insight into the determinants of household behaviour.

1. Introduction

The vast majority of poor people reside in rural areas in developing countries. A substantial number of these households are largely dependent on agriculture and thus their livelihood strategies rely on the use of natural resources (i.e. soil and water). These resources have come under pressure as a result of demographic growth, limited fertility maintenance and changing institutional arrangements that have resulted in more restricted carrying capacities. Since the poorest people tend to reside in areas where the pressure on natural resources is highest, this poses a formidable challenge as regards policies aimed at sustainable development.

A large range of possible technical, economic and institutional solutions have been proposed over the past few decades, but their (usual partial) modes of implementation have not been able to halt the processes of resource degradation or reduce poverty. While it is generally acknowledged that sustainable development hinges on the dimensions of ecological sustainability, economic feasibility and social acceptance, trade-offs occur between these dimensions, and win-win-win situations are more the exception than the rule. In particular, less favoured areas – characterized by unfavourable agro-ecological conditions, relatively high population densities, poor market access and/or institutional arrangements – are apparently not conducive to sustainable development (Pender *et al.*, 1999).

Agricultural potential is low due to fragile and unstable agro-climatic conditions that constrain the quantity and quality of the natural resource base. In many parts of Africa, the limits of carrying capacity are reached at low levels of population density. Poor market access is related to relative isolation, often linked to a scarce physical infrastructure. Institutional constraints refer to the rules of the game that govern the interaction amongst and between individuals and groups (i.e. property rights, division of labour, exchange networks, and informal safety and insurance arrangements).

Linkage between resource degradation and poverty is often labelled as the poverty-environment nexus. In this nexus, resource-use decisions by individual farm households, as well as those made by the communities in which they live, provide the key to understanding how resource degradation and poverty are interlinked. Farmers use soil and water resources to engage in activities that permit them to attain their goals and aspirations. However, they are constrained by the agro-ecological opportunities, the existing social and physical infrastructure, access to markets and services, and by the particular institutional arrangements regarding land use and exchange that are in place.

Providing an adequate analytical framework for understanding the poverty environment nexus represents a major challenge to science. The key interactions between households, communities and the environment are outlined in Figure 1. Major resource-use decisions are taken by agricultural households (1), who depend on a dynamic stock of natural resources (2). The linkages between the natural resource base and the agricultural households (4) are essentially addressed using bio-economic programming approaches. These models consists of a set of biophysical process models (including crop-soil interaction models and crop-growth simulation models), that define the relationship between the natural resource base and household decision making, and *vice versa*.

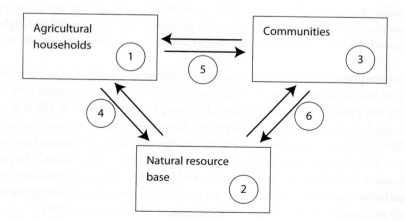

Figure 1. Linkages between households, communities and natural resource base.

The third realm refers to the community (3). The relationships between the community and the households (5) become apparent in local markets, institutions and interactions between households. Understanding these arrangements requires reliance on econometric modelling approaches that capture behavioural motives for resource-use decisions. The last set of interactions relates to the linkages between community-level activities and the natural resource base (6). Further analyses of these linkages can be based either on aggregation (in programming models) or through collective action analysis (in econometric approaches).

In order to improve the lives of these poorest people of the world living in less-favoured areas there is a need for a combination f appropriate technology, an institutional setup that assists households to cope with existing market and government failures, and a set of policy measures that induces adaptive behaviour that leads to both increased household welfare and improved management of the natural resource base (Deininger, 2003).

In the remainder of this chapter two methodological approaches for addressing agricultural households' resource-use decisions are compared with respect to their usefulness in quantifying the relationship between household behaviour and agricultural resource degradation for policy analysis. Using examples from representative studies on two East African countries - Kenya and Ethiopia - we draw some conclusions regarding the relevance, usefulness and specific recommendation domains for each of these modelling approaches.[20]

The organisation of the chapter is as follows. We start with a discussion on the structure of the agricultural household decisions regarding agricultural resources like soil and water. Hereafter, we discuss the econometric approach to agricultural household modelling and how it addresses household choices in relation to soil and water management practice. In the next section, we present optimisation-based simulation models in order to understand the implications of policy change on household choices in relation to soil and water management. We outline the major strengths and weaknesses of each approach drawing on examples from empirical studies in Ethiopia and Kenya. In the last section, we compare both approaches and draw conclusions concerning the possibilities for combining them in a unified analytical framework

2. Agricultural household decisions

Two decades have passed since the publication of the seminal work on agricultural household models by Singh *et al.* (1986). Their analysis unified different strands of thought on agricultural household behaviour into a single conceptual framework that offered possibilities for quantitative analysis. Some important lines of thought that contributed to the concept were the mathematical household theory of Nobel Laureate Gary Becker

---

[20] For a comprehensive review of agricultural household modelling research in relation to the poverty – agricultural resource nexus; see: Kruseman (2000).

(1965), the notion that production and consumption are linked in a livelihood strategy developed by Chayanov in the 1920s (published in English in 1966), and pioneering work on quantitative micro-economic analysis of farm households in developing countries in the 1970s (Barnum and Squire, 1979).

Agricultural household modelling has been widely applied to a wide variety of domains, giving rise to the emergence of two rather different approaches. The first approach used the concept of the agricultural household model to build a theoretically robust econometric model that sheds light on the determinants of agricultural household behaviour. The second approach constructs a simulation model based on agricultural household theory in order to explore the effects of policy, environment, and /or economic incentives on household behaviour. The former approach relies heavily on behavioural assumptions using empirical survey evidence, while the latter approach uses biophysical simulation to address policy questions regarding resource use and degradation, often considering future scenarios regarding actual and potential production options.

Recently, some attempts have been made to better incorporate institutional arrangements into the framework of both types of agricultural household modelling (Holden *et al.*, 2002; Kruseman 2004b). Particular attention is usually given to two types of institutions: (i) tenure arrangements that provide regulated access to land, and (ii) safety nets that permit rural households to generate a minimum income through engagement in public work programs.

The basic structure of a household model, following Singh *et al.* (1986), usually assumed that households maximize their consumption utility, subject to budget and resource constraints, a production function, and a set of constraints capturing context-specific conditions. The model contains a number of balance equations capturing the household level resource equilibrium. There are essentially two areas of choice: (i) the utility function capturing the consumptive behaviour of household decisions, and (ii) the production function capturing the productive side of decision-making procedures.

For practical purposes, the basic model can be expanded in different directions. In order to better capture biophysical processes related to the degradation and/or conservation of agricultural resources, and the underlying institutional arrangements, the production side can be expanded by including resource stocks, investments and linkages between resource quality and productivity. In order to improve insight into the decisions regarding preferences for consumption of goods, services and leisure, household characteristics and institutional arrangements can be included. This also permits the analysis of utility maximization in an inter-temporal context, since the goals and aspirations of the household often evolve over time as household size and composition changes. Since the livelihood strategies of poor rural households are often geared towards managing uncertainty in the face of socio-economic, institutional or agro-climatic risk, a further specification of risk preferences and implications of risk for technology choice and resource allocations can be included in both model specifications.

3. Econometric analysis

Econometric analysis of agricultural household models is frequently used to understand the exogenous determinants of endogenous choice. Resource-use decisions are thus explained from a set of fixed farm and household characteristics. Reduced-form equations are considered an appropriate procedure for dealing with interactions, since the coefficients capture the sum of both direct and indirect effects. This approach, however, deserves careful use. If we derive the first-order conditions for the agricultural household model and combine and collapse the resulting equations, the end result is a system where the dependent variables consist of the choice variables of the household (production structure, investment, consumption, resource allocation) with all the exogenous factors on the right-hand side(household characteristics, farm characteristics, institutional characteristics). However, we should be very careful regarding causality and attribution in the inter-temporal context.

Another problem is that the key equations that capture decision-making processes include quasi-fixed inputs and determinants of wealth. Past decisions that lead to current wealth and already available soil and water conservation (SWC) structures are – in principle - based on the same set of independent variables. Total cumulative investment and wealth are thus part of a set of inter-temporal dependent variables that need to be addressed in a dynamic modelling framework.

We provide an example from a recent study in Kenya where the effects of tenure security on soil and water conservation are addressed (Kabubo-Mariara *et al.*, 2006). In Kenya, conservation and sustainable use of the environment and natural resources represent an integral part of national planning and poverty-reduction efforts. However, weak environmental management practices are a major impediment to agricultural productivity growth.

The basic econometric model that is estimated refers to the probability of investment in soil and water conservation at plot level. This decision is related to farm household characteristics (like age, family size, farm area, education, gender) and village characteristics (location, access), including tenure arrangements influencing access to arable land and the asset base of the household. In this approach, soil quality (degradation) is an endogenous variable, subject to choice within the farm household resource allocation process. Further expansion of the model permits the inclusion of parameters for farmers' awareness of resource degradation, the implications of having access to extension services, and the impact of membership of special interest groups or community organizations promoting sustainable land management practices.

Field research focussed on the determinants of farmers' willingness to invest in improved Natural Resource Management (NRM) practices. Only investments taken up during the last five years are considered. The econometric approach estimates the likelihood of NRM investments as a function of farm, household and village characteristics. Past investments can be considered for both dichotomous choice investments and continuous variable investments. The independent variables capture both the direct and indirect effects, while the residual terms capture the awareness for NRM practices, the willingness to listen to extension messages, as well as the presence and membership of NRM groups.

Estimation results from this study show that tenure security is positively correlated with the adoption of investments in SWC, as well as the intensity of adoption. This supports the literature, which argues that tenure security favours long-term conservation investments more than short-term investments. Previous studies also showed that farmers with long-term tenure security are more likely to invest in costly and more durable conservation measures (Kabubo-Mariara *et al.*, 2006; Gebremedhin and Swinton, 2003). Farmers' attitudes towards the adoption of new technologies also depend on the profitability and riskiness of available technologies.

The impact of tenure security factors is captured by five types of variables based on a factor analysis of a large number of factors that influence land tenure security: owned plots (often inherited); family land that can be sold or bequeathed with or without permission; land registered in family name; rented-out land and the right to rent out land without permission. The first three derived variables represent the strongest rights to land and, consequently, the coefficient of these variables exhibits positive and significant coefficients for most adoption models, thus confirming the importance of tenure security in the adoption of SWC investments. The differences in coefficients for permanent and seasonal investments support the finding that tenure security favours long-term conservation investments above short-term investments (Gebremedhin and Swinton, 2003). The negative and mostly significant coefficients of the latter two variables show that weak security of tenure will discourage SWC investment, particularly investments such as grass strips, terracing and tree planting.

The lessons that can be learned from this study, where agricultural household theory and the theoretical implications of reduced form estimation are applied, are the following. Firstly, it is possible to identify the determinants of household choice variables even in the face of limited information regarding exogenous parameters. Secondly, the econometric analysis could capture the state of soil quality to some extent, but certainly not the effects of household choice on environmental externalities. Thirdly, to a very limited extent it is possible to disentangle the direct and indirect effects of exogenous parameters (land tenancy) on household choice variables (for instance, changes in cropping pattern or input use). Using such an econometric model it proved to be possible to address the determinants of household choice, but only if there is sufficient variation in the exogenous and endogenous variables, e.g. between farmers, land rights and SWV practices.

A similar study from the Tigray region in Northern Ethiopia using an econometric model based on the same principles, also allows for the inclusion of variation due to weather conditions (Fitsum Hagos *et al.*, 2007). It was possible to analyse the relationship between investments in water harvesting for supplementary irrigation in periods of water stress, identifying its determinants and the derived income effects from water harvesting investments (comparing cases with and without water harvesting). Since rainfall patterns are erratic and in some years rainfall is concentrated in a short period of time, the effective growing season for agriculture becomes too short to yield an adequate level of production. Therefore, cross-section data may be easily over- or underestimating potential returns.

As part of the campaign for land rehabilitation, the Ethiopian authorities supported investment in irrigation infrastructure throughout the 1990s. This infrastructure consisted of river diversions and micro-dams within a major rural development program called Sustainable Agriculture and Environmental Rehabilitation in Tigray (SAERT).[21] Areas with low rainfall dependent upon rain-fed agriculture are prone to drought risk, and the rationale for water-harvesting investment is that it diminishes this risk. Where stream diversion is possible it has often been applied, but only limited effects in terms of water utilization can be expected. In the past few years there have been experiments with new options for water-harvesting using ponds. These are deep dug-out ponds in a corner of a farmer's field, lined with plastic or cement and meant to harvest excess precipitation that would otherwise run off the fields into the streams. Contrary to traditional irrigation programs, the water is primarily used for supplementary irrigation at otherwise rain-fed fields. In addition, there has been a major reliance on shallow wells in areas where ground water levels are not deep.

The year that the data were collected proved to be exceptionally good in terms of rainfall; therefore, the added value of water harvesting could not be straightforwardly determined (Fitsum Hagos *et al.*, 2007). To examine whether households are benefiting from ponds and whether that leads to increased per capita consumption expenditures, means separation tests could be used. The sample test reflects the mean difference in per capita expenditure for households with and without ponds. It was also examined whether households have different input use, since increased access to water may encourage the use of purchased farm inputs, particularly fertilizer. The results showed that households with ponds have significantly higher fertilizer use. This finding is further explored using a bio-economic model framework as presented in the next section.

---

[21] River diversion is an ancient technique practised in many hilly and mountainous semi-arid parts of the world. Although evidence is scarce, there are indications that this type of irrigation predates 500 BC in Tigray (Tesfay *et al.*, 2000).

Gideon Kruseman

4. Optimization approaches

In optimisation or programming models, structural relationships between resource-use practices and production and welfare outcomes are explicitly modelled. Obviously this poses several problems, since the specification of household-specific production functions and even more so the household-specific utility function is difficult. Nevertheless, there are examples where this was successfully done (Kruseman, 2000; Kruseman, 2004a; Holden *et al.*, 2002) This approach is commonly known as integrated bio-economic household modelling, because it includes a more or less detailed specification of the underlying bio-physical processes.

As an example we rely on a study conducted in Northern Ethiopia which contrasts nicely with the econometric model on water harvesting mentioned in the previous section. The main methodological difference is that the previous study was an econometric analysis of the determinants of investments and its impact on income, while the model simulations in the alternative study are concerned with simulations of the effects of policy incentives on soil degradation (Kruseman, 2004b). Both studies deal with an area with low rainfall and predominantly rain-fed agriculture, extreme poverty due to poor market access, low agricultural potential and high population density.

The bio-economic model used in the latter study was developed to better understand the long-term dynamics of poverty and soil degradation. This requires explicit modelling of biophysical processes as part of the production function. Given the fragile resource base, problems of moisture stress and limited plant macro-nutrients tend to coincide. There are few available studies addressing both these aspects in a dynamic context. The bio-economic modelling framework provides opportunities to understand these interactions and feedback processes, using a detailed sub-model of soil-plant interactions.

The building blocks for the biophysical model are readily available. It is based on the general analytical framework developed by De Wit (1992), linking nutrient applications to crop production via nutrient uptake. Using this analytical framework as a starting point, soil processes relevant for crop production are adapted with the model of soil-plant interaction developed by Wolf and Van Keulen (1989). Finally, a relatively simple crop growth model is used, based on nutrient uptake responses with a plateau based on water-limited yields. Production options are derived making use of existing Technical Coefficient Generators (Hengsdijk, 2003), but reversing the processing order: instead of finding appropriate input levels with a target yield, the relevant yield with respect to a certain level of input application is calculated. The framework of soil-plant interactions is thus expanded to take into account non-equilibrium production systems under the conditions of moisture and nutrient stress, in line with the current conditions of many farm households in the region. Although parameterisation of such a model is difficult, it is considered the only way to understand the processes of soil degradation and to identify a wide set of appropriate strategies to combat soil degradation.

The second complexity involved in agricultural household simulation models concerns the specification of the utility function. For subsistence households, utility is related primarily to food consumption. Although leisure is often included in the utility function, in the present study it is omitted, since the vast majority of the population is Orthodox Christian and are obliged to celebrate numerous religious festival days where it is forbidden to engage in activities related to agricultural production. These days account on average for more than 25% of each month. This implies that we can treat leisure as an exogenous parameter.

Consequently, consumption of food and services remains a key feature of utility. As argued elsewhere, a non-separable model should be used when there are numerous market imperfections, and thus an indirect utility function is not appropriate (Kruseman, 2001). Procedures for estimating direct utility functions are based on revealed preferences and the theoretical knowledge of the properties of such a direct utility function (Kruseman, 2000). Using consumption data from a household survey that includes both quantities and values of different consumed commodities, it is possible to estimate a direct utility function using non-linear econometrics. A key aspect of this utility function is that data permit the calculation of substitution elasticities between different cereals of high and low preference, given different relative prices.[22] Finally, differences in risk were also considered, accounting for different levels and distributions of rainfall with corresponding probabilities of occurrence with their impact on expected utility (see: Kruseman, 2000 for a comparative study in West Africa). The data used for the estimation was obtained from a household survey (conducted by IFPRI/ILRI/MU in the late 1990s).

The simulation results from the bio-economic household model refer to typical households. Compared to econometric estimates, this has a serious drawback, since the effects of household idiosyncratic aspects are not taken into account. Only general circumstances facing all households in the community under consideration are included. In addition, institutional arrangements are only considered by differentiating the subjective discount rate with regard to investments in SWC measures. These subjective discount rates tend to reflect differences in perceived tenure security. To incorporate dynamics, the utility function is given an inter-temporal character with a subjective discount rate of 25 percent (default value). The model thus uses the net present value of discounted utility over a finite time horizon. The rate of time preference has been estimated at between 25 and 50% in the case of Ethiopia (Holden *et al.*, 2002).

---

[22] A negative exponential utility function is assumed and formulated in such a way (with the possibility of estimated non-zero minimum consumption levels) that the functional form becomes flexible and is able to capture a wide range of consumption patterns. The model has been successfully tested for robustness (Kruseman, 2001).

The outcome of the simulation model indicates that investment in ponds is often prohibitive for small farmers (see Kruseman 2004a and 2004b for further details).The popularity of the ponds is due to their public promotion within the framework of food-for-work programs that provide a safety net for food-deficit small holders. We can analyse the potential benefits of pond construction using a bio-economic modelling framework (Kruseman, 2004a). Figure 2 shows that ponds stabilize household income (albeit at a low level), but also contribute to improved soil quality parameters (which more than compensate for the land loss related to pond construction). This implies that (larger) farmers with less risk and a longer time horizon will be more inclined to reap the benefits of this improved technology.

The programming model elicits options for technology choice, given farm household preferences and perceptions. Ponds may reduce risk by controlling the occurrence of water shortfalls in the growing season. On the one hand, pond construction leads to loss of land that can no longer be cultivated. This has a small effect on income and hardly any effect on soil quality. On the other hand, the use of ponds reduces vulnerability to the worst state of nature and improves expectations regarding water availability under rain-fed conditions, yielding a positive impact on both income and sustainability. The model results indicate that income levels increased by 10 percent and soil nutrient depletion is halted or even reversed.

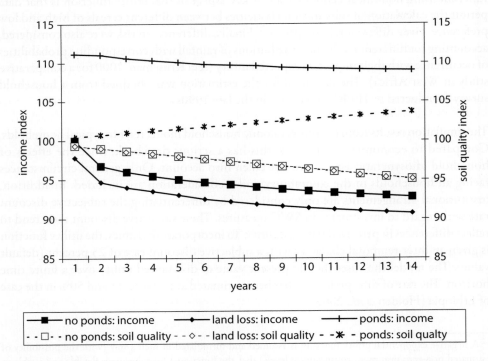

Figure 2. Long-term effects on income and soil quality with and without ponds (Kruseman, 2004b).

The latter effect is mainly due to fertilizer becoming an economically feasible input at lower levels of moisture stress risk.

The model can also be used to explore alternative institutional arrangements. Here, only food-for-work safety bets were considered. The majority of households in the study area depend on *food-for-work* and *cash-for-work* programs to supplement their income. These programs are important to enable coping with the risk related to inter-annual rainfall variation. Other options refer to public investments for reducing transaction costs related to poor market access and subsequent weak exchange networks. This may even be the result of food-for-work projects that improve road infrastructure facilities. Further improvements are, of course, possible by specifying better and different land tenure regimes (now only captured by differences in stipulated discount rates).

5. Discussion and conclusions

In the preceding sections we have discussed the advantages and limitations of agricultural household modelling for a policy analysis of the management of the agro-environmental resource base. Although there is a general consensus on the importance of joint poverty alleviation and sound management of the environment to permit sustainable intensification of rural areas and to allow viable household livelihood strategies, there remains a lack of consensus on how to achieve these goals. This lack of consensus is partly due to a poor understanding of what is commonly called the poverty – environment nexus.

We argued that an adequate understanding of the motives for household resource allocation decisions is central to providing useful policy advice. The generally accepted notion is that households use the available knowledge (regarding technology) and resource endowments to attain their goals and aspirations. In the process of pursuing these goals, farmers are constrained by agro-climatic and environmental conditions, the availability of public infrastructure, markets and services, and last but not least the prevailing institutional arrangements. The relative importance of these different aspects of decision making is subject to discussion.

Agricultural household models provide a mathematical representation of the decision-making procedures. Such models need to capture the logic of the household decision-making process, considering both behavioural attitudes (i.e. preferences, motivations, risk attitudes, etc) and the material conditions and options for adjusting resource use. Both aspects are critical, but different modelling approaches give more or less attention to each of them.

Mathematical programming models offer a detailed insight into the material options for technology change, but usually require a large number of interlinked equations where consistency has to be guaranteed by using a set of balance equations (for production and consumption, investments and savings, and labour use and labour demand, respectively). The adequate estimation of the structural equations for the utility function (driving consumption decisions) and for the production functions (and related institutional constraints) still poses major methodological problems.

The second issue is the discussion on whether to use a structural model or a reduced-form model. This is the central theme of this chapter. As we argued in this chapter the reduced-form equations are especially useful for econometric estimation of the exogenous determinants of household decision-making outcomes. They capture the direct and indirect effects of these determinants and circumvent the pitfalls of having to formulate correctly the problematic structural equations of the household model.

The econometric approach has the advantage that it can rely on reduced-form equations, but these do not provide insight into the mechanisms by which underlying exogenous determinants lead to specific household decisions. The most important implication is that extreme caution has to be maintained when extrapolating results beyond the data set analysed and beyond the scope and range of the parameters addressed. We could argue that reduced-form models are suitable for finding the determinants of the current situation, but have limited use in exploring future policy options.

The advantage of simulation models is that they have more potential for exploring policy options that fall outside the range of the current situation. Another advantage is that they offer better scope for combining data from different sources and as such include the bio-physical processes in order to calculate the environmental externalities. There are, however, some serious drawbacks to the use of simulation models.. Bio-economic models are not very suitable models when there is a great deal of heterogeneity amongst the target population in terms of resource endowments, institutional arrangements and household characteristics in general, since they commonly work with stylised (typical) households.

The combined use of both approaches is, however, quite possible. If household choice variables are measured and information is available about the household exogenous parameters that play an important role in resource allocation decisions, it is possible in principle to use the outcomes of econometric models to calculate the environmental externalities using a bio-economic optimisation procedure. Simulated data related to resource degradation processes proved to be generally fairly consistent with household survey data (Kruseman and Van Keulen, 2001).

For policy analysis, it depends on the type of questions posed as to which analytical framework is most appropriate. Econometric analysis based on household surveys offers great scope for identifying the determinants of household behaviour, while circumventing the pitfalls of having to fully specify the underlying material relationships regarding utility and production functions. Econometric analysis also offer more possibilities for addressing the effects of institutional arrangements (property rights, exchange networks) on resource-use decisions.

On the other hand, simulation models can explore a wider range of technological options and explicitly take into account shocks and risk-coping strategies of households for dealing with shocks. Another advantage of bio-economic models is that they provide some answers to questions at the interface between economics and biophysical sciences, for example, policy options for simultaneously improving welfare and sustainability.

The case studies presented from Kenya and Ethiopia illustrate that combining results from both econometric analyses and bio-economic modelling may provide parts of the puzzle. Given the different types of data requirements, it will not always be feasible to apply both approaches. Econometric models permit the unravelling of the effects of some institutional arrangements on household choice, as demonstrated in the case study on land tenure security in Kenya. Policy changes that induce new (and combined) institutional arrangements appropriate for improved land management can be derived using optimisation models in the Ethiopia study. One of the major challenges to the development economics profession in the near future is therefore the design of an integrated set of methodologies incorporating both econometric analysis and policy simulations with limited data needs. Analyzing the merits and shortcomings of both approaches is a first step in that direction.

References

Barnum, H.N. and L. Squire, 1979. A model of an agricultural household: theory and evidence. Johns Hopkins University Press, Baltimore, USA.

Becker, G.S., 1965. A theory of the allocation of time. Economic journal 75: 493:517.

Chayanov, A.V., 1966. The theory of peasant economy, Homewood, USA, 317 p.

Deininger, K., 2003. Land policies for growth and poverty reduction. World Bank Policy research report, Oxford University Press and World Bank, Washington, D.C., USA, 239 p.

De Wit, C.T., 1992. Resource use efficiency in agriculture. Agricultural Systems 40: 125-151.

Gebremedhin, B. and S. Swinton, 2003. Investment in soil conservation in northern Ethiopia: the role of land tenure security and public programs. Agricultural Economics 29:69-84.

Fitsum Hagos, F., G. Kruseman, Z. Abreha, V. Linderhof, A. Mulugeta and G. GebreSamuel, 2007. Impact of small scale water harvesting on household poverty: Evidence from northern Ethiopia. PREM Working Paper 07-01. IVM-VU (Institute for Environmental Studies, Vrije Universiteit), Amsterdam, the Netherlands, 14p.

Hengsdijk, H., 2003. An introduction to the Technical Coefficient Generator for land use systems in Tigray. Report No. 241. Plant Research International, Wageningen, the Netherlands, 34p.

Holden, S., B. Shiferaw and J. Pender, 2002. Policy analysis for sustainable land management and food security: a bio-economic model with market imperfections. Paper presented at the conference on Policies for sustainable land management in the East African Highlands. UNECA, Addis Abeba, April 24-26, 2002

Kabubo-Mariara, J., G. Mwabu and P. Kimuyu, 2006. Farm Productivity and Poverty in Kenya: The Effect of Soil Conservation. Journal of Food, Agriculture and Environment 4: 291-297.

Gideon Kruseman

Kruseman, G., 2000. Bio-economic household modelling for agricultural intensification. Wageningen University. Mansholt Studies No. 20, Wageningen, the Netherlands.

Kruseman, G., 2001. Household technology choice and sustainable land use. In: N.B.M. Heerink, H. Van Keulen and M. Kuiper (eds.) Economic policy reforms and sustainable land use in LDCs: Recent advances in quantitative analysis. Physica-Verlag, Heidelberg, pp. 135-150.

Kruseman G. and H. Van Keulen, 2001. Soil degradation and production function analysis in a bio economic context. In: N.B.M. Heerink, H. van Keulen and M. Kuiper (eds.) Economic policy reforms and sustainable land use in LDCs: Recent advances in quantitative analysis. Physica-Verlag, Heidelberg. pp. 21-48.

Kruseman, G., 2004a. Long-term bio-economic modelling for the assessment of policies for sustainable land use in less-favoured areas' (mimeograph), Washington D.C.: IFPRI., 79p.

Kruseman, G., 2004b. Village level incentives and disincentives for investment in sustainable land management. (mimeograph), Washington D.C.: IFPRI., 45p.

Pender, J., F. Place and S. Ehui, 1999. Strategies for sustainable agricultural development in the East African highlands. Environmental and production technology division working paper 41. Washington DC: IFPRI, 86p.

Singh, I., L. Squire and J. Strauss, 1986. Agricultural household models: extensions, applications, and policy. The Johns Hopkins University Press, Baltimore, USA, 335p.

Tesfay, G., M. Haile, B. Gebremedhin, J. Pender and E. Yazew, 2000. Small scale irrigation in Tigray: management and institutional considerations. In: M.A. Jabber, J. Pender and S.K. Ehui (eds.) Policies for sustainable land management in the highlands of Ethiopia. summary of papers and proceedings of a seminar held at ILRI, Addis Ababa, ethiopia, 22-23 May, 2000. ILRI working paper no. 30. Addis Ababa: ILRI, pp. 27-30.

Wolf, J. and H. van Keulen, 1989. Modeling long-term crop response to fertilizer and soil nitrogen. Plant Soil 120: 11-38.

Strategies for enhancing food security

Ethics, stability and hunger

Per Pinstrup-Andersen

Abstract

Failure to prioritize action to eliminate poverty, hunger and malnutrition presents a serious ethical, economic and stability problem for both developing and developed countries. This chapter discusses the ethical aspects associated with the failure to convert plans, strategies, and goals, including the Millennium Development Goals and the World Food Summit goals, into action. It then proceeds to analyze whether poverty, hunger and lack of social justice contributes to armed conflict and terrorism. Preliminary results from on-going research show that the probability of armed conflict is much larger in the poorest than in the less poor developing countries and suggest that many failed or fragile states are caught in a poverty-conflict trap. The chapter concluded that development strategies for low-income developing countries should pursue the dual goal of eliminating poverty, hunger and inequality and reducing the probability of armed conflict and terrorism and suggests five priorities for action. While the main responsibility for such action rests with each developing country, high-income countries should provide appropriate support in the form of fair trade policies and developing assistance. Since most countries in armed conflict are failed or fragile states with widespread poverty and hunger, development assistance aimed at achieving the dual goal should focus on such countries.

1. Introduction

Is the elimination of hunger an ethical obligation of societies and would the fulfilling of such obligation also improve international stability by reducing armed conflict in low-income countries? Gandhi stated many years ago that 'To a people famishing and idle, the only acceptable form in which god can dare appear is work and promise of food and wages' (Fischer, 1983). More than 40 years ago, President John F. Kennedy stated at the World Food Congress, 'We have the means, we have the capacity to eliminate hunger from the face of the earth in our lifetime. We need only the will.'

Promises have also been plentiful. Consider, for example, Henry Kissinger's statement at the World Food Conference in 1974: 'Within a decade no man, woman or child will go to bed hungry.' Thirty-three years later, more than 800 million people go to bed hungry. Thousands of similar statements have been made throughout the years. Based on the rhetoric, it appears that policy makers believe that poverty and hunger should be eradicated or at least severely reduced. Governments of virtually every country have agreed to a number of targets and goals to do just that. Recently, in connection with the Millennium Summit, countries agreed to a target of reducing by half the percentage of the world population that suffers from poverty and hunger by the year 2015. The commitment to the target was reaffirmed by all UN members at a UN Summit in 2005. Similarly, at the World Food Summit in 1996, the governments of 186 countries agreed to a goal of reducing by half the number of hungry people between 1990 and 2015. Five years later, at a follow-up conference, the same countries reconfirmed the goal. Unfortunately, action has not followed rhetoric. Hunger affects more than 800 million people and that number has been constant during the last fifteen years.

Malnutrition affects about one-fourth of all preschool children in developing countries – or about 135 million. While the number of underweight preschool children is falling, the number of wasted preschool children has not changed during the last ten years. More than 10 million children die every year; the overwhelming reason is poverty, with more than half dying from hunger and nutrition related factors.

The difference in preschool child mortality rates between high and low-income countries is astounding and is increasing. Thus, the rate in Sub-Saharan Africa was 18 times the rate in high-income countries in 1990, increasing to 28 times in 2003. In South Asia, the rate was 13 times higher than in high-income countries in 1990, increasing to 16 times in 2003 (UNDP, 2005). In 2003, 179 of every 1,000 live-born children in Sub-Saharan Africa died before their fifth birthday. The average mortality rate covers considerable variation among countries and between poor and rich within each country. Thus, more than one-fourth of all live-born children in Sierra Leone die before their fifth birthday, while less than one in 10 die in Lesotho and Eritrea (UNDP, 2005). In Mali, 25% of the children born into the poorest sector of the population die before they reach five years of age, compared to 15% for those born into the richest quintile (UNDP, 2005). Six of every 1,000 children born in high-income countries die before the age of five years.

Lack of access to clean water, primary health care, education, and a variety of goods and services that the non-poor take for granted is commonplace among the poor and disenfranchised. In spite of the targets agreed upon at the World Food Summit and the Millennium Summit, the number of hungry people is roughly the same today as it was in 1990, and with business as usual, extrapolations based on existing data indicate that the number will stay roughly constant between now and 2015. More than half of the developing countries failed to reduce the number of hungry people during the 1990s and if the most successful country, China, is excluded from the statistics, the number of hungry people has risen since 1990 – a trend that appears likely to continue until 2015 (Pinstrup-Andersen, 2002). Several international conferences, including the World Food Conference in 1974, the Nutrition Conference in 1990, the World Food Summit in 1996, and the Millennium Summit in 2000, have specified hunger alleviation goals to be achieved within specified time periods. None have been achieved.

The lack of action is accompanied by much rhetoric, plans, targets, and promises. The most cynical of these promises and agreements is undoubtedly the United Nations' Declaration that freedom from hunger is a basic human right. It is cynical because it is not empowered with any means to enforce it. Other recent efforts such as the poverty reduction strategy programs (PRSPs) and the program proposed by the Millennium Development Goals Task Force stress the development of more plans and strategies. So far, these plans have resulted in very little action.

This chapter argues that failure to take appropriate action is unethical and contributes to national and international instability.

2. What ethical standards?

But what ethical perspectives guide the action or lack of action to alleviate hunger? Four such perspectives are common: utilitarianism, deontological ethics, virtue ethics, and human-rights ethics. Utilitarianism is part of the consequentialist school. What counts is the outcome or the consequences of the action taken, not the intent of the action. Whether an action is right or wrong depends on the resulting utility. Deontological ethics, on the other hand, refers to the intent rather than the outcome. The intent may relate to 'doing the right thing' in the context of meeting the motives of society, groups, or individuals, and respecting rules and rights. Virtue ethics also relate to the intent, but differ from deontological ethics by focusing on the actor rather than the action. Doing the right thing makes the actor virtuous. Human-rights ethics prioritize action towards the assurance of specified basic human rights such as freedom from hunger. As further discussed below, the four ethical standards may, but need not, conflict.

The goal of maximizing utility says nothing about whether the gainers deserve the utility, unless, of course, a social-welfare function with assigned differential weights for various population groups is designed and used. For example, a poor person or a child may be more deserving than a non-poor adult. Therefore, a greater weight could be assigned to utility captured by such more-deserving groups. By assigning different weights to income gains obtained by various population groups, we can compare the social net benefit or utility from each policy option. But how do we derive the appropriate weights? Most people would probably agree that a $1 net gain for the poor carries a higher social benefit than a $1 net gain for the rich; but how much higher?

Assigning weights to the expected utility of current versus future generations and to humans versus animals present particularly challenging questions. Whose welfare should be maximized and what is a reasonable trade-off between current and future generations, between rich and poor, between two groups of poor people, between people and animals, and between utility maximization and intrinsic values of the natural environment?

Do we explicitly or implicitly subscribe to deontological ethics when deciding among trade-offs? In other words, are there things we do not want to do, even if they are legal and would enhance utility? Yes, of course. Utility maximization takes place within legal as well as ethical constraints. But the ethical trade-offs are sometimes so socially sensitive that we do not make them explicit. For example, most people would probably agree with Singer (1993, p. 229), at least in principle, that 'If it is in our power to prevent something very bad from happening, without thereby sacrificing anything of comparable moral significance, we ought to do it.' But such agreement is not represented in policy action to prevent the annual death of 10 million preschool children of hunger, malnutrition, infectious diseases, and other preventable causes (UNICEF, 2002).

A recent study (Bryce *et al.*, 2005) concluded that 6 million of the 10 million child deaths could be avoided by known interventions in each of 42 countries at a recurrent annual cost of $5.1 billion, or $1.23 per person living in those countries. Could the richest 10% of the populations of those countries afford to pay $12.30 annually without sacrificing anything of comparable moral or material significance? Of course they could. Do they? No. There are several possible explanations for the failure to apply Singer's principle; the obvious one is that we may agree with the principle when it comes to people we know or feel we have something in common with, while the suffering people far away from us – physically and mentally – are of less significance.

But is there a significant ethical difference between the failure to take action which would save children from dying and actively killing children? From a purely utilitarian perspective, the outcome, which is what counts, is the same. From a deontological ethical perspective, however, I can argue that in the case of not taking action, I have done nothing wrong. But killing children is clearly wrong. We can hide behind this argument as long as the children who die are an unseen mass (Singer, 1993). But what if my neighbour's child is starving and I could take action to make the child well without sacrificing anything of comparable moral significance? I would presumably subscribe to virtue ethics and do the right thing. Of course, if the well-being of my neighbour's child enters into my utility, utilitarian ethics might lead me to the same action. But what if it is not my neighbour's child but the children of the country for which I am a policy-maker or a food policy analyst?

I agree with Wisman (2003, p. 431), when he states that 'Our potential for happiness is limited by the extent to which we find misery and injustice around us.' The question is whether 'around us' refers to the family, the local community, the country, or the world. Wisman's argument may also explain why governments of some countries forcefully remove poor and starving people from the streets and other public places. In personal correspondence, Thorbecke (2004) suggested the hypothesis or conjecture that the welfare weights assigned to individuals are inversely related to the square of the distance between the one who assigns the weights and the one to whom it is assigned; where distance could refer to geographical, income level or some other measure of social status.

Is it genocide when millions of children die because of neglect by the State? Is failure by governments to take action, as promised on various occasions, a crime against humanity? Not according to the United Nations and the International Criminal Court. The terms 'genocide' and 'crime against humanity' apply only when certain acts are committed. Failure to act to save lives is not covered, even when States have the means to act (United Nations, 1948; International Criminal Court, 1999).

Socially sensitive trade-offs are common in economic policy related to the food system. These trade-offs are frequently implicit in the action taken but are not stated. For example, should policies focus on improving the well-being of the poorest children or households even if total utility or social welfare would be maximized by focusing on the less poor? Should public funds be spent on reducing hunger among current children or on prenatal care among pregnant women, which might increase birth weight and prevent many more children from dying in the future?

But in cases where utility is the appropriate indicator, are incomes a good measure? What about personal freedom and security? China and Vietnam have experienced very impressive reductions in absolute poverty and food insecurity, but personal freedom and democracy are still very limited. In India, personal freedom and democracy is at a high level, but the reductions in poverty and food insecurity are limited.

A 'right-to-food' ethics is presented in the Universal Declaration of Human Rights: 'Everybody has the right to a standard of living adequate for the health and well-being of himself and his family, including food, clothing, housing and medical care...' (United Nations, 2004). The International Covenant on Economic, Social and Cultural Rights further states that 'The States Parties to the present Covenant, recognizing the fundamental right of everyone to be free from hunger ... will take appropriate steps to ensure the realization of this right...' (Eide, 2002).

Although the United Nations declarations related to food security and freedom from hunger as a human right give the appearance of being legally binding, they are not. Instead of 'right', a more appropriate term would be privilege or charity, because it is neither an enforceable right nor a claim. It is a privilege that can be granted or, more commonly, not granted. Contrary to enforced rights, privileges or charity can be viewed as an attempt to escape from justice (Kolm, 2004); an attempt, such as a transfer program, that can be removed at any time. One could argue that it is a moral right that would be enforced only if those with enforcement power share the underlying moral or ethical standard. Poverty transfer programs are frequently of a short-term nature. Privileges and charity tend to treat symptoms, while efforts to achieve social justice would aim to change underlying causes.

Moral imperatives such as a right to freedom from hunger can be very expensive both in terms of fiscal and economic costs and are very difficult to enforce. Tweeten (2003, p.

19) argues that 'No nation has sufficient resources to honor simultaneously the right to food, shelter, clothing, education and health care for all.' I would argue that few nations can afford a human resource base that lacks any of these basic necessities because of the resulting low labour productivity. Failure to invest in the human resource is not just an ethical problem; it is one of the main reasons for poverty and lack of economic growth in many developing countries – something that nations can ill afford. The widespread recognition that investment in human resources not only helps people out of poverty but also contributes to national economic growth removes a potential conflict between ethics and economics. Another potential conflict is gradually disappearing, as recent evidence which shows that increasing relative inequality is not a precondition for rapid economic growth is accepted (Thorbecke, 2004).

There is no penalty associated with either the failure to protect the right to food or the failure to achieve the goals that countries agreed to as part of international conferences. In fact, accountability is all but absent at both national and international levels. One could argue that the ethical standards and principles to which policy-makers pretend to subscribe are valid for rhetoric but not for action. Although 'The human right to adequate food is recognized in several instruments under international law' (United Nations, 1999, p. 1), nobody seems to enforce international law.

Lack of seriousness in the fight against hunger is also illustrated by international agreements and institutions. We have legally binding international agreements on trade but not hunger. Under the umbrella of the World Trade Organization, one country can take legal action against another if the latter breaks the rules. But trade policy that can be demonstrated to cause poverty, hunger, and child malnutrition and death, such as the current agricultural policies in the EU and US, is legal. Furthermore, international agencies such as FAO and UNICEF that are concerned with poverty, hunger, and the well-being of children, have no enforcement power over governments.

3. Society versus household ethics

One of the factors that would make government enforcement of the individual's right to freedom from hunger difficult is that the individual usually resides in a household, and therefore is subject to household decision rules. Government attempts to help individuals achieve their rights may be modified and possibly nullified by household decisions regarding intra-household allocation of food and other resources that run counter to an absolute priority on achieving freedom from hunger for all household members before any resources are allocated to other needs or wants. Similarly, when transfers are made in kind, e.g. food aid or specific food commodities, households are likely to make counter adjustments in the overall consumption basket. Thus, good intentions by the public sector and failure to take into account household decision processes and goals, may add transaction costs without adding benefits.

Low-income households are likely to have several priorities competing for scarce resources. Whether the household decisions are made by a dictatorial household head or through some bargaining process, the hungry and malnourished are usually those in the weakest bargaining position, i.e., preschool children, pregnant and lactating women.

It is difficult to distinguish between needs and wants. Attempts by government or society to assist households in meeting what society considers basic needs may run counter to the desires of the head of the household, who may use his or her decision-making power to convert public interventions aimed at meeting needs to meet expenses that are not considered basic needs by society. Using food stamps to purchase cigarettes or, if that is not possible, matching the value of the food stamps with equal cuts in food purchases from other income sources, is a case in point. Such conflict between society's attempt to deliver on virtue or deontological ethics and the household head's attempt to maximize his or his household's utility is typical of transfer programs aimed at improving the nutritional status of preschool children, pregnant women, or other household members perceived to be disadvantaged (Pinstrup-Andersen, 1993).

4. Ethics and stability goals: are they compatible

According to President George W. Bush, 'A world where some live in comfort and plenty while half of the human race lives on less than $2 a day is neither just nor stable' (The White House, 2001). There is mounting evidence showing a causal link between poverty, hunger, and unequal income distribution on the one hand, and national instability on the other (Messer *et al.*, 2001; Messer and Cohen, 2004; Messer *et al.*, 1998; Pinstrup-Andersen, 2002, 2003). While the empirical evidence of a causal link between poverty and international terrorism is weaker than the evidence of a causal link between poverty and national conflict, it is clear that at least some who support or undertake international instability and terrorism, find their moral justification in the existence of poverty, hunger, and unnecessary human suffering.

Ethnic and religious tension, colonial history, improper drawing of postcolonial country borders, as well as political rivalries are the usual purported reasons for armed conflict. Yet, while these factors clearly contribute to and facilitate armed conflicts, the underlying causes may be poverty, perceived unfairness, and hopelessness. For example, while blamed on ethnic conflict, the mass killings in Rwanda and the spillovers to neighbouring countries were in large measure a result of rapid soil degradation, hunger, and lack of access to productive land, water, and employment (Pinstrup-Andersen, 1998). Ethnic differences facilitated the killing but did not bring it about.

Thus, while a number of factors may cause or contribute to armed conflict and terrorism, there is a growing body of evidence that poverty and follow-on conditions such as hunger, malnutrition, and hopelessness are the major underlying causes. Thus, Collier *et al.* (2003)

conclude that 'the key root cause of conflict is the failure of economic development. Countries with low, stagnate, and unequally distributed per capita income that have remained dependent on primary commodities for their exports, face dangerously high risks of prolonged conflict. Collier and Hoeffler (2002) found a significant relationship between the probability of conflict and per capita gross domestic product (GDP). In the absence of economic development, neither good political institutions, nor ethnic and religious homogeneity, nor high military spending provide significant defenses against large-scale violence' (p. 53).

Nafziger and Auvinen (1997) agree and conclude that 'tension ripen into violent conflict especially where economic conditions deteriorate and people face subsistence crises. Hunger causes conflict when people feel they have nothing more to lose and so are willing to fight for resources, political power, and cultural respect. Slow growth in food production may be a source of violent conflict'. A high level United Nations (UN) Panel on Threats, Challenges and Change concludes that 'terrorism flourishes in environments of despair, humiliation, poverty, political oppression, extremism and human rights abuse; it also flourishes in context of regional conflict and foreign occupation; and it profits from weak state capacity to maintain law and order' (UNDPI, 2004, p. 47). The lack of capacity to maintain law and order may in fact translate weak states into failed states as in the case of Somalia, Liberia, and the Democratic Republic of the Congo. Such weak or failed states provide, as illustrated by the case of Afghanistan, protection for international terrorism and weak or failed states may well pose a greater danger to the interests of Organisation for Economic Co-operation and Development (OECD) countries than strong states. At the same time, strong authoritarian states may supply the terrorists which will operate in weak states as exemplified by a large number of terrorists coming from Saudi Arabia. The Center for International Development and Conflict Management at the University of Maryland concludes that 'most terrorists come from low-income authoritarian countries in conflict' (Marshall and Gurr, 2005).

Political leaders seem to agree that there is a strong causal link between poverty and terrorism. Here are quotes from some of the statements made by such leaders at the Development Summit in Monterrey, Mexico in 2002 (BBC News, 2002): 'There is a need for more development to help stamp out extreme poverty as a motivation for terrorism.' 'Poverty in all its forms is the greatest single threat to peace, security, democracy, human rights and the environment.' Several speakers supported the notion that defeating poverty would eliminate a major driving force behind international terrorism and that poverty causes violence. Thus the Peruvian President stated that 'to speak of development is to speak also of a strong and determined fight against terrorism' and the President of the UN General Assembly called the world's poorest countries 'the breeding grounds for violence and despair'.

In a television interview in November 2004, the British Chancellor of the Exchequer, Gordon Brown, is quoted as having said 'poverty is a breeding ground for discontent' and that 'frustration and a lack of hope will drive instability'. In another interview, the World

Bank President James Wolfenson suggested that 'without dealing with the question of poverty, there can't be any peace'.

However, armed conflict and terrorism are caused by a multitude of factors and some researchers conclude that poverty and hunger may not be major contributors. Thus, Lancaster (2003) concludes that 'poverty does not produce terrorists' and 'eliminating poverty is not likely to eliminate terrorism'. She further concludes that 'in some cases there does appear to be an indirect relationship between poverty and the poor governance that can lead to civil violence and state collapse' and that 'reducing poverty and improving education, health, and economic well-being of a population may, all things being equal, lead to better governance over time and fewer opportunities for terrorists or criminal elements to operate in these countries'. Krueger and Malečková (2003) suggest that the connection between poverty and terrorism is indirect and probably quite weak. Schmid (2003) argues that antiterrorism policies should be based on good governance, democracy, rule of law, and social justice. He argues that the correlation between poverty and terrorism is much weaker than the correlation between rule of law and terrorism.

In virtually all of the poorest countries, where the probability of conflict is highest, the population is predominantly rural and depends heavily on agriculture. Although urban poverty is growing, 75% of the world's poor people still live in rural areas of developing countries. It therefore seems eminently reasonable to focus a strategy to alleviate poverty and conflict on rural areas. Many are caught in a poverty trap, failing to generate enough income from farming and other rural activities to escape the human suffering expressed in high levels of hunger, malnutrition, and child mortality. Such situations are a breeding ground for desperation, discontent, and conflict. Although some countries, notably China, are paying close attention to the simmering discontent in rural areas and local outbreaks of violence and trying to remedy the situation, governments of most low-income countries still do not appear to see the quiet crisis in rural areas as a source of instability and a threat to their legitimacy.

Rural poverty is closely associated with access to natural resources such as land and water and conflict over such access is common. Water conflicts may be within national borders or spill over into neighbouring countries.

5. Preliminary findings from ongoing research

Sixty-seven of a total of 141 developing countries for which data are available, experienced armed conflict during parts or all of the period 1990-2003. A direct comparison between the countries with conflict and those without found strong correlation between each of a set of social variables and conflict. Thus, it appears that countries in conflict have lower incomes, less economic growth, lower levels of education and more serious health and nutrition problems. In view of the well-documented negative effects of conflict, such differences would be expected. In other words, conflict caused poverty. An alternative explanation would be that countries with low incomes, less economic growth, low levels of education, and serious health problems – in other words poor countries – are more prone to armed conflict. This would imply that poverty caused conflict. In reality, poverty and conflict are likely to reinforce each other, leaving countries in a poverty-conflict trap.

In order to test the hypothesis that poverty causes conflict, countries that entered into armed conflict after 1993 were compared to countries without conflict using data from the period prior to the outbreak of the conflict. The income level of the conflict countries prior to the conflict was less than half the income level for the countries not in conflict and the conflict countries experienced negative economic growth prior to the conflict as opposed to a 2% growth for the countries not in conflict. The poverty rate in conflict countries was twice that for the countries not in conflict, the poverty gap was more than twice, all the food and nutrition indicators were worse, child mortality rates were two and a half times higher, a lower percentage of the population in the conflict countries had access to safe drinking water, and the investment in agricultural research was only one-third that of countries not in conflict. Thus, since the indicators refer to the periods prior to outbreak of conflicts, it seems safe to conclude that poverty and related factors, such as hunger and malnutrition, increase the probability of armed conflict.

Poor and hungry people are not terrorists, but people with no hope and nothing to lose except their lives and those of their children provide not only a perceived moral justification for terrorism, but also the anger and hatred that drive it. They are susceptible to terrorist appeals. As stated by Jim Wallis in a conversation with President George W. Bush, 'Mr. President, if we don't devote our energy, our focus and our time on also overcoming global poverty and desperation, we will lose not only the war on poverty, but we'll lose the war on terrorism' (New York Times Magazine, 17 October 2004, p. 50). In the same article, Mr. Wallis is quoted as saying, 'Unless we drain the swamp of injustice in which the mosquitoes of terrorism breed, we'll never defeat the threat of terrorism.' The Nobel Peace Prize winner, Norman Borlaug, puts it this way: 'We cannot build world peace on empty stomachs' (World Food Prize Symposium, Des Moines, Iowa, 14 October 2004).

Failure to deal effectively with perceived and real social injustice will render current military efforts ineffective in dealing with the threat of terrorism. No society – national or international – will be secure when material inequalities and material deprivations are as extreme as they are now. We must try to understand the frustration, hopelessness, and anger of the many millions of people who are poor, hungry, and without opportunities to escape the human misery they are in. We must then tailor our efforts to assure a socially just, stable, and secure world accordingly. This is our ethical obligation and it is in our own self-interest from both an economic and a stability perspective.

The elimination of hunger is a sound economic objective for the non-poor to pursue. For those who subscribe to virtue or deontological ethics, this is clear. Elimination of human suffering is simply the right thing to do, irrespective of whether it adds to one's own utility.

Irrespective of the ethical motive, investment in the human resource through food security, education, and primary health care has been shown to have a very high economic payoff, not only for the target individuals but also for the non-poor in the societies to which they belong and internationally, through mutually beneficial trade expansion and improved national and international stability. Alleviation of hunger is both an ethical imperative and a matter of enlightened self-interest for the non-poor. Failure by the non-poor in developing countries and by rich countries to prioritize the eradication of hunger is morally wrong and a danger to their security (Singer, 2002). The enlightened self-interest is articulated by the United Nations (2001) as cited in Singer (2002, p. 7) as follows:

> In the global village, someone else's poverty very soon becomes one's own problem: of lack of markets for one's products, illegal immigration, pollution, contagious disease, insecurity, fanaticism, terrorism.

Given the strong ethical, humanitarian, economic, and security reasons for solving hunger problems, it is puzzling that decision-makers have chosen not to do so. One explanation for the lack of action to commit existing resources to help people out of hunger and related hopelessness is the lack of a shared vision of the moral rights and duties of the various population groups. Another might be that those who have the power to choose either do not expect to benefit from solving the hunger problems or they do not believe they will be harmed by a continuation of the status quo. Given certain social and political constructs they may, of course, be right – at least in the short run.

Some may have a very short time horizon for making choices, and therefore seek immediate results through the application of force instead of the more sustainable but longer-term solution of removing the underlying causes of conflict, namely poverty, hunger, and hopelessness. Furthermore, short time horizons are likely to imply high internal discount rates, making future income gains from investments in human resources less attractive than the immediate extraction of economic surplus through discrimination, exploitation, and force. The short time horizon, together with massive attention by the news media, may also explain why acute hunger problems resulting from natural or human-made disasters are more likely to be solved than the longer-term hunger problem that requires sustainable attention over a long period of time. Lack of understanding of economic relationships and the mistaken belief that the economy is a zero-sum game, in which a dollar gained by the poor is a dollar lost by the non-poor, may be other explanations for the lack of action to seek social justice.

6. Required action

Efforts to achieve food security for all, or at least meet the World Food Summit goal within a reasonable timeframe, will require action on five fronts: first, developing countries should design and enforce policies that help the poor gain access to employment, productive resources and markets. Second, they should significantly expand investments in public goods, such as in education, health, infrastructure and research. Third, the least developed countries, including virtually all African countries and many countries in Asia and Latin America, should refocus public policies and investment priorities towards rural areas with emphasis on the creation of a socio-economic and institutional environment that facilitates agricultural development and pro-poor income generation such as rural infrastructure, non-farm employment, market information and agricultural research and technology. Fourth, OECD countries should decouple domestic agricultural subsidies from quantity produced and area used in production, and they should eliminate tariffs and other import restrictions on agricultural commodities, processed foods, and textiles from developing countries and they should discontinue subsidized exports, including dumping and non-emergency food aid. Last but not least, developing countries should maintain social safety nets to deal with immediate and short-run poverty, hunger and nutrition problems

The five sets of action must be pursued simultaneously. Instituting the right policies and investment priorities in a low-income developing country, which is subjected to an inflow of subsidized exports and closed international markets, is unlikely to achieve food security for its citizens. Similarly, implementing the policy changes suggested for OECD countries without appropriate policies in developing countries will primarily benefit high-income countries, including developing countries with appropriate policies and rural infrastructure. Low-income developing countries that fail to pursue domestic policy reform will be left behind.

Globalization, including international trade liberalization, more integrated international capital markets and a freer flow of labour, information and technology has benefited hundreds of millions of people but many others have been made worse off. Effective food and agricultural policy and institutions are needed to complement and guide globalization to reduce poverty and hunger.

Insufficient access to productive resources and income-earning opportunities by low-income population groups has led to widespread inequality. While attempts to redistribute existing wealth have been unsuccessful in most cases, land reforms in some countries, most notably in South Korea, have been successful. Efforts to give private title to collectively owned and managed land in some African countries such as Somalia and Malawi, have had mixed results. It is still too early to judge the impact of ongoing land reforms in Zimbabwe and South Africa and it is not yet clear whether and how the explicitly declared goal by President Lula of Brazil to eliminate hunger in that country will be achieved. It is one of the most refreshing positions taken by any Head of State in recent times and it is my impression that it will be attempted by land reforms, new investments in public goods of particular importance to the poor along with pro-poor growth and distribution of subsidized food to target groups. Presumably, the approach to pro-poor growth will include expanded access by the poor to productive resources and employment.

Moving beyond Brazil, one of the key questions related to efforts to eliminate food insecurity is whether growth should be pursued independent of its immediate distributional effects or whether pro-poor growth, which may imply a lower overall rate of growth, should be pursued. What is clear is that attempts to redistribute wealth without growth are unlikely to be politically feasible. Existing literature is strong on the importance of economic growth for poverty alleviation but weak on the opportunities for accelerated poverty alleviation through pro-poor growth strategies. However, while some economic trade-offs may embody an ethical dilemma, investment in the human resource through the eradication of poverty, hunger and malnutrition is likely to achieve both growth and equity goals.

The importance of investments in public goods such as primary education and health care to enhance access to remunerative employment and other income-earning opportunities for both growth and equity is widely recognized, along with poor farmers' access to land, credit, appropriate technology, and well functioning markets for inputs and outputs. While micro-credit programs have been successful in strengthening poor peoples' access to credit in a number of countries, access to credit is a major bottleneck for smallholder agriculture. Poor infrastructure and related high transportation and transactions costs in rural areas of most low-income developing countries, along with poorly functioning markets, make it very difficult for the rural poor to benefit from opportunities embodied in technological developments and globalization. Without expanded investment in rural infrastructure and improved institutions, even those rural poor who have access to land, will continue to be marginalized and will remain food insecure.

7. Conclusion

In conclusion, the global food system, and especially poor people within it, suffers from social injustice and unethical behaviour by decision-makers in both low- and high-income countries. The gap between rhetoric and action to reduce poverty and hunger by the governments of the majority of the developing countries is particularly worrisome from an ethical point of view. While most of the blame for lack of action rests with developing country governments, several policies in rich countries have severe negative effects on poor countries and people. Failure to explicitly consider such international spillovers from national policies and international institutions slows down globalization and contributes to social injustice and armed conflict. Ethical goals and stability goals are compatible.

References

Bryce, J., R.E. Black, N. Walker, Z.A. Bhutta, J.E. Lawn and R.W. Steketee, 2005. Can the World Afford to Save the Lives of Six Million Children Each Year? Lancet 365(9478): 2193–2200.

Collier, P. and A. Hoeffler, 2002. On economic cause of war. Oxford Economic Paper (50) 563-573.

Collier, P., V.L. Elliot, A. Hoeffler, H. Hegre, M. Reynal-Querol and N. Sambanis (2003) Breaking the Conflict Trap: Civil War and Development Policy. World Bank Policy Research Report. Oxford, UK: Oxford University Pres

Eide, A., 2002. The Right to Food: From Vision to Substance. In: M. Borghi, L.P. Blommestein (eds.) For an Effective Right to Adequate Food. Switzerland: University Press, Fribourg, pp. 27–50.

Fischer, L., 1983. The Life of Mahatma Gandhi. Harper and Row, New York.

International Criminal Court, 1999. Rome Statute of the International Criminal Court, Articles 6 and 7 [available at http://www.un.org/law/icc/statute/romefra.htm].

Kolm, S.-C., 2004. Macro justice: The Political Economy of Fairness. Cambridge: Cambridge University Press.

Krueger, A.B. and J. Malečková , 2003. Education, Poverty and Terrorism: Is There a Causal Connection?. Journal of Economic Perspectives, vol. 17, no. 4 (Fall 2003,) pp. 119-144.

Lancaster, C., 2003. Poverty, terrorism, and national security. ECSP Report 9: 19-22 Woodrow Wilson International Centre for Scholars.

Marshall, M.G. and T. Gurr, 2005. A global survey of armed conflicts, self determination movements and democracy. Centre for International Development & Conflict Management (CIDCM) University of Maryland.

Messer, E. and M. Cohen, 2004. Breaking the links between conflict and hunger in Africa. IFPRI 2020 Africa Conference, Brief 10.

Messer, E., M. Cohen and J. D'Costa, 1998. Food from peace: Breaking the links between conflict and hunger. IFPRI Food, Agriculture and the Environment Discussion Paper 24.

Messer, E., M. Cohen and T. Marchione, 2001. Conflict: A cause and effect of hunger. In: Environmental Change and Security Project Report No. 7. Washington, DC: Woodrow Wilson International Center for Scholars, Smithsonian Institution.

Nafziger, E.W. and J.Y. Auvinen, 1997. War, hunger, and displacement: An econometric investigation into the sources of humanitarian emergencies. Working Paper 142. UN/WIDER Helsinki.

Pinstrup-Andersen, P. (ed.), 1993. The Political Economy of Food and Nutrition Policies. Baltimore, MD: Johns Hopkins University.

Pinstrup-Andersen, P., 1998. Challenging Approaches to Development Aid: The Effect on International Stability. In Global Governance 4: 381-394.

Pinstrup-Andersen, P., 2002. Food and agricultural policy for a globalizing world: Preparing for the future. American Journal of Agricultural Economics 84(5): 1189–1404.

Pinstrup-Andersen, P., 2003. Eradicating poverty and hunger as a national security issue for the United States. In: Environmental Change and Security Project Report 9. Washington, DC: Wilson International Center for Scholars, pp. 22–27.

Schmid, A.P., 2003. Prevention of terrorism: a multi pronged approach. Presentation at international Expert Meeting 'Root Cause of Terrorism', Oslo (Norway), June 9-11 2003.

Singer, P., 1993. Practical Ethics. Cambridge: Cambridge University Press.

Singer, P., 2002. One World: The Ethics of Globalization. New Haven: Yale University Press.

Thorbecke, E., 2004. Economic development, income distribution, and ethics. Paper presented at Ethics, Globalization, and Hunger Workshop, Cornell University, Ithaca, NY, 17–19 November.

Tweeten, L., 2003. Terrorism, Radicalism, and Populism in Agriculture. Ames, Iowa: Iowa State Press.

UNDP, 2005. International Cooperation at a Crossroads: Aid, Trade and Security in an Unequal World. Human Development Report 2005. New York: United Nations Development Program.

UNDPI (United Nations Department of Public Information), 2004. A more secure world: our shared responsibility. Report to Secretary's General High-Level Panel on Threats, Challenges, and Changes. UNDPI/2367 (December).

UNICEF [United Nations Children's Fund], 2002. The State of the World's Children. New York: UNICEF.

United Nations, 1948. Convention on the Prevention and Punishment of the Crime of Genocide. UN General Assembly 260A(III), 9 December [available at http://www.unhchr.ch/html/menu3/b/p_genoci.htm].

United Nations, 1999. The Right to Adequate Food. Committee on Economic Social and Cultural Rights, 20th Session, Geneva, 26 April–14 May.

United Nations, 2001. Report of the High-Level Panel on Financing for Development. UN General Assembly, 55th Session, Agenda item 101, 26 June [available at http://www.un.org/esa/ffd/a55-1000.pdf].

United Nations, 2004. Nutrition for Improved Development Outcomes: Fifth Report on the World Nutrition Situation. Geneva, Switzerland: United Nations System Standing Committee on Nutrition.

White House, 2001. Office of the Press Secretary.

Wisman, J.D., 2003. The scope and promising future of social economics. Review of Social Economy 61(4): 425–445.

Food aid and governance

Michiel A. Keyzer and Lia van Wesenbeeck[23]

Abstract

Arguing that food security management belongs to the core of government tasks and has been since Biblical times, we assess current practices in this domain and their effect on local governance, with special reference to the food aid operations of the World Food Program. Within the food security management task, we distinguish three basic elements: (i) entitlement: to provide income entitlements that enable the needy to acquire sufficient food, (ii) delivery: to ensure that the food physically gets within the reach of the needy, and (iii) taxation: to secure funding that covers the full cost of the operations. Our review of literature suggests that the taxation aspect is often neglected, and that little attention is paid to the weakening of respect for local authorities that inevitably results when they remain dependent on foreigners in fulfilling core security tasks, financially and otherwise, let alone when they no longer play any role in this sphere. Next, we present a theoretical framework that coherently deals with the three elements, and discuss its application through a spatially explicit model of optimal food aid provision for Sub Saharan Africa that is being developed within an ongoing joint research project between WFP and SOW-VU. Our main message is that foreign donors, when called upon to intervene during food crises, should always program for a post-emergency exit strategy that leaves behind local authorities with sufficient authority and means for food security management.

1. Introduction

Throughout his career, a keen interest in poverty alleviation and improvement of food security led Professor Arie Kuyvenhoven to engage in a multitude of research projects and advisory activities in this field. His research embraced sustainable land use and policies to reduce the vulnerability of farmers and other rural poor in less-favoured areas, as well as the influence of trade agreements on the poor. His wider interest in the subject is witnessed, among others, by his serving on the board of both the International Food Policy Research Institute (IFPRI) and, for over 15 years, the Centre for World Food Studies (SOW-VU), our institute. In this latter capacity Arie Kuyvenhoven often gave testimony of his religious inspiration in studying poverty and hunger. Hence, it is only fitting for us, as a tribute to his contributions, to present a paper on food security, with emphasis on food aid, and with a Biblical rooting.

The chapter proceeds as follows. Section 2 reviews the basics of food security management, by comparison of the relaxed situation in the Garden of Eden with the tight food management

---

[23] We thank Max Merbis for his comments.

in Joseph's Egypt, and the food crisis faced by Moses in Sinai, where manna from heaven brought relief. Section 3 reviews the main lines of thought on food aid provision in modern times. Section 4 presents some facts and figures on present-day food aid and describes the operation of the World Food Program (WFP). Next, section 5 reviews impact assessments of food aid. Section 6 concludes that there is a need for a more formalized treatment of the issues, and formulates a theoretical framework to accommodate the three key elements of food security management – provision of income entitlements, supply of goods, and adequate funding of food security programs. Finally, as an application it also reports on the workings of a spatially explicit model of optimal food aid provision for Sub Saharan Africa that is being developed within an ongoing research project conducted jointly by WFP and SOW-VU. Section 7 concludes.

2. Eden, Egypt and Sinai: three systems of food security

2.1. Eden

In Paradise, the hospitable Garden of Eden provided all food security needed (Genesis 2: 8-25). Adam and Eve could without any effort, pick all fruits from the trees (but one) and never suffered hunger. In historical times, the food system of Mesopotamia is thought to come closest to this ideal, with abundant and well irrigated fruit trees providing ample food throughout the year, and also allowing for easy drying and preservation of food providing for dry spells and for journeys across the territory. After being chased from Paradise, Adam and Eve had to work more but they could live peacefully with their offspring without need of any governmental agencies to protect them against threats from nature or from other tribes.

2.2. Egypt

How radically different was the situation much later, in Pharaoh's Egypt. Recall Pharaoh's Dream (Genesis 41: 1-4) of the seven fat cows coming out of the Nile followed by the seven hungry cows that eventually ate all fat cows. When called upon by Pharaoh, Joseph interprets this dream as a warning that after seven exceptionally rich harvests will follow seven years of hunger and (Genesis 41: 33-36) suggests the following measures to safeguard Egypt's food security. First, there is taxation: one fifth of all harvested grain should be stored in warehouses owned by Pharaoh. Indeed, in view of the Nile valley's seasonal flooding, stocks had to be kept high and dry, far away from the fields, presumably in caves up the East Bank. These caves had to be guarded. Since farmers had to work on the fields, special troops had to fulfil this task, justifying a standing army to be kept even during peace time. Genesis tells us that when the seven years of hunger began, the stored grain was sold to the hungry population, and in the seven years that followed, Pharaoh managed to get control over all resources of the population - initially, the population spent their cash reserves, then they sold their cattle to Pharaoh, next their land and, finally, they became his slaves (Genesis, 47: 13-25).

This narrative highlights the key role of stockholding in securing and to some extent justifying the power of the state, in this case even to the extent of leading to serfdom. Indeed, Pharaoh abuses his power: he had already taxed farmers, so why let them pay again for the stocks in time of need? Nonetheless, the contours of a more balanced food security policy emerge, as one envisages the functioning of the system once serfdom was established. Since the slaves had to be fed, and those who worked the fields had to be fed well, the later pharaohs must have distributed food from stock to their slaves, in good as well as in bad years. Able bodied slaves may have received food-for-work but young children and elderly must have been entitled to food in another manner, possibly through vouchers or stamps of some sort, to secure their access to food while preventing multiple visits to the point of distribution. Hence, the system must eventually have comprised the three basic elements of a food security scheme: (i) targeted provision of income entitlements to beneficiaries, in normal years the poor, in bad years the whole population (ii) adequate delivery, supported by stock operation and trade with the outside world, and (iii) adequate funding, through taxation or requisition in kind. We refer to (i)-(iii) as the triad of food security. People living at the rhythm of the seasons in the Nile delta were naturally led to abide by these rules, virtually natural laws, as long as food deficits could be covered from stocks or imports, which proved impossible eventually.

2.3. Sinai

Indeed, as Exodus 12: 31 tells us, the ruling regime eventually was unable to deliver when hit the Ten Plagues (described in Exodus 8-12). Pharaoh had to let go the People of Israel, thus reducing the number of mouths to be fed. However, liberty came at the price of famine and lawlessness throughout the journey. The Lord, we are told, re-established His authority over His people by rescuing them, dropping manna and quail from heaven (Exodus 16 and Numeri 11), while punishing those who failed to be satisfied with the gift. After these miracles and revelations, Moses finds his people willing to accept the Law, in the form of the Ten Commandments, establishing that under the new circumstances, individual transactions became important and property rights had to be protected through regulations, essentially demanding respect for the neighbours' life and property. The centralized planning of Egypt turned into a market economy.

Of course, with the transition and after reaching the Promised Land, the question of social equity and safety nets arose, as some families might not possess enough cattle and farmland to survive at times of hardship. The Book Leviticus (25: 35) provides some guidance in this respect, forbidding any profiteering from food sales to the hungry. Regarding equity, the same verse also orders the rich to help the needy. Furthermore, as proclaimed in Deuteronomium 15: 1-6, debt relief should be given every seventh year. All these are mere prescriptions for individual conduct. Government institutions fulfil no direct role in this respect. Moreover, as the Bible generally attributes the occurrence of natural disasters to Godly punishment for collective sins, such as disobedience, and was addressed to the People rather than to its rulers, it does not contain many precepts for public food security management per se. Indeed, under the dry conditions of the Middle East, with their limited scope for irrigation, modest harvests, and poor means of transport, beyond some public stockholding little could be done during prolonged periods of droughts or plagues. Instead, enforcing respect of the law became the key security providing mechanisms. Leviticus and later scriptures indicate that the Law carved in stone had to be protected by other security improving customs ("Hedge around the Law"), and by dedicated guardians with powers to decide in the face of circumstances: priests, judges and political leaders (the later "rules versus discretion"[24]). The many books of the later Babylonian Talmud were an attempt to codify a comprehensive set of rules and jointly with their prescribed interpretation, so as to offer security with respect to the law itself. Clearly, the mere vastness of the series compromised its use in practice, and even the text itself explicitly presents questions for further debate. Unlike in Egypt, adaptability became of essence in the new societies and required some discretion and some separation of powers to be established.

Be this as it may, in both Egypt and Sinai, the higher-level institutions based their legitimacy on their capacity to protect the people from shocks they would be unable to address individually (self-protection and self-insurance), or jointly with family, friends and neighbours (mutual protection and mutual insurance). This defined their tasks in the area of food security management. In Egypt, the natural demands of irrigated agriculture hardly supported any system of governance other than tight central control. In Sinai, decentralization became necessary, through regulations and specialized institutions to enforce the law on the basis of the respect they commanded rather than by repression.

3. Food aid in modern times

3.1. Enlightment and revolution

To date, in many parts of the world food security management still operates under regimes very similar to those of ancient Egypt, and Sinai. However, in 17th century Europe, the Age of Enlightenment led philosophers to look for rational justification, with key contributions in the writings by Hobbes, Locke and Spinoza. The crucial element in this transfer of power is the notion that the state should truly represent the citizens (Spinoza's notions of 'acting as if by one mind', 1670; Hobbes' 1651 emphasis on the inherent weakness of any leader who fails to perform basic functions). This rationalization pursued in the 18th century by Rousseau ("contrat social", 1762) and Voltaire (1759) emphasizes the nature of the state as an institution that is "of the people, by the people", and should, therefore, not only protect the people against collective threats but also be primarily funded by these people, as opposed to foreigners. Whether the rich should be forced to give to the hungry poor in need, in other words whether food security management included redistribution as opposed to

---

[24] The modern debate started with the seminal Coase (1960) article on how to deal with externalities.

being limited to providing for emergencies remained a contentious issue, and in fact still is. The French Revolution was to some extent inspired by this intellectual debate but the actual insurgence was unleashed when the Ancien Regime proved incapable, in the wake of a financial crisis, to provide sufficient food and purchasing power to the Paris population as a whole. Queen Marie-Antoinette's famous outcry 'Qu'ils mangent de la brioche' (let them eat cake) when told that there was no bread left was illustrative of the estrangement of the rulers from their people. . Soon the radical ideology dominated that allowed the poor to take from the rich by force, but this obviously did not fill their stomachs for long and eventually respect of private property returned, while right-to-food claims and forced redistribution subsided.

3.2. World Wars I and II and their aftermath

Whereas food crises are immanent in human history, concerted international actions to cope with them are relatively recent phenomenons. Massive inflows of food aid, primarily from the United States, were supplied after World War I. In 1918-1919, the American Relief Administration (ARA) under the direction of future president (1929-1933) Herbert Hoover shipped nearly 4 million tons of food and other supplies to Europe, an immense undertaking that was paralleled only by Hoover's subsequent action to address the Ukrainian famine of 1921-1923. By the summer of 1922, at the peak of operations, the US were feeding nearly 11 million people a day. The total cost of the mission were approximately $60 million (or about 0.1 percent of GDP), of which $20 million was an appropriation from the US Congress used to purchase corn and seed. In relative terms this may well have been among the largest food aid operations ever.

Also towards the end and in the aftermath of World War II, during the period 1945–1948 the allies shipped millions of tons of food to Europe. For example, from April 29 until May 8, the British Bomber Command dropped 11,000 tons of food on the hunger stricken Northern part of the Netherlands, to save 3.5 millions from starvation. This famine, and the aid campaign that helped addressing it, under the fitting name of Operation Manna, is thought to have fostered the country's persistent support for development assistance.

After 1945, the US became a major donor of food aid (in kind), to former allies as well as to former enemies, in a complex mixture of benevolence and political strategy to 'Win the Hearts and Minds of the People'. This policy has been pursued uninterruptedly until present.

It must be added that after World War II, political and commercial motives also became part of the equation. In a nutshell, subsidies to the farmer states in the Midwest became instrumental in attempts to win the primaries and to secure seats in Congress. Foreign policy was also a major driver, to establish goodwill and forge strong alliances in the post-war world, partly to avoid the spread of communism, partly to foster decolonization. Commercially, an explicitly stated motive was to win the taste of foreign consumers for US-food products, especially in the mid 1950s, when economic life had recovered from the war, and US-food exports faced declining demand as agricultural production recovered in Europe and Asia, and these regions were beginning to become food exporters themselves. Against this background, the US Congress passed in 1954 the Agricultural Trade Development and Assistance Act, Public Law 83-480, known as PL480 or 'Food for Peace Act'. PL480, jointly with the 1949 Agricultural Act provided the legal basis for subsequent food aid shipments to developing countries.

In the 1950s and 1960s, food aid became a major component of development assistance, reaching 20 percent of ODA in the 1960s (FAO, 2006), partly because the problem of hunger was viewed primarily as one of food scarcity, but also as a means to find outlets for the large agricultural surpluses in the US, and since the mid sixties in the European Community. This resulted in shipments of around 10 million metric tons annually on average, mainly cereals. The aid was generally transferred to the government of the receiving countries, as a grant or at subsidized prices, mostly for sale on the domestic markets. In terms of the triad discussed in section 2, this aid focused on the supply side (ii), through imports from donor countries, with grants and subsidies substituting for taxes (iii) but bypasses the entitlement side (i).

It took the 1974 floods and resulting famine in Bangladesh to change this perception, when presumably as much as 80,000-100,000 people starved to death. Floods occur in Bangladesh every year, and are, therefore, like in Egypt, as it were built into the social fabric. However, in 1974 they lasted exceptionally long, and governance was particularly weak as the country had just become independent two years earlier after a war with (West) Pakistan, and very short of funds and expertise. Because of the floods, the field work was postponed, leaving without livelihood many landless labourers who used to be paid in kind (Ravallion, 1987). At the same time, prices rose due to expectations of a poor harvest, which did not materialize as, partly thanks to the floods, paddy yields came out to be even higher than in the preceding and following years. This phenomenon was not well understood at the time by the donor community that delivered massive food aid in kind, which the poor were unable to buy. Sen's (1981a,b) later assessment brought this discrepancy to international attention, emphasizing that entitlements of the poor (purchasing power) rather than availability were the critical factor that had caused famine in this case.

Bose (1990), in a comparative study of three famines in Asia during the period 1942-1945 concludes that in all cases, production decreases, if even present, could not be blamed directly for the famine. Lin and Yang (2000), in a case study on the Chinese famine during the "Great Leap Forward" in 1959-1961, also find support for Sen's thesis, in that the fall in production was not the direct cause for the famine. However, unlike Sen, who emphasized the role of government in preventing a famine, the Chinese authors conclude that government actually caused the famine as its procurement activities strongly favoured the urban population. In a review of literature, Ravallion (1997) mentions that price increases in rural areas come out as the prime cause for famines in many empirical studies. In relation to food crises in Sub-Saharan Africa, Keyzer *et al.* (2003) also find that, rather than adverse climatic conditions per se, entitlement failures and poor policies are the main causes of famines. Indeed, it appears that lavish provision of food aid may cause prices to become too low in rural areas, rather than too high, and through this discourage production. Indeed, while Sen's writings on entitlements have been very influential in academic circles, the two major donors, the US and until recently the EU have kept on providing food in kind. We have argued that all elements of the triad (entitlements, delivery, taxation) are important and need to be operated jointly and with great care. Hence, the common divide in policy debates between those who advocate entitlements as "clean" and those who foreign support aid in kind as "practical" cannot do justice to the subtlety of the issues at stake.

4. Present-day food aid and the role of WFP

Worldwide, over the last 25 years, some 10 million tons of cereals were on average shipped annually as food aid, peaking at 15 million tons in 1992 (FAO/WFP). Figure 1 shows the pattern by region.

As to the source of food aid, the amount purchased in the country itself (local procurement) and in other developing countries (triangular purchases) increased by about 67 percent in the last ten years (71 percent for Sub Saharan Africa) but remains relatively low rising from 22 percent in 1996 to 33 percent in 2005 worldwide, while remaining almost constant in SSA. In fact SSA has ranked high in terms of relative share of local and triangular purchases (Figure 2). This is the direct result of the EU emphasis on local and triangular purchases laid down in Council regulation 1292/96 that followed the review of the EU food aid policy.

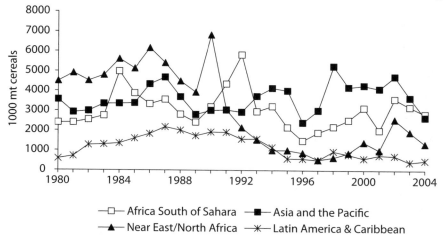

Figure 1. food aid flows by destination, cereals, 1980 – 2004. Source: FAOSTAT.

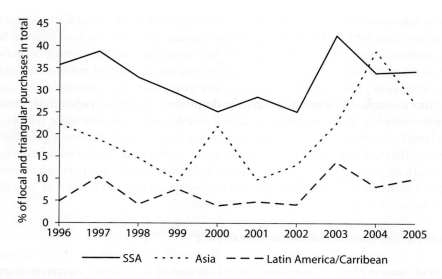

Figure 2. share of local and triangular purchases in total food aid by region. Source: INTERFAIS.

Yet, most of local and triangular purchases originate from few countries only; in 2007, in value terms, Uganda contributed 20% (mainly triangular), Ethiopia 18% (mainly local), Kenya 15% (both) and South Africa 14% (triangular) (WFP, 2007c).

With respect to the nature of the aid, a major shift has occurred over the last 20 years (Figure 3) from project and program aid to emergency relief in response to sudden natural

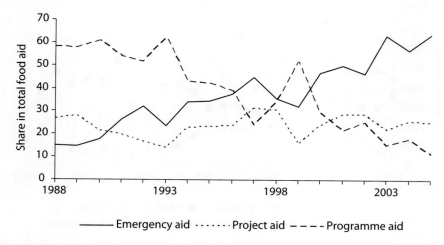

Figure 3. Composition of food aid, 1988 – 2005. Sources: WFP (2006a) and FAO (2003).

disasters, slow-onset disasters such as crop failures, and complex emergencies such as (civil) wars. In 2005, 64 percent of total food aid was emergency aid, as opposed to a mere 15 percent in 1988. For SSA the 2005 percentage of emergency food aid in total even reached 76 percent.

These shifts also reflect that over the years, the donor community has persistently lowered its expectations of program and project food aid contributions to development, witness the share of food aid share in total aid dropping from over 20 percent in the 1960s to 5 percent in 2005 (Barrett and Maxwell, 2006).

4.1. Role and mode of operation of WFP

Be this as it may, emergency food aid is there to stay as basic element of relief operations. Nowadays, the delivery of most of emergency food aid is coordinated by the UN's World Food Program (WFP). In 2005, 35 million people received 8.2 million tons of emergency food aid through WFP coordinated operations (Figure 4). WFP often cooperates with other UN organizations, in particular with the UN's High Commission for Refugees (UNHCR), to provide for millions of refugees and internally displaced persons who have fled war or natural disaster (Figure 5)

Naturally, WFP-operations directly impact on the achievement of Millennium Development Goals 1 (MDG-1) for reduction of hunger and malnutrition. In addition, WFP refers to food aid as contributing to the other MDGs indirectly by improving the nutritional status of special groups such as women, children, and those affected by HIV/AIDS.

Regarding the financing demands of these MDGs, we may note that for 2005, WFP estimated its total operating expenditure at USD 2.9 billion to reach 96.7 million people (WFP, 2006b). This amounts to an average expenditure of USD 30 per person worldwide. However, if Africa is singled out, average per person expenditure turns out to be almost USD

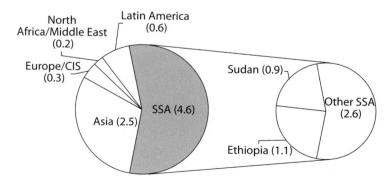

Figure 4. Emergency food aid shipments in 2005, by region in million mt. Source: WFP, 2006a.

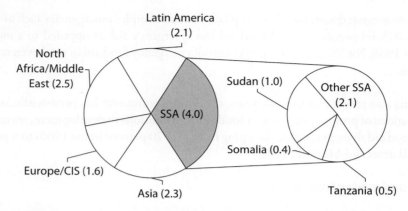

Figure 5. Refugees and IDPs in 2005 supported by WFP. Adapted from UNHCR, 2006.

77 per person. This clearly illustrates that the financing requirements to reach the MDGs as presented in the Millennium Reports (USD 6.5 for MDG-1, 17 for MDG-2, and 28 for MDG-4 to 6), must grossly underestimate the actual MDG-financing needs (see also Keyzer and Van Wesenbeeck, 2006).

WFP is the co-ordinating agency that brings food to those in need under emergencies. Its activities focus on procurement, shipment and distribution. In case of an emergency, WFP joins UN assessment teams to make an inventory of needs and possibilities for shipping food to the target groups. To cover immediate expenses, two sources of short-terms finance are open to WFP. First, WFP-country directors can borrow up to USD 500,000 from the Immediate Response Account, a centrally held special account that is funded multilaterally, and usually is able to provide the requested resources within days, as loans that are later to be repaid to the IRA once dedicated funds for the emergency at hand have been received. Second, in December 2005, the Central Emergency Response Facility (CERF), a special UN-wide fund for humanitarian assistance has been established. The fund is projected to have financial reserves totalling USD 500 million in 2008, and is replenished via donations from public and private donors (in 2007, the Netherlands ranks third on the donor list, with a total donation of USD 53.4 million or 20% of total donations made up to May 2007). CERF, like IRA, provides the financial assistance needed to keep an operation afloat before dedicated funds have been received. In addition, this fund facilitates the continuation of 'under-funded' activities, i.e. operations for which specific appeals have not triggered an adequate donor response. Allocation of CERF funds (mainly grants) to UN organization, specialized agencies and the international organization for migration is done by the Emergency Relief Coordinator, on behalf of the Secretary General of the UN. In 2006, WFP received USD 108 million from CERF, for instance for assistance to Darfur and to maintain aid to refugees in Kenya.

However important these quickly available financial resources may be to allow a timely response to emergencies, their size is modest relative to the total WFP-budget, of about USD 3 billion in 2006, up from USD 130 million in 1971. This only confirms that WFP is in fact an executing agency that mostly acts on demand by its donors.

Besides its emergency operations, WFP also assists in recovery, with funds collected on the basis of dedicated calls to donors. In all these operations, the urgency of the needs tends to dwarf the depth of preparations, and WFP has increasingly been criticized for lacking any sort of comprehensive analytical framework to guide its operations (see DISI, 2006). The practice under emergencies is to send in missions that conduct rapid appraisals so as to gain an impression of the people affected, while the follow-up focuses on target groups through dedicated actions, such as school feeding projects, food-for-work programs and food provided especially to those suffering from HIV/AIDS, as a temporary supplement to the diet to help the affected groups in rebuilding their productive lives. However, as is also concluded in the internal evaluation of the Darfur operation, the "machinery" of WFP appears to be deficient in monitoring the effects of such operations on the nutritional status and asset position of the recipients, as the focus lies on the delivery of the goods rather than on its effects (WFP, 2007b). Similarly, efforts at using food aid to promote sustainability of agriculture, as well as investment in education and infrastructure by lifting basic nutritional constraints seem to lack theoretical guidance and empirical foundation. Hence, the estimation of entitlement needs, first element of the triad, is essentially based on expert judgment. Regarding delivery, second element of the triad, the price effect of WFP actions on local markets does not receive much attention, even though the needs assessment indirectly account for it by seeking to eliminate deficits. Finally, the last element, local taxation, hardly enters consideration. Consequently, WFP can only expect that ongoing aid projects will be terminated once it retires, rather than being adopted and pursued by local authorities. More generally, WFP does not avail of any explicit strategy to leave an adequate governance of food security upon its disengagement. It is as if the dismal situation usually encountered on the ground when foreign aid is called in had erased the donors' memory of better times when traditional stockholding and food distribution by local clan or village leaders provided a functioning institutional arrangement that could deal with drought and food scarcity most of the time.

Clearly, food aid is a subsidized export of sorts but one that is admitted under WTO-regulations, provided it is sanctioned under the Food Aid Convention (FAC), an international agreement, signed in 1999, "to improve food security and the ability of the international community to respond to emergencies". To avoid the use of food aid as a balancing outlet for food production – providing much when prices are low and little when they are high – minimum requirements for the provision of food aid were set. However, it appears that in practice low prices have corresponded to food aid exports in kind that greatly exceeded these minimum requirements, whereas tight supply corresponded to downward renegotiation of these levels (Huff and Jimenez, 2003). Over time, attempts have been made to reform the

FAC in line with its stated objective, but this has so far only resulted in a widening of the array of commodities that could be supplied to fulfil the quota, and to the acceptability of contributions in cash rather than kind. This neglect of elements (i) and (iii) of the triad is in part attributable to the fact the FAC was established as a "by-product" of the Uruguay Round of multilateral trade negotiations, to avoid trade disputes in this sphere. However, judging by the topics currently under discussion for a new FAC, the prospects for a more balanced coverage seem dim, as the list of "key issues" for negotiation so far includes besides the representation on and housing of the Food Aid Committee such issues as the nature and level of commitments as well as the monitoring and enforcement of these commitments, (Hoddinott and Cohen, 2007), all exclusively referring to element (ii) delivery through imports, and without reference to local stockpiling.

5. Impact assessments

Most of present-day literature on food aid is concerned with assessment of the impact on local production and prices, for prevailing mode of delivery (e.g. cash, kind, school-feeding, food for work), type of commodity, timing of delivery and methods for targeting. Basically, this research has been testing, on the basis of survey data, a subset of the eight propositions by Maxwell and Singer (1979) on the impact of food aid.[25] The present section reviews contributions along these lines.

5.1. Impacts on food market

For Mozambique, Donovan (1996) studies the effects of yellow maize inflows of food aid on the domestic production of the consumer-preferred white maize and concludes that, partially because of the strong preference for white maize, yellow-maize inflows only had a minor and short-lived depressing effect on the price of white maize. This conclusion holds for the war period of 1990-1993 as well as for the post-war period of 1993-1995, although effects were slightly more pronounced in the second period than in the first. The chapter uses a VAR-technique supplemented with an assessment of the process of market integration within the country. By contrast, concentrating on the post-war experience, Tschirley *et al.* (1996) conclude that the continuous inflow of yellow maize as food aid may have discouraged production of traditional white maize. They point to the co-movement of prices of white and yellow maize in non-emergency years as evidence that the untargeted inflow of yellow maize as food aid depressed prices of yellow maize and, via substitution effects of white maize as well.

---

[25] The eight propositions are: (1) food aid may lift credit constraints to investment, (2) food aid may disproportionately benefit the most vulnerable, (3) food aid may assist governments in stabilizing food prices, (4) food aid is at least partly additional, (5) food aid has a disincentive effect on local food production, (6) the allocation of food aid reflects political and military interests rather than need, (7) food aid leads to greater dependence rather than greater self-reliance, (8) food aid often is a second-best option.

For Tanzania, Tapio-Biström (2001) presents a model of the agricultural sector that allows for producer risk and also explicitly accounts for market segmentation, both geographically and in terms of the distinction between official, controlled markets and unofficial markets and finds a positive relation between food aid inflows and production. In this study, no distinction is made between program and emergency food aid due to lack of time series data. Nonetheless, the author assumes on the basis of evidence from secondary sources that only a minor fraction of food aid was used for emergency, while the bulk was sold on the open market... She finds that segmentation between official and unofficial markets makes it possible for food aid delivery to promote local production, along arguments similar to those of Von Braun and Huddleston (1988), who stress that in many countries grain markets obey a dual structure with output sold at fixed (floor) prices to procurement agencies while consumers pay subsidized, low prices, for given rations. Clearly, this refers to the pre-structural adjustment situation, in which the procurement prices did not have to suffer when the shops selling to consumers were supplied from food aid. Nowadays such a separation is rare. Yet, in some countries where coastal cities are supplied from foreign sources, while farmers are living in relative isolation in rural areas, undernourishment among the urban poor can still be alleviated through imported food aid, without directly harming the rural areas. By the same token, however, rural development will require such aid to become sourced domestically as soon as rural infrastructure permits.

Many more studies of price impacts have been published, see e.g. Barrett (2002) and Tapio-Biström (2001)) for surveys. From an econometric viewpoint, all have the shortcoming that food aid never is truly exogenous, since it is usually provided almost inevitably in circumstances that impact directly on production and prices as well, such as droughts, unrest, speculative hoarding (Abdulai *et al.*, 2005). Straight reduced form estimations neglect these circumstances. At the same time, it is virtually impossible in such a context to find reliable instrumentation variables that might cure this problem.

5.2. Impacts on labour markets

Another branch of investigations, less related to emergency evaluates whether food-for-work programmes (FFW) are more effective than free distribution. Advocates of FFW stress its contribution to the provision of public goods, usually road infrastructure, and its effectiveness as well targeted safety net for the poor, since the well-to-do will never participate. The negative side is that FFW may discourage labour participation in more productive jobs. Along these lines, Osakwe (1998) formulates a small open economy model in which food intake raises labour productivity, in accordance with the efficiency wage relation (Dasgupta and Ray, 1986). The author finds that when food aid is used to finance infrastructure development projects, it has no labour disincentive effects in the food industry and improves food security but has an ambiguous effect on aggregate welfare. When food is given to unemployed workers without any obligation to work, these workers stay away from the food industry and overall food security suffers but the effect on aggregate welfare can be of either sign. Quisumbing (2003) examines the effects of food aid on child nutritional status in rural Ethiopia and finds a gender effect: whereas food received as direct transfer mainly benefits girls' nutrition, FFW primarily helps boys. Holden *et al.* (2006) focus on Northern Ethiopia and conclude that effects are ambiguous and critically depending on the quality of program design and implementation, which unfortunately prove not to be positive on average.

5.3. Cash versus kind, targeting

In general, there is a growing recognition that many of the 'failures' of food aid can be attributed to errors in targeting and timing of interventions. Targeting is often found to be more effective with food aid in kind. For example, Basu (1996) constructs a simple model with a threshold on consumption below which all income will be spent on food, and finds that aid in cash is less helpful, since it raises food prices and through this harms those who are not included in the food aid program. Similarly, Arndt and Tarp (2001) in a computable general equilibrium model find that aid in kind is superior because it is less easily diverted from the poor.

More generally, on the issue of targeting, the overall conclusion of empirical research seems to be that it generally is deficient and driven by other aims than the improvement of nutritional status. Clay *et al.* (1999) use survey data for Ethiopia and cannot find a significant association between household food security and food aid receipts, while Barrett (2001) evaluates the performance of US-donated food aid under the PL480 programs using time-series data and concludes that it did not stabilize food availability per capita, as the response to crises was too slow, too much politically motivated and the quantity provided was too small. However, Yamano *et al.* (2005) find for Ethiopia, based on three representative national surveys held in 1995-1996, that food aid does have a positive effect on child growth and that it mitigates the effects of weather shocks but they also note that many villages that were exposed to such shocks did not receive food aid. Del Ninno *et al.* (2005) compare the experiences of four major recipients of food aid (India, Bangladesh, Ethiopia and Zambia) and conclude that the effects on production depend on the adequacy of targeting, its importance relative to local production, its substitutatibility to local crops, and the extent to which it is accompanied by investment programs in agriculture. Finally, Donovan *et al.*, 2006) assess current procedures used in emergence needs assessments and advocate improving these to avoid depressing effects of food aid on production.

5.4. Policy simulation exercises

Until the nineties', Bangladesh has been operating a system of public food grain provision that comprises a mix of food-for-work programs, open market operations to stabilize grain prices, and rationing (through ration cards issued to individuals with which food in specialized shops can be bought), modified rationing for the population in rural areas, and rationing to priority groups such as the military, and the police, patients in hospitals, and students in student hostels (e.g. see Ahmed, 1988). As extensively discussed in Section 3 above, consensus has emerged that in Bangladesh entitlement failures have been a major cause of food crises. Using the general equilibrium model developed for the third Five-Year Plan (1985-1990), Keyzer (1987) illustrates that the best policy to help the poor would be to increase the ration subsidy, which requires inflows of foreign aid as balance of payments support to supplement government finance. In a simulation run without balance of payments support but direct inflows of food grains as aid, calorie consumption is higher, but incomes of especially the poorest people are lower, and most importantly, domestic savings decrease, implying less investments and continued future dependency on aid. More recent studies for Bangladesh based on model simulations include Fontana *et al.* (2001) who with a CGE model study the effect of decreased food aid imports with and without compensating balance of payments support as two of the policy scenarios. Model simulation outcomes are similar but effects are smaller. Dorosh *et al.* (2002) also find that continuous inflows of grain as food aid into the Bangladesh economy discourages rice production, while arguing that reduced inflows of grains should be compensated by aid in cash to allow the government to continue supporting the poor through income transfers.

For Ethiopia, Gelan (2006a,b) formulates a CGE-model with food aid represented as an endowment injected into the economy as a perfect substitute to local food grains and without any costs to the recipients. The simulations compare food aid in kind, the cash equivalent of the costs the donor made to deliver the food aid (value of food aid on the market plus transportation costs), and the cash equivalent of the food aid received at local market prices. Here also food aid in kind is found to lower prices and to discourage local production.

All the contributions above take a static approach, neglecting that food security naturally impacts on expectations and through this on recipient behaviour, especially since individuals will try to avoid hunger by all means. Coate (1989) formulates a formal two-period model, with consumption decisions made in period 1 affecting the probability of survival in period 2. The model treats the possibility of famine as a catastrophic risk by including a threshold level on consumption below which the probability of survival is less than one and is decreasing as consumption drops further below the threshold. Another major innovation in this model is that it explicitly identifies the donor as an agent who seeks to minimize the prevailing level of mortality, subject to a budget constraint. It is found that, if food is exported from the region, cash relief is optimal as long as traders do not collude and are relatively efficient in transporting the food to the region. If food is imported into the region, then cash relief is to be preferred if traders are more efficient than the relief agency and do not have market power. These analytical results are, however, obtained at the expense of theoretical rigor, with all incomes taken to be in cash and hence non-responsive to any price changes. Nonetheless, the key point is that whether to opt for food aid in cash or in kind depends on the market conditions and circumstances and is definitely not a matter for ideological debate.

5.5. Entitlement and taxation through insurance

While the literature on food aid distribution under uncertainty is remarkably scarce, the branch on improving distribution of entitlements under uncertainty and associated taxation has been blooming in recent years, particularly through papers on semi-commercial crop insurance, with indemnification generating entitlements and premiums-cum-aid playing the role of taxes in the funding of the arrangement. To reduce the monitoring costs, the proposed arrangements are making use of index-based indemnification schedules depending on easily measurable variables on prices and rainfall, rather than on individualized post-harvest damage assessments. Pilot studies have been carried out in a selection of developing countries, including Mexico (Skees *et al.*, 1999), Morocco (Skees *et al.*, 2001), India (Kalavakonda and Mahul, 2005; Veeramani *et al.*, 2005), and Malawi (Hess and Syroka, 2005), with mixed results so far. Remarkably, the delivery side is absent from these investigations, presumably because the current proposals are only targeting very small fractions of the rural population.

Dercon and Krishnan (2003) is one of the few papers directly linking insurance to food aid. For rural Ethiopia, the authors conclude that food aid inflows have reduced traditional risk sharing, by reducing the risk of being non-insured. This may not be a negative development, as it is increasingly being recognized that the poor often have to be very risk averse, which leads them to opt for low-risk, low-return livelihood strategies that leave them in poverty (Dercon, 2004, Carter and Barrett, 2006, FAO, 2006). In such a situation, food aid or insurance might encourage them to engage in more activities whose returns combine higher expected value with higher variance.

5.6. Summarizing

To summarize, it appears that the regression-based impact assessments essentially serve to inform the donor whether the money was well spent and how to improve the mode of delivery, so as to avoid disturbing the operation of commodity as well as labour markets, while maintaining adequate targeting and meeting the needs. These assessments tend to apply reduced forms that neglect endogeneity, and, therefore, might tend to attribute to the food aid delivery impacts that should in fact be attributed to the forces that triggered the operations, such as political unrest, drought, and poor governance. Since food aid is increasingly distributed under emergencies when conditions are exceptional and data collection is not a first priority, this is to a large extent inevitable. At the same time it must be noted that for many indicators the cost of data collection is minor and hardly competes with additional deliveries of food to the needy.

Econometric exercises coexist alongside with simulation studies as policy makers and WFP-staff need more than ex post assessments alone. To plan effectively, they can also benefit from use of policy simulation models that may inform their choices. We have seen that such models have been developed in the past but it would seem that modern techniques could be relied upon to construct decision support tools that are better tailored to the planners' needs. Such tools should have the capacity to capture the key elements of food security managements. First, there is uncertainty itself. Providing food security is a risk management task that needs to take into consideration that the future is unknown. It has to plan under uncertainty, and design prevention as well a coping mechanisms, such as stockholding, procurement in other locations, and indemnification through insurance arrangements. Individual, mutual and public strategies have to be considered jointly. Second, all elements of the triad should receive due attention: the entitlements to avoid vulnerable groups being left out or receiving inadequate entitlements, while paying due attention to aspects of domestic redistribution; the delivery to see to it that the entitlements are sufficient to buy the necessary food without price hikes causing additional undernourishment for non-participants; and finally, the tax-side that should be developed in sufficient detail to show ways for the beneficiaries to contribute and eventually take charge of the arrangement, so as to give local government institutions a chance to assume their responsibilities again and develop sufficient support among the population.

6. Towards a decision support tool for WFP

We conclude from the previous sections that the effective management of the triad of entitlements, delivery and taxation is a subtle operation that has challenged policy makers ever since the expulsion from the Garden of Eden. In Egypt, a very fertile soil, a most reliable rhythm of the seasons and a tight central control could not prevent calamities leading to famine. In Sinai and in the Promised Land, famine could not be eliminated, despite greater decentralization and reliance on free markets. Currently, excessive emphasis on questions such as cash versus kind have unnecessarily polarized and simplified the debates, leading to unwarranted attempts at formulation of unified rules, where adaptive tailoring the solution to the specificity of the prevailing conditions should be the guiding principle that calls for dedicated analytical tools.

In this section, we present a theoretical framework for modelling the triad, and also describe the present status of a spatially explicit model for food aid deliveries in SSA, which is currently under construction in a co-operative venture between our institute (SOW-VU) and WFP.

6.1. Modelling the triad: theoretical framework

A theoretical framework may serve food security management in two ways. One is to describe how the various components of food security interact logically, in time and space, and how financial aspects relate to physical flows and satisfaction of needs. The other is to provide the underpinnings for the applied models that operate a food security decision support system. Though abstract, the theoretical framework has the major advantage that it sketches

Michiel A. Keyzer and Lia van Wesenbeeck

the scene without requiring construction of a database that in this field of investigation is bound to be incomplete and unreliable.

As theoretical framework, we use a single period welfare program in Negishi format with interdependent utilities representing empathic consumers, as a special case of the model specified and analysed in Ginsburgh and Keyzer, 2002, ch. 9, p. 330-336; we refer to this source for proofs and technicalities. Empathy drives consumers to care for the well-being of others in an uncertain future. This makes them willing to exhibit solidarity.

The model comprises donors as well recipients, characterized as consumers indexed i, $i = 1,...,I$. Each consumer maximizes a single period utility function with planning decisions made at the beginning of the period and uncertainty revealed at the end before coping decisions are made. Care for the own uncertain future finds reflection in the curvature of this function. Each consumer I may also seek to avoid utility of any other consumer h dropping below some threshold. We represent this by supposing that consumer i attributes a non-negative empathy weight ρ_{ih} to the fate of consumer h, $h \neq i$. On the supply side of the economy, producers indexed j, $j = 1,...,J$, operate a convex technology under decreasing returns to scale, buying inputs at the beginning of the period and obtaining outputs (crops) subject to uncertainty, at the end of the period. The resulting welfare program reads:

$$max_{xi0, xis, vj \geq 0} \sum_i \alpha_i \left(\sum_s P_s \left(u_i \left(x_{i0}, x_{is} \right) + \sum_{h \neq i} \rho_{ih} \, min \left(u_h \left(x_{h0}, x_{hs} \right), \bar{u}_h \right) \right) \right)$$

subject to

$$\sum_i x_{is} \leq \sum_j q_{js} \left(v_j \right) + \sum_i \omega_{is} \qquad (p_s)$$
$$\sum_i x_{i0} + \sum_j v_j \leq \sum_i \omega_{i0} \qquad (p_0)$$

(1)

where subscript s refers to the possible states of nature, i to the various consumers donors as well as recipients, h is an alias for i, when looked at as possible a recipient of aid; P_s is the probability of occurrence of state s, α_i the welfare weight of consumer i; $u_i(\cdot)$ is the strictly concave, differentiable utility function of consumer i, with arguments x_{i0} for beginning of period consumption and x_{is} for end of period consumption in state s; $\bar{u}_h$ is the threshold on utility below which assistance is required; $q_{js}(v_j)$ is the strictly concave production function of firm j in state s, with inputs v_j bought at the beginning of the period, ω_{is} are the state-specific endowments of the consumers, and ω_{i0} the endowments at the beginning of the period. Finally, Lagrange multipliers associated to the commodity balances (in brackets) are the market clearing prices p_0 and p_s. All vectors $(p_0, p_s, q_{js}, v_j, x_{i0}, x_{is}, \omega_{i0}, \omega_{is})$ are of dimension K, the number of commodities in the economy; commodities carry subscript k.

Thus, in this model care for the own fate finds reflection in the first utility term in the objective of (1), with risk aversion expressed via the curvature of the utility function, whereas empathy enters via the second term and is taken to remain effective only as long as utility does not exceed the threshold. The associated budget constraint for consumer i with utility above threshold is now:

$$\sum_s p_s^T x_{is} + p_0^T x_{i0} \le \sum_j \theta_{ij} [\sum_s p_s^T q_{js}(v_j) - p_0^T v_j] + \sum_s p_s^T \omega_{is} + p_0^T \omega_{i0} + T_i \qquad (2)$$

where T denotes the transpose, θ_{ij} is the share of consumer i in the profits of firm j, and T_i a cash equivalent transfer from the donor minus the tax to fund arrangements. Hence, (net) donors will have negative and (net) recipients positive T_i-value. Determining this transfer is a somewhat technical but computationally straightforward matter, essentially amounting to treating the well-being of the recipient as a public good, to which every recipient consumer contributes by Lindahl-pricing of consumptions x_{i0k} and x_{isk} at discounted and probability weighted prices

$$\phi_{ih0k} = \alpha_i \rho_{ih} \delta_h \sum_s P_s \frac{\partial u_h(x_{i0}, x_{is})}{\partial x_{i0k}}; \phi_{ihsk} = \alpha_i P_s \rho_{ih} \delta_h \frac{\partial u_h}{\partial x_{isk}}, \text{ where } \delta_h = 0 \text{ if utility is above}$$

threshold and 1 otherwise; receipts obtained are treated as given. Hence, $\phi_{ii0} = 0$ and $\phi_{iis} = 0$. This defines the transfers:

$$T_i = \sum_h \phi_{hi0}^T x_{i0} + \sum_{h,s} \phi_{his}^T x_{is} - \sum_h \phi_{ih0}^T x_{h0} - \sum_{h,s} \phi_{ihs}^T x_{hs} \qquad (3)$$

Model (1)-(3) provides a theoretical framework to represent the entitlement via budgets (2) and commodity prices from (1); the delivery via program (1) and the taxation via transfers (3) with marginal utilities obtained from (1).

We remark that premium payments for self as well as mutual insurance are also coming out of this model, and that a recipient of aid might also be a donor. Agents above the threshold ($\delta_h = 0$) receive no assistance since the donor derives no marginal utility from helping them. Whether all utility levels $\tilde{u}_{hs} = u_h(x_{h0}, u_{hs})$ will, for all possible states s in the solution reach these threshold depends on the empathy-coefficient ρ_{ih} as well as on the own thriftiness of the recipients and their willingness to take precautions (the curvature of the utility function).

A food crisis can be represented in this framework by assuming that in the state s that materializes, the value of endowments of some consumer i is very low, either because the quantity produced is deficient, say, with assets destroyed by disaster or war, or because the price of the endowments is very low, as in Sen's Bangladesh case discussed earlier (Sen, 1981a,b) with demand for the labour power of the landless and rural wages falling dramatically.

Regarding market imperfections, we note that budget equations (2) suppose that the consumer is perfectly able to smoothen consumption over time, by borrowing or lending, or by maintaining stocks. The opposite situation with zero borrowing can be accounted for via separate budgets for every state s, and state-specific welfare weights α_{is}. In this case, the individual will need more aid, as few coping options are left. This confirms that, improving the borrowing and lending operations may be as important as looking after food security directly, since it is key to the entitlement side of the issue.

Note that in this model selfish agents can free ride on others who have more empathy. It remains an open issue whether solidarity should be purely based on voluntary transfers driven by empathy as is done here in the Sinai-tradition, as opposed to assuming some right-to-food not based on donor empathy. Because of this the model can depict lack of empathy, illustrating for instance that it may be unsafe to rely fully on foreign donors, if only because their empathy tends to be triggered by images of disasters and, therefore, come too late for many of the victims of the disaster at hand.

Furthermore, the model illustrates that for the efficient solution of (1)-(3) to be attained, many institutions should be in place, in particular strong law enforcement to protect property rights so as to avoid misappropriation of aid flows. This calls for a well respected leadership that can act without excessive repression. Therefore, local government authorities should remain conscious of the donor's upcoming departure and made responsible for taking over the tasks temporarily assumed by the donor's "intensive care". Like any good doctor, the donor on its part should help the patient in regaining autonomy. While it would be naïve to assume that dictatorial or fully corrupt regimes can be forced to change course in this way, we would argue that the current donor practice of channelling aid through NGOs in such cases may act as a further undermining of the legitimacy of local institutions, possibly also those with agents acting in good faith. In short, in all three elements of the triad good governance is both product and input of food security management.

6.2. Implications for donor presence

So far, the discussion suggests that the donor should keep distances and not be very visible in the recipient country. It also points to a dilemma: in view of the subtleties of the issues, the donor should make sure to be well informed, and often almost act as substitute for the market's Invisible Hand when local market infrastructure fails. This means that quite a few data have to be collected and that well-calibrated models should be made operational so as to avoid various failures in entitlement, delivery and financing aspects of the operation.

Besides guiding the operation, this analytical support will also help offering the necessary transparency to the donors, short of which their willingness to pay might erode, in terms of the model because of the uncertainty on their part that financial contributions actually help. Specifically, and admittedly somewhat in the margin of our main argument, we remark in relation to WFP that the organisation should go beyond its current practice of offering well documented reports on its physical deliveries, and, in addition, provide comprehensive analytical accounts of its management of financial resources, for example give perfect clarity as to the relation between its operations in local currency and the donor contributions in foreign exchange. Currently, funds donated in USD are used for local purchases of goods and services, obviously via the prevailing local exchange rate. Yet, in accounting, the official UN exchange rate is applied to convert costs in local currency back to a USD financial report, which obviously leads to accounting differences. At a practical level, this means that

WFP should, in good as well as bad years, hold a permanent account in domestic currency from which local purchases can be made. Moreover, continued presence is also needed in situations where WFP seeks to provide assistance beyond emergency relief, in reconstruction and in prevention of disasters. This might obviously conflict with any principle of leaving as soon as the situation permits but it does not have to. WFP could maintain presence but its role and status should be a technical one, in support of the reconstruction of local institutions as much as of local infrastructure.

6.3. Applied model: optimal food aid provision in SSA

Applied policy simulation models are key ingredients in performing this technical role, as a component of decision support systems. We end with a brief overview of such a model, as recently designed in a joint research project of WFP and SOW-VU.

Within theoretical framework (1)-(3), the most demanding component is obviously welfare program (1) that covers the delivery elements of food security management. When moving to application of such a framework, the first step is to make the index sets of consumers i, producers j, states s, and commodities k explicit. At this stage, the spatial dimension becomes critical, because circumstances (weather conditions, population densities, soil qualities etc.) vary strongly across space in SSA as much as elsewhere, and especially in SSA because the food aid may have to be procured very far from the deficit areas, and transport infrastructure is deficient. Therefore, the model developed in this project is SSA-wide and distinguishes over 250,000 separate sites, points on grid of 5 arc minutes resolution (about 10 x 10 km).

In its present formulation the model considers the ex-post solution of (1)-(3) in any particular year, after uncertainty (essentially about production levels) has been revealed, and seeks to identify optimal coping strategies through routing of commercial as well as food aid flows across the continue. It generates consumption, production and trade of food for each cell on a, distinguishing between local food (produced by farmers for own consumption), commercial food (produced for sale), and food aid. At every cell on the grid, consumers buy each commodity, if available, depending on the prevailing market clearing price and on their purchasing power. Food aid in cash is represented as a general increase in purchasing power of (a sub-group of) consumers, food aid in kind as a specific increase in the power to 'buy' the commodity food aid only. The model can generate efficient transport flows consistently over the entire grid, for a given configuration of freight costs, inclusive of tariff and non-tariff barriers and within the capacity constraints imposed by the available transport infrastructure. The model uses dedicated algorithms and software is now fully calibrated to represent the year 2000.

Regarding the data underlying this model calibration, the estimated calorie intake in 2000 is derived from data on women's and children's weights in Demographic and Health Surveys (DHS, 2006). Calories are obtained from three sources: locally produced goods, commercially traded food and food aid. Figures 6a and 6b indicate the share of commercially traded goods and food aid in total consumption, respectively. Figure 6a confirms the generally shared view that Africa is not a well-integrated continent and people still are largely dependent on food produced in the immediate surroundings; commercially traded goods are only important along the West-African coast, along the south-east coast and in the most densely populated areas in Ethiopia. In Figure 6b, the particularities of the year 2000 are shown, when there was a major drought in the Horn of Africa. Hence, the share of food aid in total consumption was close to 1 in those areas. Other major areas of concentration of food aid are parts of Angola, and refugee camp sites scattered across the continent.

Turning to model simulation, the issue under study in the exercise reported is where to obtain the food from and how to route it at lowest cost to its destination. Figure 7 depicts the optimal flows of food aid, where the recipients are predefined by the database for the year 2000, but the routes followed and the locations of procurement or import are determined endogenously by the model.

7. Concluding remarks

Emergency food aid does not solve the poverty problem, but it helps the needy when nothing else works. It is always given under circumstances of upheaval and unrest, when no time is to be lost, and when careful data collection and policy analysis should not stand in the way. Nonetheless, precisely because it deals with the most vulnerable segments of the world's population, food aid provision deserves permanent attention as a research topic, to inform decisions better, to avoid capture of the subject and pro and cons of its implementation by various interest groups. The policy choices to be made should not be ideological. While those dispensing aid should be given a free rein, without excessive administrative burden,

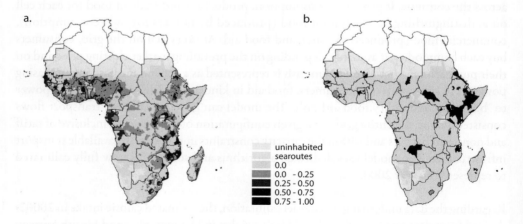

Figure 6. a: Share of commercial consumption; b: Share of food aid in total consumption.

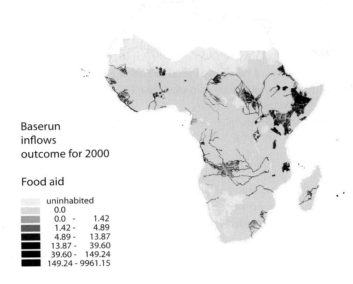

Baserun
inflows
outcome for 2000

Food aid

	uninhabited
	0.0
	0.0 - 1.42
	1.42 - 4.89
	4.89 - 13.87
	13.87 - 39.60
	39.60 - 149.24
	149.24 - 9961.15

Figure 7. Inflows of food aid.

the food aid planners in the background should be given the means to act on the basis of sound empirical data within a state-of-the-art analytical framework.

At each emergency, the approach to be adopted should be chosen afresh but quickly, on the basis of a good understanding of the causes of the crisis as well as on the quality of the locally available physical infrastructure and the state of the social institutions. For this, the planners would seem to need decision support tools that are more up-to-date than those they rely on at present.

At the core of food aid provision lies the issue that it involves foreign intervention, which needs local presence to be well informed and effective, and at the same time has to recognize that dealing with present and future food crises is primarily the responsibility of local authorities. The fact that outsiders are taking up this responsibility may be unavoidable and even natural for extreme events such as tsunamis but even then whenever possible the local authorities should remain in charge, at least in the perception of their constituencies, short of which local governance will be undermined. This is not to deny that in some places incumbent regimes are more part of the problem than of its solution. Yet, in our view the lesson learnt since Egypt and Sinai must be that reconstruction after disaster relates to social as much as to physical infrastructure, and that the donor community, when disengaging from a place, should leave behind a local governance with sufficient political capital that pays due attention to every element of the triad of food security management.

References

Abdulai, A., C.B. Barrett and J. Hoddinott, 2005. Does food aid *really* has disincentive effects? New evidence from Sub-Saharan Africa. World Development 33(10): 1689-1704.

Ahmed, R., 1988. Structure, cost and benefits of food subsidies in Bangladesh. In: P. Pinstrup-Andersen (ed.) Food subsidies in developing countries. John Hopkins University Press. Baltimore (Chapter 15).

Arndt, C. and F. Tarp, 2001. Who gets the goods? A general equilibrium perspective on food aid in Mozambique. Food Policy 26: 107-119.

Barrett, C.B., 2001. Does food aid stabilize food availability? Economic development and cultural change 49(2): 335-349.

Barrett, C.B., 2002. Food security and food assistance programs. In: B.L. Gardner and G.C. Rausser (eds.), Handbook of Agricultural Economics. Elsevier Science, Amsterdam.

Barrett, C.B. and D.G. Maxwell, 2006. Towards a global food aid compact. Food Policy 31(2): 105-118.

Basu, K., 1996. Relief programs: when it may be better to give food instead of cash. World Development 24(1): 91-96.

Bose, S., 1990. Starvation amidst plenty: the making of famine in Bengal, Honan and Tonkin, 1942-1945. Modern Asian Studies 24(4): 699-727.

Carter, M. and C.B. Barrett, 2006. The economics of poverty traps and persistent poverty: an asset based approach. Journal of Development Studies 42(2): 178–199.

Clay, D.C., D. Molla and D. Habtewold, 1999. Food aid targeting in Ethiopia. A study of who needs it and who gets it. Food Policy 24: 391-409.

Coase, R.H., 1960. The problem of social costs. Journal of Law and Economics 3: 1-44.

Coate, S., 1989. Cash versus direct food relief. Journal of Development Economics 30: 199-224.

Dasgupta, P. and D. Ray, 1986. Inequality as determinant of malnutrition and unemployment: theory. The Economic Journal 96(384): 1011-1034.

Del Ninno, C., P. Dorosh and K. Subberao, 2005. Food aid and food security in the Short- and Long-run: Country experience from Asia and Sub-Saharan Africa. World Bank, Washington D.C..

Dercon, S., 2004. Insurance against poverty. Oxford University Press, Oxford, UK.

Dercon, S. and P. Krishnan, 2003. Risk sharing and public transfers. The Economic Journal 113: C86-C94.

DHS, 2006. Demographic and health surveys, Measure DHS+, Calverton, USA, Calverton, USA: Macro International Inc

DISI, 2006. Comprehensive Food Security and Vulnerability Analysis: An Internal Review. Prepared by Development Information Services International (DISI) for WFP, Rome.

Donovan, C.,1996. Effects of monetized food aid on local maize prices in Mozambique. East Lansing: Michigan State Unversity, Department of Agricultural Economics.

Donovan, C., M. McGlinchy, J. Staatz and D. Tschirley, 2006. Emergency needs assessment and the impact of food aid on local markets. MSU International Development Paper 87, Department of Agricultural Economics, Michigan State University, East Lansing, Michigan, USA.

Dorosh, P., Q. Shahabuddin, and M.A. Aziz, 2002. Bumper crops, sagging prices and poor people: implications for food aid programs in Bangladesh. Market and Structural Studies Division Discussion Paper no 43, International Food Policy Research Institute, Washington, D.C.

FAO, 2003. Information note on food aid flows. Committee on commodity problems Sixty-fourth Session, Rome, Italy, 18-21 March 2003.

FAO, 2006. The state of food and agriculture 2006. FAO. Rome.

Fontana, M., P. Wobst and P. Dorosh, 2001. Macro policies and the food sector in Bangladesh: a general equilibrium analysis. Trade and macroeconomics discussion paper no 73, International Food Policy Research Institute, Washington, D.C.

Gelan, A.U., 2006a. Does food aid have disincentive effects n local production? A general equilibrium perspective on food aid in Ethiopia. Food Policy (in press).

Gelan, A.U., 2006b. Cash or food aid? A general Equilibrium Analysis for Ethiopia, Development Policy Review 24(5): 601-624.

Ginsburgh, V. and M.A. Keyzer, 2002. The structure of applied general equilibrium models. MIT press, Cambridge (Mass).

Hess, U. and J. Syroka, 2005. Weather-based insurance in Southern Africa: the Malawi case study. World Bank Agricultural and Rural Development Discussion Paper 13, World Bank, Washington DC.

Hobbes, R., 1651. Leviathan. Reprinted 1965 by Clarendon Press, Oxford.

Hoddinott, J. and M.J. Cohen, 2007. Renegotiating the Food Aid Convention: background, context and issues. IFPRI discussion paper 00690. IFPR, Washington D.C.

Holden, S., C.B. Barrett and F. Hagos, 2006. Food-for-work for poverty reduction and the promotion of sustainable land use: can it work?. Environment and Development Economics 11: 15-38.

Huff, H.B. and M. Jimenez, 2003. The Food Aid Convention: Past Performance and Future Role within the new Global Trade and Development Environment. Contributed paper presented at the International Conference Agricultural policy reform and the WTO: where are we heading? Capri (Italy), June 23-26, 2003.

INTERFAIS. Database maintained by WFP. Available at: http://www.wfp.org/interfais/

Kalavkonda V. and O. Mahul, 2005. Crop insurance in Karnataka. World Bank Policy Research Working Paper 3654. World Bank, Washington D.C.

Keyzer, M.A., 1987. Consequences of increased foodgrain production on the Bangladesh economy. In: A.J.J. Talman and G. van der Laan (eds) The computation and modelling of economic equilibria. Elsevier Science, Amsterdam. P. 59-83.

Keyzer, M.A., M.D. Merbis, M. Nubé, B.G.J.S. Sonneveld and R.L. Voortman, 2003. Food crisis management in Sub-Saharan Africa: a bird's eye view of the continent. Study commissioned by the Ministry for Development Cooperation, The Hague. SOW-VU, Amsterdam

Keyzer, M.A. and C.F.A. van Wesenbeeck, 2007. The Millennium Development Goals, how realistic are they? De Econimist 154(3): 443-466.

Lin, J.Y. and D.T.Yang, 2000. Food Availability, Entitlements and the Chinese Famine of 1959-61. The Economic Journal 110(460): 136-158.

Maxwell, S.J. and H.W. Singer, 1979. Food aid to developing countries: a survey. World Development 7: 225-247.

Osakwe, P.N., 1998. Food aid deliveries, food security and aggregate welfare in a small open economy: theory and evidence. Bank of Canada Working Paper 98-1, Ottawa, Canada.

Quisumbing, A., 2003. Food aid and nutrition in rural Ethiopia. World Development 31(7): 1309-1324.

Ravallion, M., 1987. Towards a theory of famine relief policy. Journal of Public Economics 33(1): 21-39.

Ravallion, M., 1997. Famines and Economics. Journal of Economic Literature 35(3): 1205-1242.

Rousseau, J.-J., 1762. Du contrat social, ou principes du droit politique. Marc Michel Rey, Amsterdam.

Sen, A., 1981a. Poverty and famines: an essay on entitlement and deprivation. Clarendon, Oxford, UK.

Sen, A., 1981b. Ingredients of famine analysis: availability and entitlement. Quarterly Journal of Economics 96(3): 433-464.

Skees, J.R., P. Hazell and M. Miranda, 1999. New approaches to crop yield insurance in developing countries. EPDT Discussion Paper 55. International Food Policy Research Institute, Washington DC.

Skees, J.R, S. Gober, P. Varangis, R. Lester and V. Kalavakonda, 2001. Developing rainfall-based index insurance in Morocco. Policy Research Working Paper 2577. World Bank, Washington D.C.

Spinoza, B., 1670. Tractatus Theologico-Philosophicus.

Tapia-Biström, M.-L., 2001.Food aid and the disincentive effect in Tanzania. Department of Economics and Management, University of Heksinki, Helsinki, Finland.

Tschirley, D., C. Donovan and M.T. Weber, 1996. Food aid and food markets: lessons from Mozambique. Food Policy 21(2): 189-209.

UNHCR, 2006. 2005 UNHCR statistical yearbook. UNHCR, Geneva.

Veeramani, V.N., L.J. Maynard and J.R. Skees, 2005. Assessment of the risk management potential of a rainfall based insurance index and rainfall options in Andhra Pradesh, India. Indian Journal of Economics & Business 4(1): 195-208.

Voltaire, 1759. Candide, ou l'Optimisme. Reprinted and translated by Penguin books, 1947.

Von Braun, J. and B. Huddleston, 1988. Implications of food aid for price policy in recipient countries. In: J.W. Mellor and R. Ahmed (eds) Agricultural price policy for developing countries, John Hopkins University Press, Baltimore, MD, pp. 253-264.

WFP, 2006a. Food aid monitor, available at http://www.wfp.org/interfais/index2.htm

WFP, 2006b. WFP annual report 2005. WFP, Rome.

WFP, 2007a. How to use food aid. Overview of uses of food aid. Available at: http://www.wfp.org/food_aid/introduction/index.asp?section=12&sub_section=1

WFP, 2007b. Evaluation of EMOP 10339.0/01: Assistance to populations affected by conflict in greater Darfur, West Sudan. WFP, Rome.

WFP,2007c. Food Procurement Map. Available at: http://www.wfp.org/operations/Procurement/food_pro_map_06/foodmap.html

Yamano, T., L. Christiaensen and H. Alderman, 2005. Child growth, shocks, and food aid in rural Ethiopia. American Journal of Agricultural Economics 87(2): 273-288.

Food security policies in West Africa: a mix of options to be adjusted to local and regional conditions

Michel Benoit-Cattin

Abstract

'Food security [is] a situation that exists when all people, at all times, have physical, social and economic access to sufficient, safe and nutritious food that meets their dietary needs and food preferences for an active and healthy life' (FAO, 2002). From the standpoint of aggregate country-wide trends, food requirements increase quantitatively as the number of consumers grows and these requirements gain diversity as incomes rise and urbanisation increases. In West Africa, the rural population is still growing despite a rapid rural exodus and the subsequent increase in urbanisation. West African countries face the same challenges but have varied geographic, agro-ecological, historical and economic contexts. These are reflected by different food security problems and thus by different combinations of conceivable options.

1. Introduction

Among the many definitions of food security, the one put forward by FAO can be retained: 'Food security [is] a situation that exists when all people, at all times, have physical, social and economic access to sufficient, safe and nutritious food that meets their dietary needs and food preferences for an active and healthy life' (FAO, 2002). Whether at household or country level, food security depends on the one hand on a technical capacity for stable, sustainable self-supply and on the other on the economic capability of earning income to purchase what is not produced.

The food challenge results from the fact that resources of agricultural origin are seasonal, uncertain and geographically dispersed while individual requirements are daily and constant, with requirements for diversity and quality that increase with income and urbanisation. The matching of requirements and resources involves storage, processing and trading operations at greater or lesser distances. Permanent structural matching should be sought but with attention paid to its conjunctural execution in the face of shocks that are foreseeable to varying degrees. At the level of households, unstable incomes and/or uncertain food crops link poverty and food insecurity. Reducing poverty certainly involves reducing food insecurity and vice versa but the two objectives differ with regard to the ways and means of achieving them. I focus here on reducing food insecurity at the country level.

With regard to the aggregate trend at country level, food requirements increase quantitatively with growing numbers of consumers and diversify with the improvement of their incomes and with urbanisation. Ways and means of reducing food insecurity at both country and household level have frequently been put forward in both the international and national literature. The recommendations, options and orientations concern the technical aspects of agriculture, institutional measures, policy guidelines or investment priorities. The proposals are usually fairly all-purpose and the various aspects (technical, institutional and political) are not sufficiently articulated. They are sometimes presented in a markedly one-off manner with technical or institutional analysis of a local experience.

Most of these proposals are addressed here as options. They are organised and discussed with reference to the context of West Africa, with the emphasis laid on conditions of implementation and on questions remaining to be addressed by the research sector. The options usually do not exclude each other but tend more to be synergistic proposals. Selecting and ranking them is not simple and can only be performed with reference to a given context. This chapter is not a bibliographical analysis but a personal synthesis based on experience and reflection. The reader can find further technical and organisational references in the *Mémento de l'agronome* (CIRAD, 2002) and the accompanying CD. Analyses with a more political and economic slant can be found in *World Development Report 2008: Agriculture and Development* (World Bank, 2007), a fully up-to-date publication that also mentions a large number of bibliographical references.

As regards the population context, it is worth remembering that the demographic changes in West Africa, as in sub-Saharan Africa as a whole, are only just beginning and rural population growth is continuing in spite of a rapid rural exodus and the subsequent increase in the urbanisation rate. In addition, the AIDS epidemic is calling into question in a lasting manner the social and economic coherence of demographic structures; this further complicates forecasting and is another factor of insecurity.

The different available technological options should be examined, to identify options for improving the productivity and resilience of production systems through water management, mechanisation, varietal improvement, agricultural conservation measures, integrated approaches and integration of farming and post-harvest operations. The institutional options that address these technical changes are in a new so-called 'post structural adjustment' context characterised by government withdrawal, decentralisation and privatisation. The institutional options proposed require ad hoc legislative and regulatory frameworks. The countries in the region will benefit from coordination to learn from each other's experiences and to harmonise their regulations as much as possible.

Various public investment options can be mentioned for infrastructure and concern irrigation, transport, trade, storage, exports and markets. The importance of public services is emphasised for information and communication technologies (ICT), research, education and health. The combination of these broad groups of options in each national context constitutes a food security policy. This in turn is closely linked to other sectorial and economic policies. The selection and development of the options presented requires a combination of options in each country that is adapted to the agro-ecological and economic situation and strong coordination and harmonisation between countries.

Countries have varied geographic, agro-ecological, historical and economic contexts in the face of these common challenges. This results in differing food security issues and hence different combinations of possible options. The technical options that can be envisaged, the institutional options likely to enhance them and certain public investment options are examined in turn. It is considered that a food security policy consists of the combination of these three main sets of options applied in each national context. This food policy is in turn closely connected to other sectorial and economic policies at country or sub-regional level. This linking results in problems of performance and other research questions.

2. Technical references

In every case in West Africa, it is first of all necessary to increase the quantity and regularity of the availability of food products in order to reduce both structural and conjunctural food insecurity. To achieve this, farming systems should be more productive and more resilient as the land and water resources used are becoming increasingly limited, if only as a result of the increase in the rural population. These resources tend to suffer degradation when subjected to constant pressure.

Although climate is a permanent instability factor for production and hence for food security, the global warming predicted is a new and supplementary uncertainty. Climate change, whose pattern and local consequences are still poorly defined, should trigger the mobilisation of efforts in both local and international research to contribute to the invention of agricultures for the 21st century.

In West Africa today, a distinction can be made between at least three large zones mainly on the basis of rainfall. The Sahel zone is in the north at the edge of desert areas, the forest zone runs along the coast in the south and the Sudan-Sahel or savannah zone lies between the two. The Sahel zone is mainly pastoral and produces a surplus of livestock, milk and gathered foodstuffs but suffers from droughts. Household food security relies on a degree of local grain production, the size and productivity of herds and flocks and the prices of animal products and grains on the local markets. Fairly even crops of rice and maize are produced in the forest zone, together with roots, tubers and plantain, and some perennial crops (cocoa, coffee, rubber-tree latex, etc.) exported beyond the region are subject to unstable world prices. Overall, food security is fairly well assured by local production and available income. The northern part of the Sudan-Sahel zone depends mainly on rain-fed millet and sorghum crops exposed to erratic rainfall. The food insecurity of rural households is more marked among those with no income from animal farming or migration. Security is better assured further south as surpluses of maize and sorghum can be produced, sometimes complemented by cotton (or groundnut) production whose returns fluctuate increasingly with unstable world prices displaying a downward trend.

One of the first technical options for increasing and stabilising agricultural production under the various ecological conditions is the mastery of water (Tiercelin, 1998). This is fairly undeveloped in the sub-region and mainly involves a few large hydro-agricultural development operations (in the Senegal river valley and the Niger delta in Mali)(Kupper and Tonneau, 2002). Significant technical improvement to existing facilities can still be envisaged in terms of both intensification and diversification. A heavyweight option consists of extending these development operations while the maintenance and rehabilitation of existing facilities is already costly. Another option with potentially broader scope concerns lowlands in both the savannah and forest zones (Lavigne-Delville, 1998). Priority has been awarded to rice in all cases involving water management. Other options consisting of alternative or complementary crops could be examined.

As regards the improvement of agricultural labour efficiency and productivity, it is important to rehabilitate the promotion of animal draught power for cultivating operations and the transport of inputs, crops and people (Le Thiec, 1996). Mainly for reasons of funding, this has been associated more with the strongly supervised promotion of cash crops (cotton and groundnut) than with the development of food crops. The same observation could be made concerning the fertilisers recommended and distributed in close relation to the promoting of cash crops, which only indirectly benefit the grain crops in a rotation.

Conventionally, the development and distribution of improved varieties remains the basis of improvement of the productivity, hardiness and quality of agricultural crops. However, *in situ* agronomic appraisals show that water availability, soil fertility and weed control are weightier limiting factors than the potential of the varieties grown (traditional or improved) (Matlon, 1984). Breeding work should nevertheless continue and biotechnologies can be used, without overestimating their potential impact.

Indeed, the potential of improved species can only be sustained if increasingly scarce land is exploited in a more sustainable manner rather than in 'mining' style. Expanding cultivated areas for reasons of population growth is mainly - and increasingly - performed by reducing the duration of fallows. The reconstitution of soil mineral and organic fertility is becoming steadily poorer. The issue is that of identifying, developing and extending techniques that can be referred to generally as conservation agriculture (Ashburner *et al.*, 2005). This is a basic component of the sustainable management of natural resources. The range of techniques is broad and includes land development using stone contour barriers, vegetation strip barriers, the introduction of trees for land improvement managed as hedges or enclosures or grown in fallows, mulch-based direct seeding, etc. Many more or less one-off operations have been run by research bodies and numerous NGOs and should serve as the basis for research/

development actions on a larger scale capable of significantly changing farmers' practices in the region.

The role of ligneous plants and livestock should be reconsidered in the heart of these practices. In particular, animal farming plays a key role in savannah agriculture for traction, transfer of fertility and the conservation of organic matter and also as food and for assuring the incomes of small farmers. Animal husbandry, and especially the rearing of large ruminants, allows the 'horizontal' transfer of fertility between village grazing zones and cultivated areas (Berger, 1996). This maintains or even improves productivity in synergy with other conservation farming practices. The sale of small ruminants generates funds for the owner in case of an unforeseen need for cash (illness, social obligation) without compromising food reserves by sales or production capacity by mortgaging land or equipment.

The development of integrated pest and disease management programmes should continue to safeguard crops and also to conserve land and water and producer and consumer health. Damage is better controlled when pesticide application is reduced and there are less harmful effects for humans, domestic animals and the environment. As a complement, the improvement of post-harvest techniques can also reduce losses and also improve product quality. These techniques cover storage, primary processing and the packing of produce (Appert, 1985).

To complete this overview of the technical options that can be envisaged for reducing food and nutritional insecurity, I would like to stress the need to make use of farmers' innovations. Farmers have used the options mentioned as well as others in order to innovate. These innovations should be identified and their effectiveness appraised. The factors affecting their implementation should be understood so that their extension can be enhanced. From this point of view, multidisciplinary research of the participatory 'farming systems' type should be rehabilitated and promoted (Matlon *et al.*,1984, 1986). The development, promotion and extension of these various technical options require certain prior or synergistic institutional innovations.

3. New institutional arrangements

The current context can be described as post-structural adjustment and is characterised by the withdrawal of governments, decentralisation and privatisation. In particular, decentralisation is opening up new prospects for land and natural resource management ((Le Roy *et al.*, 1996; Magrin, 2006). To be sustainable, the technical possibilities for productivity gains mentioned above need secure access to land and the other resources. This security feature often requires a new legislative and regulatory framework and involves public policies negotiated between the new local authorities resulting from decentralisation and the duly organised social groups concerned. Political declarations of intent have been strong and renewed in these areas in the sub-region but actions for implementation are still fairly limited. Analysis of the conditions of implementation of these policies and their effects in terms of effectiveness and equity are needed for both the states that have initiated measures and those that have a 'wait and see' approach. This will enable certain readjustments in the formulation of options and in recommendations for implementation. Much hope is placed in producers' organisations in policy statements and the actions of certain donors and many NGOs, in particular to compensate the weakness of states and markets (Mercoiret, 1994; Rondot and Collion, 2001). Producers' organisations have been relied on to distribute inputs, collect produce and guarantee loans. Experiments have been conducted on the management of grain stocks. Numerous complementary and competing initiatives have been promoted by various operators seeking to mobilise farmers - with the latter displaying various degrees of conviction and interest. An appropriate legal status and the development of sustainable funding procedures (contributions and fiscal charges) are needed for these organisations to be viable.

Here again, it is necessary to conduct comparative analyses of the conditions of the promotion and management of producers' organisations and of their impact in terms of effectiveness and fairness. The privatisation of agriculture also involves the upstream and downstream aspects. Upstream, the privatisation of seed production and distribution raises problems of access related to patentability, especially when biotechnologies are used. The research sector (national and international) has responsibility in this domain and a key role in limiting processes that lead to the exclusion of those most lacking in resources.

Upstream again, privatisation has shown its limitations in the supply of fertiliser and the obtaining of loans. The public authorities, operating in synergy with producers' organisations and the public sector segment concerned, have a role to play here.

As for loans, the shortcomings of public banks specialised in funding for agriculture have not been counterbalanced by so-called microfinance bodies. On the contrary, microfinance develops all the better when it is underpinned by a solid formal banking system (Wampfler *et al.*, 2003). Farming is too risky to stimulate the availability of medium and long-term credit, whether by commercial banks or by decentralised financial systems, unless the latter are effectively controlled by farmers. Microfinance is more effective in providing access to small credit operations for poor households that are in difficulty. Furthermore, as it is related to the promotion of work that generates income, it allows them to repay their debts and leave the vicious circle of (more or less accidental) financial problems, food problems, debt, etc. Microfinance can thus play a significant role in conjunctural food insecurity.

Microfinance must in all cases flourish without too much risk for both borrowers and lenders within a legislative framework, as has been done for the WAMU Support project for the regulation of saving and credit associations (PARMEC/WAMU). (Projet d'Appui à la Réglementation sur les Mutuelles d'Epargne et de Crédit /West African Economic and Monetary Union).

As regards the privatisation of the downstream segment, the trend is for the integration of producers via trade and processing enterprises, with these replacing the marketing companies and public boards that started their decline during structural adjustment. These contractual relations can help to soften market risks for producers but they are not without risk of domination, hence the need for control by the public authorities in partnership with professional organisations.

Another pathway for the regulation of these oligopolistic processes is the promotion of 'contestable markets'. The existence of an efficient free market side by side with the integrated sectors is a powerful way of controlling them and preventing certain market imperfections. As both producers and consumers have the choice between an integrated sector and a competitive sector, the risks of imperfection are reduced in each of the latter. The promotion of such markets requires the setting up of material structures and information systems.

Both upstream and downstream privatisation involves the establishment or strengthening of certain professions (traders, transport services, processors). Their promotion and structuring should be planned in phase with and on the same lines as those of producers' organisations.

The same comparative analyses of the implementation conditions of the corresponding measures and their impacts in terms of effectiveness and fairness should be encouraged. The promotion of institutional options - like that of the technical options - often requires a greater or lesser degree of public expenditure in investment and in the functioning of public services.

4. Re-launching basic public investment

The implementation of the technical and institutional options mentioned above requires the re-launching of a whole range of public investments in infrastructure and public services. The rehabilitation and enlargement of large irrigated systems is doubtless the main investment in infrastructure specifically devoted to farming. However, such investment does not have solely agricultural justification in any country in the world. Their economic interest may come from the supply of hydroelectricity or be part of a public works policy. They can also be justified by regional development or food sovereignty considerations.

The infrastructure making the greatest contribution to the performances of the agrofood system consists doubtless of transport, roads, railways and river transport. It makes it possible to both reduce the price of farm inputs and the cost of movement of produce. It obviously makes a more or less direct contribution to all kinds of economic activities that are not directly related to the agrofood system. Commercial infrastructure (storage, exports, market facilities) acts in synergy with the infrastructure above and is related to industrial infrastructure (free zones, etc). This can be an excellent field for the increasingly talked about public/private partnerships. Indeed, as sources of public funding are increasingly rare, such partnerships are more necessary than ever for financing investments in infrastructure in West Africa. The regional banking sector is rich in unproductive funds and could support and accompany these private initiatives.

Good information for all the private players in the agrofood system, as in the rest of the economy, requires substantial investment in information and communication technology. Public/private partnerships are essential in this field too. In particular, the ongoing expansion of the cellular phone networks in rural areas is very promising.

Agricultural research and animal health can be mentioned in the category of public services that make a direct contribution to the performance and sustainability of agriculture. As seen in section 2, a lot of technical options are available. Maybe the priority is on their transfer and dissemination through more participative programs. For renewed programs on plant improvement and protection, they have to be promoted and conducted at the regional level in the context of thematic networks mobilising regional and international expertise. Education and human health are part of other sectorial policies but certainly contribute to reducing food and nutritional insecurity.

Mention of the main public expenditure options leads to the political options that in turn depend on the situation in each country, its past, and its regional and international positions.

5. Policy options

A food security policy has been defined here as a combination of technical, institutional and public expenditure options. Examination of these various groups of options brings up more general political options concerning relations between the public and private sectors or agricultural policy options that take non-food crops into account, economic policy options (the role and financing of investments), regional development options (large projects) or sectorial policy options (education and health).

It will be remembered that in the 1960s the ideological and political contrasts between the new independent countries were strong. However, this did not prevent them from going through the same major development phases to reach political formulations that are fairly homogeneous today in spite of the variety of conditions.

The agro-ecological diversity of the sub-region has been mentioned in passing. From the angle of geographic and economic diversity, three groups of West African countries can be identified as follows:
- the 'coastal' group of countries;
- the Sahel group;
- all the other countries, described as being intermediate.

This variety, combined with agro-ecological diversity, means that there is little point in wishing to identify sets of technical, institutional and economic recommendations with standard priorities. In the coastal countries, on the one hand fisheries resources and the substantial availability of tubers make it fairly easy to make up for possible grain deficits and on the other their maritime access to international markets makes it possible to obtain supplies more cheaply, which also has the disadvantage of setting up competition with national production. In contrast, the Sahel countries are more protected from foreign competition by their remoteness from major ports and poor roads, but they have fairly few alternatives in case of grain deficit and imported cereals are generally the only option - a costly one when this is not funded by international aid bodies. It seems difficult to make any generalisations about the intermediate countries as they tend to combine the handicaps of the coastal and Sahel countries, that is to say precarious domestic production and strong competition from abroad. Furthermore, the availability and source of foreign exchange play a major role in the food security expectations of the population. When the source of foreign currency is non-agricultural, such as mining operations (uranium in Niger, phosphate in Togo and Senegal, gold in Ghana and Mali, oil in Nigeria, etc.), the positive impact on food security is obvious, especially via food imports easily funded by foreign currency earnings from the export of mining production. However, these facilities may go against food sovereignty. Yet, when foreign currency earnings are from cash crops (cocoa in Côte d'Ivoire and Ghana, cotton in Burkina Faso and Mali, groundnut in Senegal, etc.), the question of the possible choice between food crops and cash crops and hence the problem of the optimum allocation of land and agricultural labour is raised. However, the impact of such division should be seen in relative terms, especially as it has been seen in some regions that cash crops like cotton have positive external effects on food production (Raymond and Fok, 1995).

Beyond their agro-ecological, geographic and economic diversity, the countries of West Africa have all experienced the same major phases of development for more than half a century. It is true that they were colonised by two major countries, Britain and France, together with Portugal. It is true that they may have chosen one side or another, including the non-aligned group, during the cold war, but with hindsight one can identify several common features that lastingly marked the minds and behaviour of farmers, officials, private operators, experts and scientists, especially with regard to the perception of the respective roles of the state, markets and civil organisations.

First of all, the land development promoted by the colonial powers - each in its own way - followed a preliminary period of administered development that was socialist or capitalist in style, under the direct control of the newly independent states. These development models were sometimes effective for export-focused agriculture even if they resulted in cumulated macroeconomic imbalances, although these were caused more by poor quality governance than inappropriateness of the models. In fact, the latter were all called into question in the 1980s by the so-called structural adjustment policies whose main aim was to correct macroeconomic imbalances. In this context dominated by the 'Washington Consensus', fairly radical institutional and economic measures were implemented as conditionalities, with controversial results. The two main orientations concerning agriculture were market liberalization and State withdrawal. Public monopolies were dismantled and private traders took control of the markets, both local and international. Henceforth, producers and consumers were confronted with unstable and unpredictable prices. These policies were all the more successful as they allowed for local and national features, which has not always been the case with the one-size solutions often proposed to the countries in West Africa.

Today, the West African context can be characterised by a determination to achieve regional economic and monetary integration within the framework of the Economic Community of West African States (ECOWAS) and, at continental level, within that of the New Partnership for Africa's Development (NEPAD). These ambitions both facilitate and form constraints for the new rules of international trade currently being negotiated within the framework of relations between the European Union and ACP (Africa, Caribbean, and Pacific) countries or at the WTO (World Trade Organisation).

As regards food security, the trade rules being negotiated will take the ECOWAS sub-region along the road towards a customs union, with facilitated intra-zonal trade and a degree of external tariff protection to prevent competition with local production, and especially food production. Easier and less costly movement of food products made possible in particular by investment in infrastructure should improve food security. On a conjunctural basis, with meteorological uncertainties, surpluses in one place can more easily make up for deficits in another; this will result in narrow price fluctuations and hence finally encourage farmers to grow more for the market. As a complement to this, stabilising the price of imported foodstuffs - rice in particular - will help to stabilise domestic prices and maintain their level. The availability of food supplies and stabilised prices make a fundamental contribution to the food security of households.

If customs union is progressing well at the regional level, monetary union is a more long-term aim, given the initial situation. Eight countries belonging to WAMU, the West African Economic and Monetary Union (Benin, Burkina Faso, Cote d'Ivoire, Mali, Niger, Senegal, Togo and Guinea Bissau) all use the CFA franc, and most of them since 1945. Six countries (Gambia, Ghana, Guinea, Liberia, Sierra Leone and Nigeria) set up the West African Monetary Zone (WAMZ) in 2000 with the aim of a common currency by 2009 and the prospect of the merging of the two monetary zones in ECOWAS. In matters of food security, the WAMU countries would be able to obtain foreign currency to overcome any food shortage by pooling their foreign exchange reserves.

Overall, convergences undeniably exist in the formulation of the main lines of policies and in the perception of the external constraints that weigh on them. However, the range of political options is not limited if they incorporate the distinctive agro-ecological, geographic, historical, demographic, economic and monetary features mentioned above.

If major trends exist in questions of food security, these can be found in basic economic and social behaviour. Each person, each household and each country is exposed to a degree of food insecurity and has his/her/its own perception of this. Anticipation and the covering of the risks perceived require negotiated public policies to share roles between the state, the markets and civil organisations. It is therefore necessary to go beyond market/state antinomy in analysis of the functioning of agricultural markets in the West African countries and replace it by a complementary vision of these two institutions that includes civil organisations. Furthermore, more allowance should be made in food security policies for the relationships between food policy in the strict sense and agricultural policy, and especially with regard to the contribution of cash crops to the improvement of food security. These policies should form part of a balanced development plan at the national level.

6. Concluding remarks

At country level, ecological diversity is combined with economic and geographic diversity. Some countries are landlocked, others have a seacoast, some are exporters of agricultural products, others have non-agricultural incomes from abroad, some are in the franc zone and others are not. It will be understood that their starting positions will not be the same and neither will the resources for the implementation of the options described.

The institutional options proposed require ad-hoc legislative and regulatory frameworks. The countries in the region would find it beneficial to coordinate their individual experiences and harmonise their rules as much as possible.

The countries are also all concerned with regional integration policies and are involved in international negotiations with various partners (EU, WTO). In this context, it is necessary to improve information bases and analysis and negotiation capacities in each country and in their regional organisations. More generally, meeting the food security challenge faced by West African economies requires capacity strengthening at various levels: governments, elected officials, civil organisations and the scientific community. Research in agriculture and in economics by regional networks contributing to international efforts on the subject is needed to address numerous points in the different fields covered.

In all the situations in West Africa and the rest of the continent, reducing food insecurity is probably the main effective criterion for appraising the relevance and effectiveness of the development policies implemented and of overall governance. From a structural point of view, a trend for the improvement of a country's capacity to cover its food requirements by means of domestic production or imports is seen as an aggregate in the movement of its balance of trade structure and more particularly in the movement of the agricultural trade balance. From a conjunctural point of view, improved capacity for facing droughts, floods or pest outbreaks results in better anticipation of problems and their real scale and by less recourse to food aid.

The selection and implementation of the options described requires a choice in each country of the combination of options that match its agro-ecological and economic situation and strong coordination and harmonisation between countries. All this can only be envisaged and effective in a context of good governance and on condition that peace and security are assured in the sub-region.

References

Appert, J., 1985. Le stockage des produits vivriers et semenciers. Maisonneuve et Larose, Paris, France, 225 p.

Ashburner, J., P. Djamen, B. Triomphe, J. Kienzle and F. Maraux, 2005. Regards sur l'agriculture de conservation en Afrique de l'ouest et du centre et ses perspectives: contribution au 3ème Congrès mondial d'agriculture de conservation, Nairobi, octobre 2005. FAO, Rome, Italy, 101p.

Berger, M., 1996, L'amélioration de la fumure organique en Afrique Soudano-Sahélienne. In Agriculture et Développement. N° hors série. CIRAD, Montpellier, France, 58p.

CIRAD, 2002. Mémento de l'agronome. Montpellier, France, 1691p.

FAO, 2002. The State of Food Insecurity in the World 2001. Rome, Italy, 58p.

Kupper, M. and J.P. Tonneau (eds.), 2002. L'office du Niger, grenier à riz du Mali. Cirad, Karthala, Paris, France 252p.

Lavigne-Delville, P., 1998. L'aménagement des bas-fonds en Afrique de l'Ouest. In: Tiercelin J.R. (ed.) Traité d'irrigation. Lavoisier, Paris, France, pp.560-582.

Le Roy, E., A. Karsenty and A. Bertrand, 1996. La sécurisation foncière en Afrique: pour une gestion viable des ressources renouvelables. Karthala, Paris, France, 388 p.

Le Thiec, G., 1996. Agriculture africaine et traction animale. CIRAD. Montpellier, France, 355p.

Magrin, G., 2006. La décentralisation règlera les problèmes de l'Etat en Afrique. In: L'Afrique des idées reçues. Belin, Paris, pp. 383-389

Matlon, P., 1984. Technology evaluation: five case studies from West Africa. In: P. Matlon, R. Cantrell, D. King and M. Benoit-Cattin (eds.) Coming full circle: farmers' participation in the development of technology. IDRC, Ottawa, Canada pp 95-118.

Matlon, P., R. Cantrell, D. King and M. Benoit-Cattin (eds.), 1984. Coming full circle: farmers' participation in the development of technology. IDRC, Ottawa, Canada, 176 p

Matlon, P., R. Cantrell, D. King and M. Benoit-Cattin (eds), 1986. Recherche à la ferme: participation des paysans au développement de la technologie agricole. CRDI, Ottawa, Canada, 217 p.

Mercoiret, M.R., 1994. L'appui aux producteurs ruraux. Ministère de la Coopération, Karthala, Paris, France, 463p.

Raymond, G. and M. Fok, 1995. Relations entre coton et vivrier en Afrique de l'Ouest et du Centre. Le coton affame les populations ? Une fausse affirmation. In Economies et Sociétés, Série Développement agro-alimentaire, A.G. n°22, Cahiers de l'ISMEA, Paris, France, pp 221-234.

Rondot, P. and M-H. Collion, 2001. Organisations paysannes: leur contribution au renforcement des capacités rurales et à la réduction de la pauvreté-compte rendu des travaux, Washington, D.C., 28-30 juin 1999. Département développement rural, Banque Mondiale, Washington, D.C. USA, 82p.

Tiercelin, J.R. (ed.), 1998. Traité d'irrigation. Lavoisier, Paris, France, 1011p.

Wampfler, B., C. Lapenu and M. Roesch (eds.), 2003. Le financement de l'agriculture familiale dans le contexte de libéralisation. Quelle contribution de la microfinance ? Résultats du programme de recherche et actes du séminaire international, 21-24 janvier 2002, Dakar, Sénégal. CIRAD, Montpellier, France, CD Rom.

World Bank, 2007. World Development Report 2008: Agriculture for Development. Washington DC, USA, 512p.

Cultivated land change in transitional China: implications for grain self-sufficiency and environmental sustainability

Shuhao Tan, Nico Heerink and Futian Qu

Abstract

Grain self-sufficiency is an important agricultural policy goal in China. To feed its large population, however, very limited land resources that are used intensively are available. Moreover, grain self-sufficiency is under threat due to the continued conversion of cultivated land into construction land or land for other uses and the increasing pressure on the remaining cultivated land. In the future, increases in grain productivity probably have to rely on even higher levels of input use. This may further worsen the quality of the natural resource base, through soil degradation, water scarcity, water pollution, etc., and further reduce the efficiency of fertilizer application. This raises the question whether China can maintain its successful grain self-reliance policy in the near future and what will be the impact of this policy on environmental quality. This chapter uses provincial data to examine the impact of cultivated land conversion on agricultural productivity and environmental quality. We find that the center of grain production is gradually moving northwards towards more fragile and water scarce areas, putting more pressure on the environment. Land conversion has caused large losses in ecosystem service values in the 1990s, but large-scale ecological restoration programs implemented in recent years more than compensate for such land conversion losses. Ecological restoration programs, however, are concentrated in regions with relatively low land productivity, while land conversion for construction usually takes place in areas with relatively high land productivity. Newly cultivated land is especially located in areas that are only marginally suited for agricultural production, and that are likely to have much lower productivity levels. Because the stock of potentially cultivable land is almost exhausted, China's grain self-sufficiency policy can only be maintained in the near future by preserving the available stock of arable land and increasing its productivity in a sustainable way.

1. Introduction

Grain self-sufficiency is an important agricultural policy goal in China. The Chinese government has set itself a target of at least 95 percent grain self-sufficiency under normal conditions (State Council, 1996). To feed its large population, which reached 1.3 billion in 2005, very limited land resources are available. Out of the total land area of 960 million hectares, only 13.5% can be used as arable land. As a result, only 0.10 hectare per capita is available for agricultural production (NBS, 2006).

China's population is expected to stabilize at a level of 1.6 billion by 2030 (Zhong *et al.*, 1999). In the coming years this growing population will put an even more severe burden on the country's agricultural sector. Sustaining the agricultural production base and improving agricultural productivity is generally considered by researchers and the government in China as the most effective way to guarantee an adequate level of food production in the long-term. Strengthening the agricultural production capacity and developing a modern agricultural sector while sticking to the policy of food self-reliance have therefore become national policy priorities in recent years (State Council, 2005 and 2007). Proposed measures include intensifying the conservation of arable land and improving the ecological environment, accelerating the construction of irrigation and water conservation facilities, stimulating the use of environment-friendly fertilizers and pesticides, and promoting agricultural science and technology.

In the future, the agricultural production capacity in China may be greatly restricted by further reductions in the available arable land area and the resulting pressure on the remaining arable land. This may further worsen the quality of the natural resource base, through soil degradation, water scarcity, water pollution, etc., and further reduce the efficiency of fertilizer application. This raises the question whether China can maintain its successful grain self-sufficiency policy in the near future and what will be the impact of this policy on environmental quality. An important issue in this respect is the conversion of cultivated land into land used for construction and other purposes. A recent study by Shao and Xie (2007) shows that 9.3% of the arable land has been taken out of cultivation during the period 1996-2004, while 2.9% has been added to the stock of cultivated land over the same period.

This chapter uses provincial and regional data to examine the impact of cultivated land conversion on agricultural productivity and ecosystem services. Various studies have addressed the cultivated land conversion issue in China. Many of these studies examine the link between cultivated land conversion and food self-sufficiency[26] (for instance Fu *et al.*, 2001; Liu and Wu, 2002; Zhao *et al.*, 2002; Yang and Li, 2000; Zhu, 2006; Deng *et al.*, 2006). Other studies examine the environmental effects of cultivated land conversion (Tan *et al.*, 2005; Chen and Wang, 2005), while Ash and Edmonds (1998) examine the relationship between land resources and both agricultural production and the environment. In this chapter we go one step further by taking also the stock of uncultivated land that may be cultivated in the future into account in the analysis. The size of this stock can evidently have a major impact on the arable land size and its quality in the near future. In addition, we use more recent data that have been estimated with greater precision than the data used in most of the earlier studies. This is especially relevant because land use structure has changed greatly during the last decade as a result of rapid urbanization and industrialization and increasing welfare levels. A third innovative aspect of our study is that we use a method

---

[26] The term food security is commonly used in China to refer to food self-sufficiency. As the use of this term deviates from the commonly accepted definition of food security ('access of all people at all times to enough food for an active, healthy life'), we will stick to the term food self-sufficiency (or self-reliance) in our study.

developed by Costanza *et al.* (1997) to estimate the value of the change in ecosystem services caused by cultivated land conversion.

The structure of the chapter is as follows. Section 2 presents the current situation and recent trends in cultivated land and grain production in China. Section 3 examines the impact of the conversion of cultivated land on environmental sustainability using the change in ecosystem service value as a measure of sustainability. Section 4 examines the impact of cultivated land conversion on agricultural production, taking into account differences in productivity between land taken into agricultural production and land taken out of agricultural production. The last section summarizes the main findings and presents some policy implications for balancing the cultivated land area and maintaining grain self-sufficiency and environmental sustainability.

2. Cultivated land and grain production in China

Only a small share of the total land area in China can be used for cropping. As can be seen from Table 1, 130 million hectares (that is, 13.5 percent) of the total available land is used as cultivated land. Irrigated land makes up 55 million ha (or 42.3 percent) of the cultivated land. The land that is not cultivated consists to a large extent of forest land (18.2 percent of the total land area) and useable grass land (32.6 percent).

The total sown area is larger than the cultivated area, because more than one crop is planted per year in some regions. As shown at the right-hand side of Table 1, the total sown area equals 155.5 million hectares. As a result, the multiple cropping index (calculated as the sown area divided by the cultivated land area) equals 1.20. About two-thirds of the sown area is planted with grain (cereals, soybeans and tubers). The three major grain crops, rice,

Table 1. Composition of total land and total sown area, 2005.

Composition of total land area			Composition of total sown area		
Type	Area (mln ha)	Share of total land area (%)	Type	Area (mln ha)	Share of total sown area (%)
Total land area	960.0		Total sown area	155.5	
Cultivated land	130.0	13.5	Grain	104.3	67.1
Irrigated land	55.0	5.7	Oil-bearing crops	14.3	9.2
Forest land	174.9	18.2	Cotton	5.1	3.3
Useable grassland	313.3	32.6	Vegetables	17.7	11.4
Other	341.7	35.6	Orchards and others	14.1	9.0

Source: NBS, 2006.

maize and wheat, constitute just over one half (50.2 percent) of the sown area. Other major crops are vegetables, oil-bearing crops (rapeseed, peanuts), fruit and cotton.

As may be expected, most paddy fields are located in areas with relatively high levels of rainfall (see Table 2). Around three-quarters of the paddy fields (75.7 percent) are located in areas with precipitation levels of more than 1,000 mm per year. On the other hand, 79.2 percent of the dry land is located in areas with precipitation levels of less than 1,000 mm. Most cultivated land is flat (slope less than 5°). Yet, 0.8 percent of the paddy land and 1.6 percent of the dry land is located on land with slopes of more than 25 percent.

Agriculture in China is characterized by high external input use and high labor intensity, as compared to other parts of the world. As shown in Table 3, the chemical fertilizer use equals 327 kg per ha. This level is more than twice the average level for Asia as a whole and more than three times the global average. Use of tractors and harvest machines in China is very low. Only in Africa, the level of agricultural mechanization is (slightly) lower. The size of the cultivated land per person working in agriculture is by far the lowest in China, confirming the great scarcity of land in China. These figures clearly show that China has achieved its agricultural growth by adopting, in the words of Hayami and Ruttan (1985), a biological pattern of agricultural intensification.

Figure 1 shows the trend in the cultivated land area in China since 1988.[27] Between 1988 and 1995 the cultivated land area showed a small decline. But after 1995, it has decreased at a pace of more than one million hectares per year. Particularly in the period 1999-2003 the speed of decline was very fast.

Table 2. Natural conditions of cultivated land, 2000.

Slope (degrees)	<5°	5°-8°	8°-15°	15°-25°	25°-35°	>35°	Total
Paddy field	86.6%	3.8%	5.6%	3.2%	0.7%	0.1%	100%
Dry land	80.9%	4.8%	7.7%	5.0%	1.3%	0.3%	100%
Precipitation (mm)	<250	250-400	400-800	800-1000	1000-1600	>1600	Total
Paddy field	1.1%	0.3%	12.2%	10.7%	60.8%	14.9%	100%
Dry land	5.1%	9.0%	54.3%	10.8%	17.8%	3.0%	100%

Source: Zhang *et al.*, 2003.

[27] Note that the cultivated land area for 2005 listed in Table 1 comes from a different source (NBS, 2006) and cannot be compared with the estimates shown in Figure 1.

Table 3. Factor use in agriculture in different parts of the world, 2000.

Area	Cultivated land area per person active in agriculture (ha)	No. of tractors used per 1,000 ha	No. of harvest machines per 1,000 ha	Chemical fertilizer use (kg/ha)
World total	1.1	19.0	3.0	99.5
Asia	0.5	14.6	3.8	146.8
Africa	0.9	3.1	0.2	21.4
North America	12.6	22.3	3.2	96.9
South America	3.5	11.2	1.3	78.9
Europe	8.8	37.3	4.0	82.2
Australia	19.8	7.2	1.1	51.4
China	0.2*	5.3	0.3	327.2

Source: Sun and Shi, 2003.

* Own estimate: Sun and Shi (2003) report a value of 0.1.

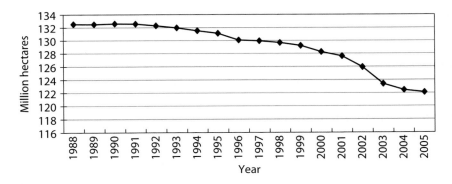

Figure 1. Cultivated land area, 1988-2005 (Heilig, 1999; Ministry of Land and Resources, 1998-2000; Ministry of Land and Resources, 2001-2005).

Grain production, on the other hand, steadily increased during the 1980s and 1990s. As can be seen from Figure 2, the production level in 1998 was about 1.26 times the grain production level in 1987. From 1999 to 2003, grain production declined by more than 80 million tons. This decline coincided with a rapid decrease in cultivated land area, as shown in Figure 1. Since 2004, grain production has recovered and is now at a level close that reached in the second half of the 1990s. This can to some extent be attributed to the agricultural direct subsidy policy implemented from the beginning of 2004.

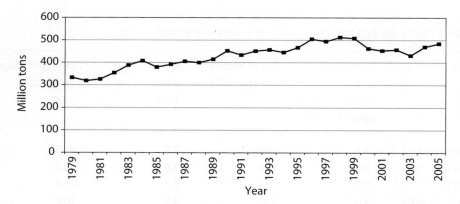

Figure 2. Grain production, 1979-2005 (Ministry of Agriculture, 1980-2006).

The geographical spread in grain production over North China and South China (including the Central-Eastern provinces)[28] is shown in Figure 3. It demonstrates that grain production has gradually moved northwards. In the beginning of the eighties around 60 percent of grain production took place in the South. In recent years, this share has declined to around 50 percent, with grain production in the North exceeding production in the South for the first time in 2005. A similar shift in grain production was found in an analysis of county-level data over the period 1980-1997 by You (2006). Our data show that this trend has continued after 1997.

Within the Northern region, grain production grew most in the Northeastern sub-region, as can be seen from Figure 4. The share of grain production in national production in the Northeast grew from around 10-11 percent in the early eighties to more than 15 percent in recent years. But grain production in the Northwestern and Northern sub-region also increased by around 2.5 percentage points each during the same period.

Figure 5 shows the trends in grain production in the South. The decline in grain production was largest in the Center-East, the traditionally strong grain production area. The share of this region's grain production in national production declined from around 35 percent in the 1980s to around 27 percent in recent years. A similar process took place in the Southeast, which saw its share of grain production decline by around 4 percentage points during the same period. Both the Center-East and Southeast are regions where rapid economic development has taken place since the mid 1980s. The share of grain production in the Southeast, on the other hand, remained more or less stable at a level of 15 percent.

---

[28] In this chapter, North China comprises the sub-regions Northeast (Liaoning, Jilin, Heilongjiang), North (Hebei, Henan, Shandong, Shanxi, Beijing and Tianjin) and Northwest (Inner Mongolia, Shaanxi, Ningxia, Gansu, Qinghai and Xinjiang). South China includes the sub-regions Southeast (Fujian, Guangdong, Guangxi and Hainan), Southwest (Sichuan, Chongqing, Guizhou, Yunnan and Tibet), and Center-East (Hunan, Hubei, Jiangxi, Jiangsu, Anhui, Zhejiang and Shanghai).

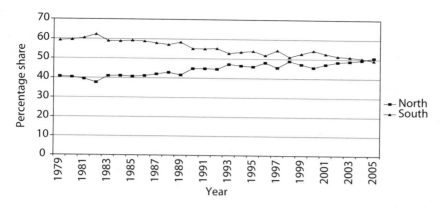

Figure 3. *Geographical shift in grain production, 1979-2005 (Ministry of Agriculture, 1980-2006).*

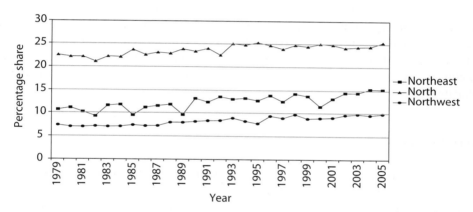

Figure 4. *Grain production in North China, 1979-2005 (Ministry of Agriculture, 1980-2006).*

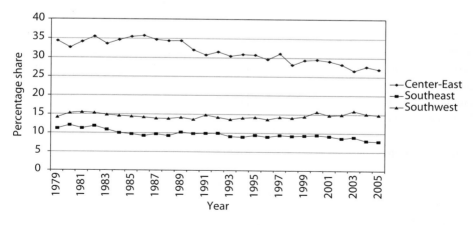

Figure 5. *Grain production in South China, 1979-2005 (Ministry of Agriculture, 1980-2006).*

These trends show that grain production has moved away from the traditional grain growing areas with relatively good production conditions towards Northern regions, where rainfall levels are much lower. Hence, levels of grain production in the near future may be more erratic due to the more fragile ecological environment in the latter regions.

According to the land use change survey that was held in 2002, a land reserve of 88.7 million ha (or 9.33% of the total land area) is available. This land reserve consists of unused wetlands, wasteland and bare land. However, only 7.34 million ha of this land reserve can be exploited as cultivated land due to natural limitations such as water availability and ecological environmental conditions. Just a small share of this, 0.33 million ha, can be reclaimed under currently prevailing technological and economic conditions. The remainder, 7.02 million ha, is cultivable in the near future.

Table 4 shows how this land reserve is distributed over the six major regions. The largest part of the land reserve is located in the Northwest (almost 40 percent) and Southwest (almost 30 percent). Only about 5 to 6 percent of land reserves are available in the Center-East and Southeast. About two-thirds of the cultivated land reserve is located in the Southwest, as climatic and agro-ecological conditions in this region are more favorable than in the Northwest.

Table 4. Available land reserves in China, 2002.

Region	Land reserve		Reclaimable land		Cultivable land reserve	
	Area (mln. ha.)	Share (%)	Area (mln. ha.)	Share (%)	Area (mln. ha.)	Share (%)
China	88.7	100	0.33	100	7.02	100
Northeast	8.6	9.7	0.02	7.3	0.53	7.6
North	9.8	11.1	0.07	22.0	0.61	8.8
Northwest	34.7	39.1	0.06	19.9	0.68	9.7
Center-East	5.6	6.3	0.00	1.2	0.12	1.7
Southeast	4.5	5.0	0.03	8.6	0.32	4.6
Southwest	25.6	28.8	0.13	41.0	4.75	67.6

Calculation based on Tang, 2006.

3. Impact of cultivated land change on ecosystem services

Cultivated land conversion can have important environmental effects. It causes a change in ecosystem services and the values associated with such services because different land use types provide different ecosystem functions. Ecosystem services refer to the goods (such as food or raw materials) and services (such as waste assimilation or soil conservation) that are derived directly or indirectly by human beings from ecosystems. In order to derive the values of ecosystem services provided by different land use types, non-marketed services may be valued by estimating the 'willingness-to-pay' of individuals for these services (Costanza *et al.*, 1997). Using the principles and methods of Costanza *et al.* (1997), Xie *et al.* (2003) estimate ecosystem services values of different ecosystem types in China from a survey among 200 Chinese ecologists (see Table 5). Their estimates indicate that wetland and water have the highest ecosystem service values. Forestland also has a much higher ecosystem service value than farmland, while grassland has a value that is only slightly higher. Only deserts have a much lower ecosystem value than farmland.

Table 6 shows the sources of land taken into cultivation and the destinations of land taken out of cultivation between the end of the 1980s and the end of 1990s, and derives the change in ecosystem service values related to these land conversions. The largest share of the land taken into cultivation, 55.7 percent, came from grassland. The remainder of the land mainly came from forestland (28.7 percent) and unused land (10.8 percent). Using the per unit ecosystem values presented in Table 6, we estimate a total loss of ecosystem values worth at least 4.57 billion US dollars due to land conversion into new cultivated land. The conversion of forestland into farmland made the largest contribution to this ecosystem value loss.

Table 5. Ecosystem service values of different Chinese ecosystem types ($/ha per year).

	Forest land	Grassland	Farmland	Wetland	Water	Desert
Gas regulation	387	89	55	199	0	0
Climate regulation	299	100	98	1,891	51	0
Water supply	354	89	66	1,714	2,254	3
Soil formation and conservation	431	216	162	189	1	2
Waste treatment	145	145	181	2,011	2,011	1
Biodiversity	361	121	79	277	275	38
Food production	11	33	111	33	11	1
Raw materials	288	6	11	8	1	0
Amenity	142	4	1	614	480	1
Total	2,417	801	764	6,936	5,084	46

Source: Xie *et al.*, 2003.

Table 6. Change in ecosystem service value for cultivated land conversion, 1989-1999.

	Area (million ha)	Per unit ecosystem service value (US\$ha^{-1}yr^{-1})	Total value (million US\$yr^{-1})
Land taken into cultivation by source:			
Grassland	3.45	801	2,765
Forest land	1.78	2,417	4,296
Unused land *	0.67	764	513
Water and wetland	0.29	6,010	1,729
Other land **	0.01	46	1
Total loss from the conversion			9,303
Minus: gain from the conversion	6.20	764	4,737
Net loss of ecosystem service value			4,566
Land taken out of cultivation by destination:			
Construction	2.14	0	0
Ecological restoration ***	2.53	1609	4,071
Destroyed by natural disasters ****	1.19	632	752
Agricultural structural adjustment	0.83	764	634
Total gains from the conversion			5,457
Minus: loss from the conversion	6.69	764	5,111
Net gains of ecosystem service value			346
Total loss of ecosystem service value			**4,220**

* Farmland value is used as approximation for the value of unused land. The actual ecosystem service value of unused land is higher as it is mainly located in Heilongjiang, Jilin and Inner Mongolia.

** Desert land value is used as an approximation for the value of other land.

*** Land taken out of cultivation for ecological restoration becomes either forest land or grassland; the average value of the ecosystem service value of these two land use types is used as an approximation of the per unit value of ecological restoration land.

**** Cultivated land destroyed by natural disasters is assumed to have the same per unit value as farmland value minus its food production value and raw material value.

Calculations based on Tian *et al.*, 2002; Xie *et al.*, 2003; and Ministry of Land and Resources (1990-2000).

During the same period, the land taken out of cultivation for ecological restoration slightly exceeded the size of the cultivated land that was converted into construction land (38 versus 32 percent, respectively). Almost 1.2 million ha (18 percent of the converted land) was destroyed by natural disasters, while more than 0.8 million ha was set aside for agricultural structural adjustment (that is, producing horticultural products, fruit and livestock with modern techniques). The resulting change in ecosystem value during this period was slightly positive (estimated at 0.35 billion US dollars), because land used for ecological restoration has a much higher ecosystem service value per ha than cultivated land while construction land is assumed to have a zero ecosystem value. Taking the estimated ecosystem values of land taken into cultivation and land taken out of cultivation together, we estimate that the total loss of ecosystem value caused by land conversions during the period 1989-1999 was equal to 4.22 billion US dollars. The main source of the loss was the conversion of forest land into cultivated land (estimated net loss 2.94 billion US dollars), which exceeded the loss caused by the conversion of cultivated land into construction land (net loss 1.63 billion US dollars).

Similar calculations have been made for the change in ecosystems value caused by the cultivated land conversion between 2000 and 2005. The results are shown in Table 7.[29] During this period a number of major ecological restoration programs such as the Sloping Land Conversion Program and the Beijing-Tianjin Sandstorm Control Program were implemented. In these programs, large areas of arable land in West China and North China were (and are still being) converted into forestland and grassland in order to prevent soil erosion and dust storms. As can be seen from Table 7, land conversion for ecological restoration dominated the conversion of land during this period. It was responsible for 67 percent of the land taken out of cultivation. As a result, there was a substantial gain in ecological service value during this period, estimated at 2.68 billion US dollars.

The estimated loss of ecosystem service value during the entire period 1989 - 2005 equals 1.54 billion US dollars. The estimated loss of land conversion into construction land during this period is equal to 3.36 billion US dollars. Table 8 shows how the latter losses were distributed over the six major sub-regions during the years 1999 – 2004. The table shows that losses caused by conversion into construction land are much larger in the Center-East and the North than in the other four sub-regions. In both the Center-East and the North, the ecosystem value losses caused by expanding construction land have been increasing over time. The other four sub-regions show relatively large losses in 1999 and 2004 and smaller losses in the period in between those years.

The estimates presented until now focused on average ecosystem service values for different types of land use. Besides that, differences in agro-climatic conditions between land taken into cultivation and already cultivated land may also have important effects on the

---

[29] Unfortunately, no information on the sources of the land taken into cultivation is available for this period.

Table 7. Change in ecosystem service value for cultivated land conversion, 2000-2005.

	Area (million ha)	Per unit ecosystem service value (US$ha⁻¹yr⁻¹)	Total value (million US$yr⁻¹)
Land taken into cultivation:	2.04		
Loss from conversion *		1,500	3,060
Minus: gain from conversion		764	1,559
Net loss of ecosystem service value			1,501
Land taken out of cultivation by destination:			
Construction	1.26	0	0
Ecological restoration **	6.14	1,609	9,879
Destroyed by natural disasters ***	0.31	632	200
Agricultural structural adjustment	1.43	764	1,100
Gain from conversion			11,179
Minus: loss from conversion	9.16	764	6,994
Net gain of ecosystem service value			4,185
Total gain of ecosystem service value			**2,684**

* The average per unit ecosystem service value of land taken into cultivation in Table 6 is used to approximate the ecosystem service value of land taken into cultivation.
** Land taken out of cultivation for ecological restoration becomes either forest land or grassland; the average value of the ecosystem service value of these two land use types is used as an approximation of the per unit value of ecological restoration land.
*** Cultivated land destroyed by natural disasters is assumed to have the same per unit value as farmland value minus its food production value and raw material value
Calculations based on Ministry of Land and Resources, 2001-2006; Xie *et al.*, 2003.

environment. To this end, we compare the natural conditions of land taken into cultivation between 1990 and 2000, shown in Table 9, with those of the cultivated land in the year 2000, presented in Table 2. We find that newly cultivated land is relatively flat. Only 11.5 percent of the land has a slope of more than 5 degrees, as compared to 13.4 percent (for paddy land) and 19.1 percent (for dry land) of the land that is already in cultivation. But rainfall levels are much lower on newly cultivated land. As much as 91.6 percent of the land receives less than 800 mm per year, while for already cultivated land this share is 68.4 percent for dry land and only 13.6 percent for paddy fields. This shows that newly cultivated land is located in areas that are only marginally suited for agricultural production, and that are likely to have much lower productivity levels. This finding reaffirms the conclusion of Lin and Ho (2003) that newly reclaimed low-graded farmland in environmentally fragile frontier areas

Table 8. Change in ecosystem service value for cultivated land conversion into construction use, 1999-2004 (million US$yr-1).

Region	1999	2000	2001	2002	2003	2004	Total
Northeast	17	9	6	5	5	21	64
North	38	32	44	47	56	68	285
Northwest	13	9	9	13	13	14	71
Centre-East	37	46	42	60	70	81	336
Southeast	18	12	10	11	13	16	79
Southwest	34	16	14	14	18	24	120
China	157	125	125	150	175	224	956

Calculations based on Ministry of Land and Resources, 2001-2006; Xie *et al.*, 2003.

Table 9. Natural conditions of land taken into cultivation, 1990-2000 (percentage).

Slope (degrees)	<5°	5°-8°	8°-15°	15°-25°	25°-35°	>35°	Total
Share	88.5	3.9	4.3	2.5	0.7	0.2	100

Precipitation (mm)	<250	250-400	400-800	800-1000	1000-1600	>1600	Total
Share	10.9	14.8	65.9	2.1	4.6	1.7	100

Source: Tian *et al.*, 2002.

has never been able to compensate for the loss of fertile farmland in the southeastern part of the country. In addition, previous research has shown that excessive reclamation of land in environmentally fragile regions has brought damage to the natural environment, causing problems of soil erosion, desertification and deforestation (Ash and Edmonds, 1998; Yang and Li, 2000).

4. Impact of cultivated land change on the potential of grain production

Figure 1 showed the trend in cultivated land change during the period 1988-2005. To analyze the impact of land conversion on agricultural production, total cultivated area is not a very useful indicator as it does not take into account differences in crop productivity caused by precipitation, soil quality and other agro-climatic factors. As we saw above, the natural conditions of newly cultivated land differ significantly from those of the land already in cultivation. Hence, to examine the impact of cultivated land change on grain production we should not only examine the size of the cultivated land but also its productivity. An important indicator in this respect is the productivity coefficient, defined as the average yield in a province divided by the average yield in China.

Table 10 shows the productivity coefficients of each province in mainland China, and uses these indices to divide the provinces into four groups:
1. Very strong production area, with p between 1.60 and 1.81. Four out of five provinces in this group are located in the Center-East.
2. Strong production area, with p between 1.20 and 1.60.
3. Medium production area, with p between 0.90 and 1.20.
4. Weak production area, with p below 0.90. The five provinces with the lowest p's are all located in Northwest China.

By multiplying the available data on land conversion per province with these coefficients, we can calculate the so-called standardized change in cultivated land since 1999. Table 11 shows the results for the period 1999-2004 for the same six sub-regions that we distinguished before. It shows that the largest declines in cultivated area took place in the Northwest and (to a lesser extent) in the Southwest, North and Center-East. If we take the productivity of the land into account (the 'standardized' cultivated land change), we find that losses were largest in the Center-East instead of the Northwest. The land taken out of production in the

Table 10. Grouping of provinces according to cultivated land productivity, 1996-2000.

Production area

Very strong (1.60< p ≤1.81)	Strong (1.20< p ≤1.60)	Medium (0.90< p ≤1.20)	Weak (0< p ≤0.90)	
Hunan 1.81	Beijing 1.58	Chongqing 1.17	Hainan 0.70	Ningxia 0.53
Shanghai 1.76	Guangdong 1.51	Anhui 1.13	Heilongjiang 0.64	Xinjiang 0.52
Jiangsu 1.73	Jiangxi 1.45	Hebei 1.05	Shaanxi 0.63	Inner Mongolia 0.48
Zhejiang 1.72	Shandong 1.42	Jilin 1.01	Tibet 0.62	Qinghai 0.43
Fujian 1.72	Sichuan 1.37	Tianjin 0.97	Guizhou 0.59	Gansu 0.42
	Hainan 1.29	Liaoning 0.96	Shanxi 0.58	
	Hubei 1.27	Guangxi 0.91	Yunnan 0.55	

p indicates productivity coefficient; it is calculated as p = (agricultural production / cultivated area) / (total agricultural production in China / total cultivated land area in China)
Source: Zheng and Feng, 2003.

Table 11. Cultivated land change and conversion into construction land, 1999-2004 (million ha).

Region	Cultivated land change			Conversion into construction land		
	Absolute changes	Standardized changes	Change ratio (%)	Absolute changes	Standardized changes	Change ratio (%)
Northeast	-0.21	-0.18	-14	-0.08	-0.07	-13
North	-1.31	-1.20	-8	-0.37	-0.42	14
Northwest	-2.80	-1.49	-47	-0.09	-0.05	-44
Centre-East	-1.09	-1.60	47	-0.44	-0.71	61
Southeast	-0.39	-0.54	38	-0.10	-0.15	50
Southwest	-1.51	-1.47	-3	-0.16	-0.15	-6
China	-7.32	-6.47	-12	-1.25	-1.56	25

Calculated from Ministry of Land and Resources (2001-2005).

Center-East was much more productive than the land that was no longer cultivated in the Northwest. So, even though the size of the land taken out of production in the Northwest was 2.7 times as large as in the Center-East, the estimated impact on agricultural production was slightly larger in the Center-East. In total, the standardized change in cultivated land for China as a whole during the period 1999-2004 turns out to be 12 percent smaller (in absolute size) than the absolute change for China as a whole, implying that the land taken out of cultivation had a productivity which was slightly below average.

An issue that has attracted much attention of policy makers and the media in China is the conversion of land of cultivated land into land used for construction of urban housing or industries. The right-hand side of Table 11 shows the conversion into construction land for the six sub-regions during the period 1999-2004. The total size of the land converted into construction land during this period was 1.25 million ha, only around one-sixth of the total land taken out of cultivation. Ecological restoration was responsible for the lion share of the land taken out of cultivation (see Table 7). Most of the land converted into construction land during this period was located in the Center-East and the North. The productivity of this land is relatively high, especially in the Center-East. As a result, the standardized change of land converted into construction land is around 25 percent larger (in an absolute sense) than the absolute change. If we consider standardized changes in cultivated land, the land converted into construction land makes up almost one quarter of the total land taken out of cultivation.

The analysis thus far uses the province-level productivity coefficient as an indicator of the productivity of cropland. Several case studies have shown, however, that especially high-yielding cultivated land (used for growing vegetables or rice) located in sub-urban areas are converted into construction land in the Eastern coastal areas (Zhang and Tan, 1998; Gao *et al.*, 1998; Tan and Peng, 2003). Hence, the productivity losses in these areas are probably considerably larger than the losses that are calculated on the right-hand side of Table 11.

5. Concluding remarks and policy implications

Grain self-reliance is an important goal of agricultural policies in China. To feed its large population, however, very limited land resources that are used very intensively are available. This raises the question whether China can maintain its successful grain self-sufficiency policy in the near future and what will be the impact of this policy on environmental quality. An important issue in this respect is the conversion of cultivated land into land used for construction and other purposes. As shown in Figure 1, the stock of cultivated land has decreased rapidly in recent years, particularly since 1999. This chapter has examined the impact of cultivated land conversion on environmental quality and agricultural productivity, using provincial and regional data.

We find that grain production has gradually moved northwards, particularly to the Northeastern region, since the beginning of the 1980s. This shift is likely to put important constraints on grain productivity in the near future since rainfall levels are much lower and the ecological environment is more fragile in the Northern region.

To assess the impact of past land conversions on the environment, we apply the concept of ecosystem service value developed by Costanza *et al.* (1997). This should not be considered a precise measure of environmental quality change, but it provides a measurable way to compare environmental losses and gains caused by taking land into and out of cultivation, and allows researchers and policy makers to compare such changes over time and between regions and use these as a basis for decision making on future policies. We estimate that the net changes in land taken into cultivation between the end of the 1980a and the end of the 1990s caused a total loss of ecosystem values worth of 4.22 billion US dollars. The conversion of forestland into farmland made the largest contribution to this ecosystem value loss. Since the beginning of this new century, the implementation of large-scale ecological restoration programs has compensated for the losses in ecosystem service values caused by land conversion and even caused a substantial gain (estimated at 2.68 billion US dollars for the period 2000-2005) in recent years.

We further find that losses in ecosystem value caused by expanding construction land are largest in the Center-East and the North. In both regions these losses have been increasing over time. Rapid urbanization and industrialization in these regions under imperfect land markets (especially rural land markets) and in many cases inappropriate government interventions, has led to rent seeking in both conversion of available land resources into cultivated land and conversion of cultivated land into construction land. This has resulted in excess land conversion and important negative environmental effects.

Of the available land reserve in China, only 7.34 million ha can be exploited as cultivated land due to natural limitations such as water availability and ecological environmental conditions. Just a small share of this, 0.33 million ha, can be reclaimed under currently prevailing technological and economic conditions. The remainder, 7.02 million ha, is cultivable in the near future. Almost no land reserves are available in the Center-East and Southeast, where land productivity is highest. This means that further expansion of cultivated land in these regions is almost impossible. Instead, the rapid urbanization and industrialization that takes place is likely to put more pressure on converting cultivated land into construction land in these regions.

Comparing the natural conditions of newly cultivated land with those of land already in cultivation, we find that newly cultivated land is relatively flat. Rainfall levels, however, are much lower on newly cultivated land. This confirms that newly cultivated land is especially located in areas that are only marginally suited for agricultural production, and that are likely to have much lower productivity levels.

Using the productivity coefficient as an indicator of land productivity at the provincial level, we calculate standardized changes in cultivated land over the period 1999-2004 that take differences in productivity levels into account. We find that the standardized change in cultivated land during this period was 12 percent smaller than the absolute change for China as a whole, implying that the land taken out of cultivation had a productivity which was slightly below average. This was caused especially by the large declines in cultivated area that took place in the Northwest, where land productivity is relatively low. Several large-scale ecological restoration programs were implemented during the examined period, particularly in Western China. Land converted into construction land made up only around one-sixth of the total land taken out of cultivation. Most of the land converted into construction land, however, was located in provinces in the Center-East and the North with relatively high productivity coefficients. As a result, the land converted into construction land was responsible for almost one quarter of the standardized change in cultivated land.

Policy making on land conversion focuses in recent years on the need to keep a balance in cultivated land area by compensating land converted into construction land by new land taken into cultivation elsewhere. Our study shows that such a policy causes both a loss in agricultural productivity, because the newly cultivated land tends to have a lower productivity, and a loss of ecosystem services. Instead of reclaiming cultivated land from ecological fragile areas, improving land productivity may be a more effective way for pursuing grain self-sufficiency and maintaining the environmental base for agricultural production. The relatively low values of the productivity indicators in several provinces presented in Table 10, and the fact that multiple cropping indices are below one for most provinces in

North and West China, indicate that substantial productivity increases can still be reached in some regions.

This asks the policy-makers to consider several points in formulating land use policies: (1) To strictly limit the conversion of cultivated land into construction land, especially the cultivated land with a high quality (flat, fertile, irrigated and suitable for use of machinery), by increasing the intensity of urban land use, converting steeply sloped cultivated land into forest or grassland, and avoiding development of land with steep slopes; and (2) To pay more attention to land conservation and improvement technologies, especially land reclamation technologies, medium-/low-yield field improvement technologies, water and soil conservation technologies, integrated technologies for small watersheds, and fertilizer-saving and water-saving technologies. In addition, stimulating land rental markets and land consolidation programs that reduce the high level of land fragmentation may also be effective policies to improve land productivity without damaging the natural resource base of grain production.

References

Ash, R.F. and R.L. Edmonds, 1998. China's land resources, environment and agricultural production. China Quarterly 156: 836-879.

Chen, Z. and Q. Wang, 2005. A positive study on China's farmland conversion and land degradation in transition. China Population, Resources and Environment (in Chinese), 15(5): 43-46.

Costanza, R., R. Arge, R. de Groot, S. Farber, M. Grasso, B. Hannon, K. Limburg, S. Naeem, R.V. O'Neill, J. Paruelo, R.G. Raskin, P. Sutton and M. van den Belt, 1997. The value of the world's ecosystem services and natural capital. Nature, 387: 253-260.

Deng, X., J. Huang, S. Rozelle and E. Uchida, 2006. Cultivated land conversion and potential agricultural productivity in China. Land Use Policy 23 (2006): 372-384.

Fu, Z., Y. Cai, Y. Yang and E. Dai, 2001. Research on the relationship of cultivated land change and food security in China. Journal of Natural Resources (in Chinese), 16(4): 313-319.

Gao, Z., J. Liu and D. Zhuang, 1998. The dynamic changes of the gravity center of the farmland area and the quantity of the farmland ecological background in China (in Chinese). Journal of Natural Resources, 13(1): 92-96.

Hayami, Y. and V.W. Ruttan, 1985. Agricultural Development: An International Perspective. The Johns Hopkins University Press, Baltimore and London.

Heilig, G.K., 1999. ChinaFood. Can China Feed itself? IIASA, Laxenburg (Austria), (CD-ROM). http://www.iiasa.ac.at/Research/LUC/ChinaFood/data/land/land_3.htm

Lin, G.C.S. and S.P.S. Ho, 2003. China's land resources and land-use change: insights from the 1996 land survey. Land Use Policy 20 (2003): 87-107.

Liu, Y. and C. Wu, 2002. Situation of land-water resources and analysis of sustainable food security in China. Journal of Natural Resources (in Chinese), 17(3): 270-275.

Ministry of Agriculture, 1980-2006. Agricultural Yearbook of China 1979-2005 (in Chinese). China Agricultural Press, Beijing.

Ministry of Land and Resources, 1998-2000. Survey Report of Land Use Change (1996-2000). Beijing, China (in Chinese).

Ministry of Land and Resources, 1990-2000. China National Report on Land and Resources (2001-2005). Beijing, China (in Chinese).

Ministry of Land and Resources, 2001-2005. China National Report on Land and Resources (2001-2005). Beijing, China (in Chinese).

National Bureau of Statistics of China (NBS), 2006 Statistical yearbook. China Statistics Press, Beijing.

Shao, X. and J. Xie, 2007. Analyzing regional changes of the cultivated land in China. Resources Science (in Chinese), 29(1): 36-42.

State Council, 1996. The grain issue in China. White paper. Information Office of the State Council, Beijing. http://english.people.com.cn/whitepaper/15.html

State Council, 2005. Opinions of the Central Committee of the Communist Party of China and the State Council on policies to strengthen the rural work and improve the overall production capacity of agriculture. Xinhua New Agency, Beijing (January 30, 2005).

State Council, 2007. Several Opinions of the Central Committee of the Communist Party of China and the State Council on actively developing modern agriculture and pushing forward the building of a new socialist countryside in a down-to-earth manner. Xinhua New Agency, Beijing.

Sun, H. and Y. Shi (eds.), 2003. Agricultural land use in China (in Chinese). Jiangsu Science and Technology Press, Nanjing.

Tan, S. and B. Peng, 2003. Examination of cultivated land utilization to guarantee our grain security and its recent adjustment thoughtfulness (in Chinese). Economic Geography, 23(3): 371-378.

Tan, Y., C. Wu, Q. Wang, L. Zhou and D. Yan, 2005. The change of cultivated land and ecological environmental effects driven by the policy of dynamic equilibrium of the total cultivated land. Journal of Natural Resource (in Chinese). 120(5):727-734.

Tang, C., 2006. Reserved land resource survey. In: X. Huang (eds.) Research on Land Management in China (in Chinese). Contemporary China Publishing House. Beijing: 418-421.

Tian, G., Q. Zhou, X. Zhao and Z. Zhang, 2002. Spatial characteristics and ecological background of neo-cultivated land in China. Resources Sciences (in Chinese), 24(6): 1-6.

Xie, G., C. Lu, Y. Leng, D. Zheng and S. Li, 2003. Ecological assets valuation of the Tibetan Plateau. Journal of Natural Resources (in Chinese), 18(2):189-195.

Yang, H. and X. Li, 2000. Cultivated land and food supply in China. Land Use Policy, 17 (2000): 73-88.

You, L., 2006. Land use and sustainability of Chinese grain production. Paper presented at International seminar on 'Transition towards Sustainable Rural Resource Use in Rural China' in Kunming, Yunnan Province, China, 22-24 October.

Zhang, G., J. Liu and Z. Zhang, 2003. Spatial-temporal changes of cropland in China for the past 10 years based on remote sensing. Acta Geographic Sinica (in Chinese), 58(3): 323-332.

Zhang, W. and S. Tan, 1998. Cultivated land conversion into non-agricultural uses in Chinese east coastal areas: problems and countermeasures. Geography and Land Research (in Chinese), 14(4): 20-30.

Zhao Q., B. Zhou, H. Yang and S. Liu, 2002.Cultivated land security and related countermeasures in China. Soils (in Chinese), 2002(6): 293-302.

Zheng, H. and Z. Feng, 2003. The quantity and quality analysis on dynamic equilibrium of the total cultivated land in China. Resources Science (in Chinese), 25(5): 33-38.

Zhong, F., C. Carter and F. Cai, 1999. China's Ongoing Agricultural Reform (in Chinese). Financial Economic Press in China, Beijing.

Zhu, H., 2006. On food security and cultivated land resource security. Research of Agricultural Modernization (in Chinese), 27 (3): 161-164.

Markets and the role of the state

Trade policy impacts on Kenyan agriculture: challenges and missed opportunities in the sugar industry

Andrew M. Karanja

Abstract

Agricultural trade policies are increasingly becoming critical in economic, social and political arenas of most developing countries such as Kenya. The pressures from globalisation and regional integration have intensely and continuously confronted policy makers in Kenya with the 'classical food price policy dilemma' on how best to deal with producer incentives without hurting the welfare of consumers. This chapter evaluates the agricultural trade policy, and its price and welfare impacts on producers and consumers. This is undertaken by estimating both the Nominal and the Effective Rate of Protection (NRP and ERP, respectively) using sugar as a case study. Results indicate that in the early 1990s the tariff structure in Kenya was more focussed on promoting competition even in the agricultural sector with certain reversals occurring in the late 1990s aimed mainly at protecting local producers. The case study demonstrates that despite the high protection accorded to the Kenyan sugar producers they are under intense competitive pressure from regional and international markets, which they are unlikely to withstand due to various constraints. The current agricultural trade policy is also not in tandem with the emerging international and regional trade regimes, and is mainly driven by the desire to safeguard the interests of farmers without due consideration on the nationwide impacts on consumers and the poor. In most cases the policies have turned to be counter-productive, as they have impacted negatively on would be beneficiaries and the poverty reduction goal.

1. Introduction

Kenya is currently facing the twin challenge of reducing poverty and achieving sustainable economic growth. The desire to alleviate poverty arises from the realisation that economic growth, although necessary, is not a sufficient condition to guarantee poverty reduction. Kenya's GDP growth rate declined since the 1970's from about 7 percent to just over 2 percent in the 1990's and averaged at 1.4 percent in 2001-03. Poverty and food insecurity increased in the same period. However, there has been a turn-around in recent years with GDP growth expected to exceed 6 percent in 2006, compared to 5.8 percent in 2005 and 4.9 percent in 2004. As a result, the per capita growth has picked up at an average rate of 1.8 percent per year. Recent statistics also indicate that poverty levels have declined from 56% in 2000 to 46% in 2005/06. Nevertheless, poverty still remains a rural phenomenon with 49% of rural population living below the poverty line as compared to 34% of urban population. Despite these positive developments, significant development challenges remain, especially with respect to sustaining growth, addressing inequalities, and improving governance.

Broad–based growth to address some of these challenges is inextricably linked to agricultural sector performance and more specifically smallholder agriculture.

Kenya's main agricultural development strategy has been based on the promotion of smallholder farming. Smallholder-led agricultural strategy has been advocated to serve the dual purpose of increasing and intensifying resource use as well as addressing equity concerns. In implementing this development strategy a variety of policy instruments has been applied. Between 1963 and 1980, policies emphasised government intervention in nearly all aspects of agricultural production and marketing (Nyangito, 1999). These direct intervention policies related to market regulation and pricing of agricultural commodities and inputs. Indirectly, the government macro-policies in respect to interest rates, exchange rates, and trade, wages and investment decisions in public goods (mainly research and rural infrastructure) were also used to influence the direction and rate of agricultural development.

There was, however, a major policy shift from 1981, when government controls were gradually removed and initial attempts made towards market liberalisation. The liberalisation of markets was aimed at enhancing participation of the private sector in agricultural production and marketing while at the same time creating an enabling environment for market forces to determine the level of agricultural prices. Kenya has dismantled most of the quantitative import restrictions and price controls on major products and the tariff is now the main trade policy instrument. Nevertheless, the government continues to apply various tariff and non-tariff policies that affect agricultural imports and exports as well as prices of non-traded goods. In the early 1990s the tariff structure was more focussed on promoting competition even in the agricultural sector. However, there were some reversals in the late 1990s with the tax structure more focused on protection aimed at satisfying an array of interest groups representing farmers, consumers, government itself, and local lobby groups. Kenyan farmers are also subject to many other taxes, both legal and non-legal, which are administered by various agencies at the local level (Freeman *et al.,* 2003). The cumulative impact of these taxes, levies, fees, permits and licences on the incomes of the poor cannot be ignored neither the price distortions they cause.

The objective of this essay is to review and demonstrate some of the policy challenges and opportunities that confront Kenyan agriculture in a dynamically changing trade regime brought about by both regional and global changes. Sugar is used as a case study due to its importance as a locally produced and traded commodity whose trade epitomises the kind of policy dilemma confronting policy makers. The commodity supply chain has lobby groups that confront policy makers with divergent policy options that are difficult to find a compromise. The essay is organised as follows. A review of Kenya's trade relations and policies in the regional and global arena is made followed by a review of the sugar industry production and policy reforms as they relate to the emerging trade issues. Estimates of the level of protection offered to domestic sugar producers in Kenya is then made and their implications to various stakeholders. Finally, discussions and conclusions are drawn.

1.1 Kenya's shift towards regional and global trade

In the 1990s, there has also been major promotion of regionalization and globalisation policies whose key component has been reduction/removal of tariff and non-tariff barriers to trade across countries. The reduction and removal of tariff and other trade barriers have resulted in withdrawal of direct and indirect protection to domestic production thus exposing domestic production and trade to fierce regional and international competition. Also in line with these globalisation trends, most non-oil commodity agreements whose objective was to guarantee certain level of producer prices has been allowed to lapse. This has led to lower and more volatile prices for such commodities, many of which originate from developing countries (Karanja *et al.*, 2003).

Kenya became a member of the World Trade Organization (WTO) in 1995 and therefore agreed to implement the Agreement on Agriculture (AoA) whose main elements are to increase market access for agricultural commodities for member countries, reduce domestic support for agricultural commodities and elimination of export subsidies to increase export competition. As a commitment to the WTO requirements, the government of Kenya gave a tariff ceiling binding of 100% for all agricultural commodities. A summary of import tariffs for major agricultural commodities in Kenya are summarized in Table 1. The import tariffs have been generally lower than 35%. However, there has been intense lobbying to the government to provide domestic market protection particularly for cereals and sugar. Kenya has never applied contingency trade remedies (anti-dumping, countervailing and safeguard measures) and the country does not have a specific legislation on safeguard measures. The Government in most cases uses suspended duties to raise the import tariffs when there is need to protect domestic production. This has occurred for maize almost annually during the glut periods when the government raises tariffs on maize well above those announced at the beginning of the year. Dairy products, eggs and honey attracted the second highest duty of 60% while sugar had the highest duty of 100% in 2002/2003 (Table 1). However, the government has announced through the Ministry of Finance of its intension to phase out the application of suspended duties as from the 2003/2004 fiscal year.

Kenya is also a member of the Common Market for East and Central Africa (COMESA) Free Trade Area (FTA). Under the free trade area, which became effective in October 2000, nine countries, including Kenya started trading on duty and quota free terms for all goods originating from within their territories. For other non-FTA countries, Kenya is bound by the COMESA trade protocol to offer rebates on normal duty for imports ranging from 10% for Ethiopia to 80% for Uganda, Comoros and Eritrea. Kenya has been under pressure to allow duty-free sugar and other agricultural goods from the region. Most of these agricultural commodities are charged a duty of between 3-5%, which is eventually

Table 1. Import tariffs (%) on selected agricultural commodities into Kenya, 1996 -2002.

Commodity\Year	96/97	97/98	98/99	99/00	00/01	01/02	02/03
Agricultural food stuffs[1]	15	15	25 50*	30	30	35	30-maize 15-wheat 35-rice
Live animals & animal products	15	15	25	30	35	35	35 60**
Processed fruits and vegetables	15	15	30	35	35	35	35
Sugar	35	15	25	35	35	35	100
Textiles	15	15	25	30	35	35	35

[1] Includes most cereals (maize, wheat and rice).
**suspended duty for maize.
** Suspended duty for dairy produce, eggs and natural honey.
Source: Kenya Gazette, Finance Bills (1996, 1997, 1998, 1999, 2001, 2002).

supposed to be eliminated. Imports from COMESA have not gone well with producers of various commodities who have been used to protection.

Under the East African Community (EAC) customs union protocol signed in March 2004, Kenya is supposed to also grant market access to commodities coming from Uganda and Tanzania. In accordance with the customs union tariff schedule which entered into force in January 2005, three tariff bands of 0%, 10% and 25% for raw, intermediate and finished products are applicable. However, in addition to the three bands, there will be limited number of products, mostly agriculture commodities, which will still be subject to additional protection varying from 60% to 30%. The main agricultural products to be accorded extra protection include, milk and its products, wheat, maize and industrial sugar.

Apart from the tariff barriers, agricultural trade is also subject to numerous non-tariff barriers that hinder and increase transaction costs. These non-tariff barriers include customs import charges and clearance procedures, quality and safety standards as well as phytosanitary requirements. The combined effect of tariff and non-tariff barriers continue to hamper formal trade especially in regional markets thereby fuelling informal trade characterised by high transaction costs.

These commitments under WTO, COMESA and EAC have direct bearing on present and future trade policy, which Kenya can pursue. Equally, the trade policy has major and far-reaching implications on agricultural production and marketing policies. Furthermore, most of these trade protocols limit the country from making unilateral policy decisions despite the prevailing political economy of the various sectors. More than ever before, policy makers in the country are also intensely and continuously confronted with the 'classical food price policy dilemma' on how best to deal with producer incentives without hurting the welfare of consumers.

2. Methodology

In principle, quantitative analysis of price distortions intends to measure the relative deviation of observed market prices from their shadow equivalents i.e. the deviation between private and social prices. The differences in the two sets of prices can originate from two main sources; policy and market effects. Specific market characteristics and technology in a particular output or input market can lead to actual prices that deviate from their efficiency equivalents, a situation often referred to as market failure. The government may intervene through policies targeted specifically to offset the market failure or to affect the income of consumers, producers or state budget. The latter form of policy interventions that do not necessarily correct for the market failure may actually leave the society worse off. This is the so-called government failure which may affect the whole society or a section of the society e.g. consumers or producers.

In agriculture, two main government intervention policies are most common. These are market price interventions and government subsidies. Market price interventions (positive or negative) consist of measures such as foreign trade barriers (e.g. import tariffs, export subsidies) and quantitative restrictions on imports and exports. Government subsidies can be given as direct payments, support for research and extension, concessionary credit and tax-breaks for capital investments. Depending on how the subsidies are tailored they may or may not affect prices. The level of economic development in Kenya severely limits government's ability to afford significant level of subsidies thereby making market price interventions the main mechanism for agricultural sector support.

The Nominal Rate of Protection (NPR) is the simplest and easiest measurable indicator of price distortion (Sadoulet *et al.*, 1995). It is computed as the ratio of the domestic price (P^d) of a tradable commodity to its border price (P^b) equivalent:

$$\text{NRP in \%} = [(p^d / p^b) - 1] \times 100 \tag{1}$$

Thus,
- If NRP > 0%, the actual market price is above its social price equivalent implying protection of producers and taxation of consumers.
- If NRP < 0%, the actual market price is less than social price and producers are taxed and consumers subsided.
- If NRP = 0%, the is neutral structure of protection

The Effective Rate of Protection (ERP) is an extension to the nominal protection concept (Sadoulet and De Janvry, 1995). This extension includes the combined effects of price distortions on output and input markets. Thus, when applied across the sector, ERPs describe the relative incentives within the sector. The ERP is computed on the basis of the ratio of value-added in production at domestic prices (VA^d) over such value-added at border prices (VA^b). Inputs for each commodity (i) are categorised into:
- Traded intermediate factors (T) such as fertilizers, chemicals and fuels;
- Primary factors; land labour and fixed capital;
- Non-traded intermediate factors (NT) e.g. services such as insurance and transport.

The ERP is:

$$\text{ERP}_i \text{ in } \% = [(VA_i^d / VA_i^b - 1] \times 100 = \left[\frac{p_i^d - \sum_j a_{ij} p_j^d}{p_i^b - \sum_j a_{ij} p_j^d} - 1 \right] \times 100 \qquad (2)$$

Where a_{ij} are the technical coefficients measuring the number of units of intermediate factor j per unit of production output i.

Thus,
- If ERP> 0%, it implies a direct protection of domestic producers of commodity i. This results in positive incentives for producers of the commodity, since they receive higher returns on their resources.
- If ERP<0%, it implies underlying disincentives to domestic producers of commodity i. Domestic producers can therefore only remain in the production of the commodity if they become more efficient than foreign producers.
- If ERP=0, it implies a neutral structure on net incentives.

For sugar, the NRP is estimated at both the production (farm-gate) and consumption or post-farm (Nairobi) levels. The official ex-factory sugar price is used as approximate farm-gate prices. Retail sugar prices in Nairobi are taken as the consumer prices. The NRP and ERP are estimated for the main product (raw sugar) and do not include by-products such as molasses.

The sugar import (CIF) price is used as the basis for the reference (border) prices. The basic assumption is that there are minor quality differences between local and imported commodities to warrant price adjustment. The CIF prices are adjusted for port charges and domestic transport costs to arrive at the actual border prices at different locations. The railway charges to various destinations from the port of Mombasa are taken as the local transport cost. It is however important to point out that high domestic transport cost like the ones being witnessed in Kenya provides additional protection to producers upcountry while at the same time reducing the prices such producers would receive if they were exporting. In particular, transport can be significant in relation to international prices of low-value commodities such as maize. For example, the 2004 rail charges from Mombasa to Nairobi of Ksh 1,870 (US$25) per ton account for 11% of the CIF price for sugar.

The sugar farm budgets for the 2003/04 production period are used as the source of information on components of cost of production. The budgets covered small-scale farms in four sugar factories i.e. Mumias, Nzoia, Sony and Muhoroni. Various working assumptions regarding the traded proportion of each intermediate input are made and are available on request. The average market prices for various inputs such as fertilizers are taken as the farm-gate prices while the FOB or CIF price adjusted for port and transport charges are used as first approximation of border prices. Due to these assumptions and some data constraints the estimated protection rates should be taken as indicative but not in their absolute terms.

3. An overview of the sugar industry in Kenya

The sugar industry in Kenya is constantly embroiled in controversy and has heavy government involvement. The industry is also highly inefficient and only survives due to high tariff and non-tariff protection. Nevertheless, the industry remains a key economic activity in Western Kenya where it directly supports an estimated 200,000 small-scale farmers who supply over 80% of the cane to the factories. Total employment in the factories is estimated to be around 40,000 workers. Kenya Sugar Board estimates that the sugar industry supports up to 2 million people in the country (KSB, 2005). Of the total 126,826 ha under cane production in 2002, ninety percent was owned by out-growers (smallholders) with the rest occupied by nucleus estates around the factories. The smallholder farmers usually own 3-4 hectares which they operate as a family unit. They are organised into out-grower organisations and co-operatives, which are expected to contract and organize cane development, and supply the cane to the factories. Inputs, credit and payments to farmers are channelled through these organizations. However, most the out-grower organisations are plagued by mis-management and poor governance and therefore unable to perform. For instance out of the Ksh 7.5 billion (US $ 110 million) owed to Kenya Sugar Board (KSB) by May 2003, the out-grower firms owed close to Ksh 2.4 billion (US $32 million). Consequently, sugar companies (millers) have taken over some of the out-grower firms' roles such as cane development, harvesting and cane transport (Republic of Kenya, 2003). Smallholder farmers are members of the Kenya Sugar Growers Association (KESGA) through which they lobby the government on policy issues.

Currently, there are seven sugar companies in the country two of which are under receivership. The government is the majority shareholder (owning over 90% of the shares) in most of the companies except in the largest factory –Mumias- where it owns around 28% of the shares. The sugar companies (millers) are organised around the Kenya Sugar Millers Association (KESMA) through which they negotiate cane prices with farmers and lobby the government on policy matters. As is the case with out-grower companies the sugar companies are also deep in debts. By May 2003, the debt burden of most sugar factories was estimated to be over

91% of their total assets with out-grower farmers owed Ksh 1.7 billion (US$ 25 million). In addition, capacity utilization in most factories is below 55% except Mumias with over 70% capacity utilization. Low capacity utilization and outdated milling technologies have been cited as some of the major factors responsible for the high sugar processing cost in the country. The sugar industry is regulated by the Kenya Sugar Board (KSB) which is established through an Act of Parliament.

Sugar production in Kenya has been on a downward tread since 1999 but recovered in 2002 when 494,249 metric tons were produced (Figure 1). The privately-owned Mumias Sugar Company is the main producer accounting for over 52% of production and sales. Cane yields are low averaging 6.6 tons per ha as compared to 18.8 tons realised under irrigation in Sudan. The planted varieties take 18-24 months to mature as opposed to 14 months in Sudan. Delays in cane harvesting that are common in Kenya, a factor that leads to low sucrose yields and sugar recovery levels. This characteristic of cane production in Kenya has been cited as one of the causes for high cost of production.

Sugar consumption in 2005 was estimated at around 700,000 metric tonnes (Figure 1) equivalent to 22 kg per capita. Forecast by KSB estimate a total consumption of 87,000 metric tonnes in 2010 and 1,287,000 tonnes in 2015, an average increase of 8% per year. Out of the 2005 consumption, 24% was covered by imports. Although there is a clear and discernible tread where years of high sugar imports alternate with years of low imports with the levels of imports being highly and negatively correlated to local production, the levels of imports have stabilised at around 25%.

Available statistics for 2005 indicate that, 47% of the imports in that year were sourced from COMESA free trade area countries mainly Malawi, Egypt, Sudan and Zimbabwe. The rest of the sugar was imported from South Africa (41%) and other non-COMESA countries. Import volumes are however higher than the official figures mainly due to diversion of transit sugar intended for the neighbouring countries.

In 2002, Kenya applied to use of COMESA safeguard measures and was allowed to impose a import quota of 200,000 tonnes per annum for sugar imported from COMESA countries. Of this quota, 80,000 tons is for industrial sugar while 120,000 tons is for domestic sugar. Under the COMESA safeguard clause, Kenya is also allowed to impose a tariff rate of 120% (Barasa, 2004). The safeguard clause under which Kenya is protected from cheap imports from other COMESA partners comes to an end in February 2008. Kenya is currently negotiating for an extension of the safeguard clause for a further four years.

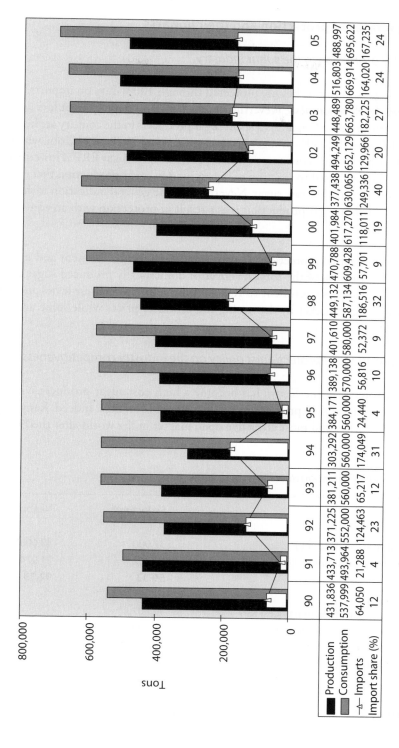

	90	91	92	93	94	95	96	97	98	99	00	01	02	03	04	05
Production	431,836	433,713	371,225	381,211	303,292	384,171	389,138	401,610	449,132	470,788	401,984	377,438	494,249	448,489	516,803	488,997
Consumption	537,999	493,964	552,000	560,000	560,000	560,000	570,000	580,000	587,134	609,428	617,270	630,065	652,129	663,780	669,914	695,622
Imports	64,050	21,288	124,463	65,217	174,049	24,440	56,816	52,372	186,516	57,701	118,011	249,336	129,966	182,225	164,020	167,235
Import share (%)	12	4	23	12	31	4	10	9	32	9	19	40	20	27	24	24

Figure 1. Sugar production, consumption and imports, 1990-2005 (KSB statistics).

4. Sugar rates of protection and their impacts

4.1. Rates protection in the sugar industry

Kenya's sugar industry is heavily protected through 100% duty on imports, 16% VAT and 7% sugar development levy. Nevertheless, the VAT and development levy are also charged on domestic produced sugar. As a result sugar producers in the country are heavily protected with NRP of around 59% in sugar production areas and 92% in Nairobi, which is the main sugar distribution and consumption area (Table 2). The sugar ERP is lower than the NPR. This may be attributed to taxation of inputs used in sugar production that erodes the high protection accorded to raw sugar. Nevertheless, the general indication is that almost 41% of value added in sugar production is as a result of protection with an equivalent taxation of consumers.

Apart from external protection of sugar producers, the government and its agencies also absorbs almost 32% of the total turnover of domestically produced sugar through taxes, levies and dividends equivalent to Ksh 4.2 billion in 2003. These levies and taxes further distort the sugar market by depressing the producer prices while some are passed on to consumers.

4.2. Implications of protectionist policy on the industry competitiveness

Due to protection, the Kenya has become a high cost sugar producer with cane prices almost equal to the CIF price of sugar in the port of Mombasa. Indeed, Kenya sugar market is arguably the second most attractive open market in the world after the EU as premium

Table 2. Sugar Protection Rates in Kenya, 2003.

Item	Ex-factory*	Nairobi
Domestic price (Ksh/tonne)	37,000	42,000
Border price (Ksh/tonne)	23,222	21,974
NRP	**59.33**	**92.13**
Intermediate inputs (actual prices) Ksh	12,607.80	
Intermediate inputs (border prices-Ksh	5,882.60	
Value added, actual prices-Ksh	24,392.20	
Value added, border price-Ksh	17,339.40	
ERP	**40.67**	

* Ex-factory refers to sugar factories in Western and Nyanza provinces.

254   *Development economics between markets and institutions*

destination of sugar. As show in Table 3, in the period 2000-2005 the average ex-factory price has been more that three times the world market price. According to International Sugar Organisation (ISO) world market prices for sugar went up in 2005 and the first half of 2006. However, in the third quarter prices went down because of growing concerns regarding the building up of world surpluses. For 2006/07 it is expected that there will be a further downward pressure on world market prices (ISO, 2007). Historically, world market prices tend to reflect production costs in the most efficient producers. The current level of world prices is still higher than the prevailing cost of production in Brazil. Furthermore, the projected world production for 2006/07 is significantly higher than consumption, world sugar stocks are expected to grow and total export availability will exceed demand. First tentative indications therefore point toward another world production surplus for the year 2007/08, thereby maintaining the downward pressure on world prices. Major exporters on the world market with very low production costs - Brazil and Australia- produce at around € 138 per tonne. The Kenyan sugar market will therefore offer an attractive market for their exports even if their exports are to be subjected to import tariff.

Table 3. Kenya sugar prices.

	2000	2001	2002	2003	2004	2005
Ex-factory price Ksh/tonne	42,675	46,778	38,943	40,548	44,064	50,898
Wholesale Ksh/tonne	52,400	48,580	39,960	41,960	46,160	57,540
Retail Ksh/tonne	55,530	53,750	44,940	46,090	51,630	64,820
Exchange rate (Ksh to Euro)	70.19	70.03	73.63	85.42	97.60	95.20
Ex-factory €/tonne	608.0	667.9	528.9	474.7	451.4	534.6
Wholesale €/tonne	746.5	693.6	542.7	491.2	472.9	604.4
Retail €/tonne	791.1	767.4	610.3	539.6	528.9	680.9
World €/tonne[1]	194.5	212.6	161.8	139.9	128.0	174.0
Ratio Kenya ex-factory/world price	3.1	3.1	3.3	3.4	3.5	3.1
New EU Sugar Protocol price (SP) in 2009/10 (€/t)						335.2
Ratio Kenya ex-factory 2005/ EU SP price						1.6

[1] ISA daily prices

Andrew M. Karanja

Furthermore, according to a recent study the average 2005 ex-factory sugar prices in Kenya was 60% higher than the new EU Sugar Protocol (SP) for 2009/10 (Republic of Kenya, 2007).[30] Even the most efficient sugar company in Kenya- Mumias- cannot profitably export sugar to the EU. According to the study, the Lisbon CIF for Mumias sugar was Euros 510 per tonne as compared to the EU reference price of Euros 335. This means that in order for Mumias to export to EU, the company has to lower its production and operational costs by 34%. The other sugar companies have higher production costs and without major changes they will be unable to even reach the new EU sugar reference price. Kenya has a low allocation of 5,000 tonnes under the EU Sugar Protocol (SP) and an allocation of 10,000 tonnes under the Agreement on Special Preferential Sugar (SPS). The low allocation means the country may not loose much revenue in the future if it's unable to export to the EU. However, the EU sugar reform will also affect the other African Sugar Protocol countries among which have huge export quotas to the EU, like Swaziland and Mauritius. Through the COMESA market these countries could redirect their exports from the EU market towards the attractive Kenyan market with its high domestic sugar prices.

Within COMESA, the greatest threat for Kenya is Swaziland which has low production costs, huge production and exports. Around 25% of Swaziland's exports are destined to the EU, but as a non-LDC, Swaziland will not qualify for sugar exports to the EU under the Everything but Arms (EBA) agreement. These exports might be diverted to Kenyan Market. Malawi, Zambia and Sudan also have sizeable sugar exports but would qualify as Least Developed Country (LDCs) to free access to EU market, which would probably remain an attractive option.

Within East African Community (EAC), both Tanzania and Uganda are still having a sugar shortage and are not able to satisfy the domestic demand. However, if the expansion plans of the (privatised) sugar industry in Tanzania are to become a reality it would imply that there could be excess supply on the medium-term. Uganda is also expanding its production. Therefore, on the longer term competition on the Kenyan market from imported sugar from Tanzania or Uganda cannot be ruled out.

It is therefore clear that the Kenyan sugar industry remains under global and regional threat despite the protection offered to sugar producers in the country. This in a way demonstrates the folly of pursuing a protectionist policy that is not complemented with deliberate efforts to introduce competitive pressure to improve internal efficiency.

---

[30] The basic elements of the EU sugar reform with regard to sugar exports of ACPs – among which the Sugar Protocol Countries - are a price cut for standard raw sugar of 36% over 4 years combined with free access for EBA sugar exports to the EU from 2009/10 onwards for LDCs. Kenya is not a LDC and will therefore not benefit from this free access. The SPS quotas will gradually be reduced with the increase of the EBA quotas and disappear in 2009/10.

4.3. Impacts of high sugar protection on welfare

As earlier indicated, the sugar consumers and industrialist in the country bear the burden through high sugar prices. For instance, by April 2003 sugar consumers in Kenya paid an average of Ksh42.57 per kg of sugar which was twice the landed cost of imported sugar in Nairobi of Ksh 21.90 per kg (excluding duty). This additional cost was estimated to be US$ 172 million in 2001 which was equivalent to US$ 6 per capita (World Bank, 2003). With the 2006 level of prices, this additional cost works out to be Ksh 16 billion (US$242 million). Based on the 1997 welfare monitoring survey, rural and urban households spend 5.6% and 4.2% of total food expenditure per month on sugar. With the current level of sugar NPR, this translates to a tax of US$3.0 and US$3.9 per month for rural and urban household, respectively. Given the fact that poor households spend a higher proportion of their income on food items, then it is clear that the poor households are heavily taxed on the account of the current sugar trade policy. It is therefore clear that sugar consumers in the country could greatly benefit from a more open trade regime that is backed by lower duty structure. The loss to the exchequer while significant would however be ameliorated by the savings made at household level and their multiplier effects.

Without protection, most of the sugar producers in Western Kenya region would find it difficult to dispose of their sugar in the local market unless measures are taken to improve efficiency in production and processing. Thus, in a scenario without protection, the 200,000 smallholder farmers and 40,000 workers employed in sugar factories would be direct losers. According to poverty mapping done by the Central Bureau of Statistics (CBS), Western Kenya, which is the main sugar belt, is a poverty hotspot with 55% (1.8 million people) of its rural population living below the poverty line (CBS, 2007). Without any credible alternatives to sugar production, the region would definitely sink further into poverty. As the days of protectionism become numbered, it is therefore prudent for all concerned to seek alternative medium and long-term interventions for the sub-sector if increases in poverty are to be avoided. Investments geared towards diversification (both on and off farm), improvements in sugar production and processing efficiency are some of the immediate alternatives that can be pursed simultaneously.

4.4. Lost opportunities in the sugar industry

Looking at the overall performance of the sugar sector in Kenya, it is evident that the industry would have been in a better position to withstand the competitive pressures if the route towards privatization was taken much earlier. A case in point is the disappointing performance of the sugar companies where the government holds majority shares. This is in total contrast with Mumias Sugar Company whose performance has continued to improve after privatization. This, therefore, makes a strong case for acceleration of the privatization and restructuring program in the sugar industry.

The prospects for biofuels such as gasohol are becoming increasingly attractive mainly due to the increase in global petrol prices and concerns for cleaner environment. This can enable diversification to occur whereby sugar industry can produce more valuable products such as ethanol for blending. However, due to limitations inherent in Kenya's energy legislation that does not allow blending, the sugar industry has not diversified into ethanol production. Furthermore, despite the potential to diversify into co-generation whereby sugar by-products are used generate electricity, this potential is yet to be realised. This is yet another opportunity that could have offered some reprieve to the industry.

5. Conclusions and policy implications

Proponents of protectionism have argued that developing countries such as Kenya should protect their main economic sectors given the pervasive protectionism prevailing in the developed countries. Nevertheless, there is there is overwhelming evidence indicating that trade protectionism comes with a high price for both developed and developing countries. Estimates of welfare gains from eliminating barriers to merchandise trade range from US$250 billion to US$620 billion annually, with about one-third to one-half of the benefits accruing to developing countries (World Bank, 2003). Agriculture markets remain among the most heavily distorted with OECD countries giving various forms of agricultural support to the level of US$ 311 billion in 2001.

For most developing countries such as Kenya, agriculture often constitutes the single most important sector of the economy where a large number of the poor derive their livelihoods, either directly or indirectly. Liberalization of global trade and especially on agriculture thus offers great potential for rapid economic growth and poverty reduction in most developing countries. While steps are made towards a more liberal trade regime, developing countries will only benefit in the long run if they adopt policies that facilitate higher productivity and competitiveness. Kenya has the necessary agricultural potential to benefit from more liberal trade especially given its position in the regional market. Most of this potential remains untapped due to various technical, institutional and policy constraints. Protectionism while serving the narrow and short-term interests of a few vocal stakeholders may prove to be an expensive and unsustainable policy option in the long term. The current, anxieties in the domestic sugar industry demonstrates very clearly the pitfalls of pursuing a protectionist policy.

In case of sugar, the results confirm that sugar producers are highly protected. Due to this protection, the sugar industry in Kenya has become inward looking, high cost and uncompetitive. The industry has become arguably one of the most lucrative open sugar destinations in the world after the EU. Despite the high domestic sugar prices that heavily tax the consumers, sugar cane producer belt in Western Kenya is rated as a poverty hotspot with 55% of the rural population living below the poverty line. This provides a policy dilemma on how best to reduce the sugar industry protection without hurting the poor producers. Investments towards diversification and enhancing efficiency and competitiveness in the industry are outright medium and long-term options which can be pursued towards resolving this dilemma.

Protection of food producers has been offered at the expense of consumers who continue to pay high prices for these food items. These high consumer prices mostly hurt the poor segment of the society and are therefore counterproductive to the poverty reduction strategy. Lower food prices reduce poverty directly – greater food intake means less hunger – and so the poor have a major stake in efforts to increase agricultural productivity (Timmer, 2002). As the results from this study indicate there are significant savings to be made especially by the poor as a result of lower food prices. These savings can be used to enhance the demand of high-value commodities such as milk, meat, poultry, tea, fruits and vegetables with a large multiplier effect to the rural and national economy. Indeed, the widespread increase in rural purchasing power brought about by the green revolution in Asia during the 1970s was key towards increasing rural employment and industrialisation. Studies done in Asia indicate that for every extra dollar of agricultural income saved there was an additional US $0.8 of non-agricultural income from local enterprises stimulated by the pending of the farm households (Delgado *et al.,* 1998). Thus, if food prices can be reduced by more widespread use of important inputs such as hybrids seeds and fertilizers augmented by better access to credit, and more conductive policy environment, there is all likelihood that the food sector can become an engine of growth in the country.

The estimated protection levels indicate high cost of farm inputs and local transport costs. The high cost of inputs maybe attributable to high transaction costs in these markets arising partly from the performance and structure of these markets. There is therefore need to review the structure agricultural input markets to identify areas where policy action can be used to improve their performance. The poor state of both rural access roads and major transport infrastructure needed to facilitate trade need attention. Marketing in the rural sector is also subject to other levies ad taxes which greatly impended market access and trade especially for the poor. Harmonisation and reduction of these taxes and levies as well as a rural trade facilitating policy are therefore required to address some of these concerns.

On the overall, the sugar case study demonstrates that the agricultural trade policy is not in tandem with the emerging international and regional trade regimes. The policy is mainly driven by the desire to safeguard the interests of farmers without due consideration on the nationwide impacts on consumers and the poor. In most cases the trade policy has turned to be counter-productive as they have impacted negatively on the would be beneficiaries. A comprehensive sector-wide approach to address these issues is therefore critical to balance their domestic and international impacts while exploring their sustainability. Further work also need to be undertaken to clearly pinpoint the sources and long-term patterns of protection or lack of it, for better policy formulation. Equally needed are multi-market studies to quantify the short term losses and long term gains arising from trade liberalization especially in a regional context.

References

Barasa, T.N., 2004. A study on the competitiveness of the sugar industry in the COMESA region- The case for Kenya. Resources Oriented Development Initiatives.

CBS, 2007. Kenya Integrated Budget Survey (KIHB), preliminary Report, Kenya Bureau of Statistics (CBS), Nairobi.

Delgado, C.L, J. Hopkins, V.A. Kelly, P. Hazell, A.A. McKenna, P. Gruhn, B. Hojjati, J. Sil and C. Courbois, 1998. Agricultural Growth Linkages in Sub-Saharan Africa. International Food Policy Research Institute (IFPRI), Research Report No. 107.

Freeman H.A, F. Ellis and E. Allison, 2003. Livelihoods and Rural Poverty Reduction In Kenya. LADDER Working Paper No. 33.

ISO, 2007. Quarterly Market Outlook; Executive Brief, May 2007-05-29

Karanja, A.M., A. Kuyvenhoven and H.A.J. Moll, 2003. Economic Reforms and Evolution of Producer Prices in Kenya: An ARCH-M Approach. African Development Review 15: 271-96.

KSB, 2005. Kenya Sugar Board (KSB) 2003 Sugar statistics

Nyangito, H., 1999. Agricultural Sector Performance in a Changing Policy Environment. In: P. Kimuyu, M. Wagacha and O. Abagi (eds.) Kenya's Strategic Policies for the 21st Century. Institute of Policy Analysis and Research (IPAR), Nairobi.

Republic of Kenya, 2003. The Sugar Sector Crises. Sugar Sector Task Force report. Ministry of Agriculture.

Republic of Kenya, 2007. National Adaptation for the Sugar Industry report, February, 2007.

Sadoulet, E and A. de Janvry, 1995. Quantitative Development Policy Analysis. The John Hopkins University Press, Baltimore, USA, 397p.

Timmer, C.P., 2002. Agriculture and Pro-Poor Growth. University of California, San Diego.

World Bank, 2003. Kenya: A Policy Agenda to Restore Growth. Country Economic Memorandum. Washington D.C.

Linking changes in the macro-economic environment to the sector and household levels: trade liberalisation and the Central American textile industry

Hans G.P. Jansen and Samuel Morley

Abstract

This chapter explores the implications of the recently-signed Central America-Dominican Republic-United States Free Trade Agreement (CAFTA) for the textile sector in Latin American economies. The textile sector is a key sector in terms of export earnings and employment for the region. Compared to previous agreements, CAFTA has widened the scope of trade relations with the U.S., and it may provide additional opportunities for growth in the sector, additional jobs, and increased vertical integration. However, an integrated approach to improve competitiveness at the enterprise level in the Central American region is necessary to achieve such effects, especially in light of reduced preferential access and enhanced Asian competition. The study produces several relevant policy implications.

1. Introduction

In August 2004, the United States (U.S.), Costa Rica, El Salvador, Guatemala, Honduras, Nicaragua and the Dominican Republic, signed the Central America-Dominican Republic-United States Free Trade Agreement (CAFTA-DR, referred to as 'CAFTA' in the remainder of this chapter). The agreement has thus far been ratified by all countries except Costa Rica. As is typically the case with free trade agreements, CAFTA has been seen as both a growth opportunity for its seven signatories and a potential threat to vulnerable sectors in each of the countries, as these sectors become exposed (even if gradually) to increased competition.

In Central America (CAM), the textile and clothing sector (often referred to as 'maquila') is the most important industrial and non-traditional export sector.[31] It has been responsible for most of the growth of manufactured exports and foreign exchange earnings, as well as for most of the employment generated since the late 1980s. The point of departure of this chapter is that despite its importance in the national economies of Central American countries, past growth in the maquila sector has not always been the result of strong comparative advantages and strong competitiveness, but rather of preferential market access provided by the U.S., following strict rules of origin requirements. Given that CAFTA contains a large number of provisions that are directly relevant to the textile trade between the U.S. and CAM, the objective of the chapter is therefore to assess the likely impact of CAFTA on the apparel value chain in CAM. In addition, the chapter tries to assess the bottlenecks and constraints

---

[31] The analysis in this chapter does not include the Dominican Republic.

to further productivity growth in the apparel industry and the resources and capabilities needed to succeed in the value chain. Research methodologies include literature review, internet sourcing, field visits and personal interviews with key players in the sector, as well as mathematical models to assess the likely macro-economic and poverty effects resulting from the impact of CAFTA on the maquila industry.

Although the chapter focuses on the effects of CAFTA, the end of the Multi-Fiber Agreement and its successor (the Agreement on Textiles and Clothing or ATC which expired on January 1 2005) has simultaneously led to increased competition – and is likely to do so even more in the future, potentially having an equal if not greater negative impact on the maquila industry than the expected (positive) impact of CAFTA.

The chapter is organized as follows: the next section provides a discussion of the origin and importance of the maquila industry in Central American countries, followed by a description of the apparel value chain in section 3. The various existing international trade agreements that shape the Central American maquila industry are discussed in section 4, with special attention to the CAFTA agreement. Section 5 provides a brief description of the current situation in the maquila industry in each of the Central American CAFTA signatory countries. In section 6 we summarize the results of simulations carried out with country-level CGE models regarding the impact of CAFTA on the maquila industry in terms of a range of macro-economic indicators such as GDP growth, employment, wages, balance of payments etc, as well the effects on poverty. We conclude in section 7.

2. Origin and importance of the maquila industry in Central America

2.1. Origin of maquila

In CAM, the term 'maquila' is often used as a near synonym for the apparel industry,[32] and in this chapter we will adhere to that convention. The maquila industry in CAM has developed relatively recently, mostly dating back to the late 1980s. The origin of capital invested in the textile maquila industry in CAM varies by country. In Honduras, Guatemala and Nicaragua, investment is largely Asian, the majority of which is Korean (about 66% in Guatemala for example). In Costa Rica, however, investment is largely American, while in El Salvador about two-thirds of investment in maquila is from national sources. The relative political stability in the region has been an important factor for investors, but the main draw has been preferential access programs granted by the U.S. which include the Caribbean Basin Initiative (CBI) and the Caribbean Basin Trade Preference Act (CBTPA).[33] About

---

[32] The exception is Costa Rica where maquila is mostly associated with the electronics industry (especially Intel, a leading producer of computer chips) which is far more important than the apparel industry in that country. However, in this chapter we will use the term 'maquila' to refer to the apparel industry also in Costa Rica.

[33] See section 4 for more information.

90% of maquila production in the Central American region is concentrated in Guatemala, Honduras and El Salvador. Maquila production in Nicaragua is growing rapidly while in Costa Rica the industry is experiencing a decline, aggravated by the fact that as of April 2007 Costa Rica had not ratified CAFTA.

2.2. Importance of maquila exports and employment

Despite its relatively recent origins, over a relatively short period of time maquila has become a leading export in every Central American country except Costa Rica (Table 1). Honduras depends on maquila for three-quarters of its total export value, while in Nicaragua maquila accounts for nearly half of total export value. Even in Guatemala and El Salvador maquila is responsible for about one-third of total export revenue. CAFTA countries' maquila exports are mostly to the U.S. All countries except Nicaragua experienced a decrease in their exports in 2005-2006 compared to 2004-2005, mainly as an effect of the completion of the Uruguay Round Agreement on Textiles and Clothing (see section 4).

Maquila is also a major provider of employment, typically involving relatively young people from rural areas. An estimated 80% of the more than 400,000 maquila employees in CAM are women (ITC, 2006). Wages can be kept relatively low largely due to an excess supply of labour. Although the maquila sector is often criticized for low wages, poor labour conditions have been more problematic, since they have led to (temporary) boycotts of certain apparel brands by consumers in the U.S.

3. Brief description of the apparel value chain

In general terms the maquila industry value chain in CAM refers to an integrated production network where basic assembly operations (mostly cutting and sewing of materials sent from U.S. plants) are undertaken and the final product exported to the U.S., frequently under tariff and quota preferences. Formally, the maquila industry can be divided into the following specialized activities that can be considered stages in the value chain:
- preparing of fibers for spinning;
- spinning of fibers into yarns;
- processing yarns into fabrics;
- design and cutting the cloth;
- sewing the cloth into finished garments (assembly); and
- finishing the item (with accessories), labelling, packaging and transporting.

Garments produced in CAM are mostly composed of fiber, yarn and textile from the U.S. (particularly in El Salvador and Honduras, in order to take advantage of preferential access, see section 4) or Asia. Maquilas in Nicaragua and Guatemala source their inputs primarily from Asia but also from the U.S. Whereas the main goal of most Asian-owned plants is to circumvent U.S. quota rather than take advantage of preferential access (and therefore

Table 1. Importance of maquila in Central America.

Country	Maquila exports to U.S. (10^6 USD)	Change on year before (%)	Growth rate 1995-2002 (%)	Imports of maquila inputs from U.S.[1] (10^6 USD)	Total export value (10^6 USD)	Maquila exports to U.S. as % of total exports	Maquila employment ('000 persons)	Maquila employment as % of total employment in manufacturing	Approximate number of maquila firms
	(1)	(2)	(3)	(4)	(5)	(6)	(7)	(8)	(9)
Costa Rica	478.6	-6.7	1.3	450.9	8610	5.6	13	8	40
El Salvador	1445.4	-18.2	193.8	205.7	4301	33.6	87	20	250
Guatemala	1717.1	-14.2	150.0	453.6	4608	37.3	142		500
Honduras	2461.8	-9.9	174.2	45.5	3066	80.3	129	27	200
Nicaragua	753.6	11.5	502.7	35.8	1653	45.6	60	30	70

[1] Total imported inputs for maquila are approximated as the sum of the imports in SITC codes #26, textile fiber, #65 which includes, yarn, thread and fabrics and SITC 84 which is clothing.

Data in (1) refer to June 30 2005 – June 30 2006; data in (4) refer to 2002; data in (6) is (1) as % of total exports in 2004; data in (7)-(9) refer to 2003 except Costa Rica which is June 2005-June 2006.

Sources: (1)-(3): IDS (2006); (4): Morley (2006); (5): World Bank database; (7): INCAE (2004) except Costa Rica which is CATECO (Camera Textil Costarricense, see www.textilescr.com); (8): IADB (2005); (9): Hernández et al. (2006) except Costa Rica which is CATECO.

they primarily source from Asian countries), U.S. owned plants (particularly those located in Nicaragua) consistently reported sourcing from U.S. mills, suggesting that previously existing ties to businesses from an investor's originating country or region affect sourcing decisions. The likely effect of CAFTA on cloth sourcing varies widely by firm, but most companies expressed openness to sourcing cloth from CAM due to CAFTA benefits.

Since the assembly step is the most labour-intensive operation in the garment production, it continues to be the operation most likely to be allocated to low-wage (mostly developing) countries, including the CAFTA countries in CAM. Within the apparel assembly sector, the following categories can be distinguished:

- 'make' or 'pure assembly' which involves only the stitching together of the pre-cut item which is typically imported: this category is also referred to as circular knit;
- 'cut and make' involving cutting the cloth and sewing the apparel item together;
- 'cut, make and trim' involving 'cut and make' and as well as trimming the item with accessories such as buttons and zippers.

In the past, maquila plants in CAM were overwhelmingly involved only in the 'make' or 'pure assembly' of the garment. Retailers (most often multinationals) purchased thread used for sewing, accessories, and cloth, cut the cloth to be assembled and sent these inputs to the maquilas for assembly. Although assembly maquilas are still the most commonly found type of maquila in CAM, the sector is increasingly moving towards 'full-package' (or 'ready to use') production – which involves the maquila sourcing and purchasing all inputs used in the production of the garment.

Full-package production typically involves purchasing all inputs, cutting and assembling the cloth, attaching accessories, undertaking any additional finishing such as ironing, packaging the garments, and sometimes shipping the garments to the client. In this way the client pays for the whole value of the apparel, rather than just the valued added in the country where the maquila is located. To the extent that supplies are locally sourced, this may lead to an increase in the traditionally few forward and backward linkages of the maquila industry to the rest of the economy, and indeed there are signs that CAFTA is already stimulating vertical integration in the Central American maquila industry. Full-package services, which are the rule in Costa Rica and becoming more common in Guatemala and Honduras, are increasingly demanded by retailers, as it allows them to focus resources on marketing and final retailing, which represent the largest part of the final value of the garment. Mexico's experience with the North American Free Trade Agreement (NAFTA) suggests that trade liberalization is important for this upgrading towards 'full-package production' to take place.

Since the apparel industry is very much 'buyer driven', the retail industry is perhaps the most important stage in the maquila value chain. The retail sector in the U.S. has become increasingly concentrated, implying more buying power for the retailer and thus increased bargaining power toward suppliers. With customer demand driving retailer demand, information regarding decisions on patterns, colours and material increasingly flows directly from retailers to textile plants. A series of logistics and business services are necessary to assure a smooth flow of goods, information and payments. 'Lean retailing' has become possible due to technologies such as bar codes, uniform product codes, electronic data interchange (EDI) and data processing as well as distribution centers and common standards across firms. Bar code equipment allows the retailer to collect 'point-of-sale' (POS) information, which enables continuous monitoring of which garments are selling and which are not and helps retailers keep track of inventories. Finally, traditional wholesalers and storage facilities are increasingly being replaced by distribution centers (DCs), which enable replenishment orders to arrive quickly to stores. Unlike wholesale storage buildings, DCs usually have a smaller floor area and the apparel is moved through automated processes.

4. International trade agreements affecting the maquila industry

4.1. CBI, CBERA and CBTPA

In order to better understand the significance of CAFTA for the Central American maquila industry, it is important to understand its preceding. The provisions of the CAFTA agreement depart from the relevant conditions of previous trade agreements between the U.S. and the five Central American countries, all of which are part of the CBI put into effect beginning January 1, 1984. The CBI granted trade preferences and other benefits to the countries of the region by the Caribbean Basin Economic Recovery Act (CBERA) enacted by the U.S. Congress in 1983. The CBERA granted unilateral preferential treatment to many products imported into the U.S. from 24 countries in the Caribbean Basin designated as beneficiaries. However, textiles and apparel, even though exempted from the world-wide quota system then in force, were not given special tariff-free access to the U.S. market; rather, they remained subject to the so-called Multi-Fibre Arrangement (MFA) which ruled from 1974 to 1994 and permitted certain countries to impose quantitative restrictions on textile and clothing in case the latter were considered a threat to their own domestic industries. There was, however, one important exception under the CBERA: under the so-called 'Special Access Program' (announced in 1986 and referred to as '807') they were exempted from the MFA provided that they were produced from inputs produced in the U.S.

Under the Special Access Program, apparel items that qualified were imported under preferential quotas called 'guaranteed access levels', rather than counting towards regular quotas. This new program applied to garments made from cloth that was produced and cut in the U.S. and then assembled in Caribbean Basin countries. In the 1970s and 1980s, U.S. apparel producers actively used '807' to outsource apparel assembly to countries with lower labour costs, especially Mexico and the Caribbean Basin countries (which include the Central American countries). The identical trade and tariff treatment of textiles from both Mexico and the Caribbean Basin countries granted by the U.S. under the CBERA changed in 1990 with the passage of the Caribbean Economic Recovery Expansion Act (CBEREA).

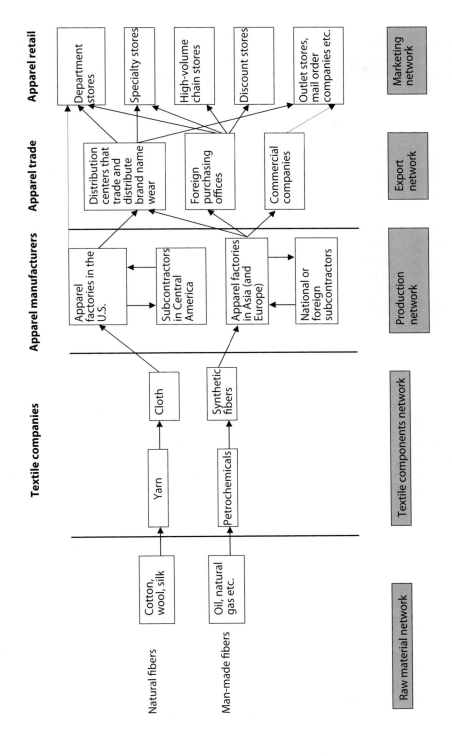

Figure 1. Networks in the maquila value chain (Based on Gereffi and Memedovic, 2003).

It reduced tariffs for the Caribbean and Central American countries by 20% over a five year period with a 2.5% floor. Thus between 1990 and the implementation of NAFTA in 1994, CAM enjoyed significant advantages over Mexico because of lower U.S. tariffs. However, until January 1 1995, textile exports from both CAM and Mexico were still subject to the MFA agreement.

NAFTA changed the position of maquila in CAM. An unintended side effect of the agreement was that the initial advantages of CBEREA beneficiary countries over Mexico were virtually eliminated because Mexican products now entered the U.S. duty-free as well as quota-free. To make matters worse for the Central American maquila industry, Mexican producers were not subject to the restrictive rules of origin on intermediate inputs. To offset this unintended and unfavourable effect of NAFTA on CAM, in 2000 the U.S.-Caribbean Basin Trade Partnership Act (CBTPA or '809') was passed. CBTPA beneficiary products include all textile and apparel products (as well as a number of other products) which were granted the same duty free access to the U.S. market and liberalized rules of origin granted to Mexico under NAFTA (often referred to as 'NAFTA parity').

The CBTPA provided a number of important opportunities for beneficiary countries in the textile and apparel sector. While the '807' program permitted the entrance – without duties or quotas – of apparel assembled in the region, if the cloth is made in the U.S. and produced from U.S. yarn, the so-called '809' program extended the '807' program and provides free trade of apparel sewn in the CBTPA region with fabric that was made from American thread and with fabric that was produced and cut in the U.S. *or the region.* In practice, the CBTPA has provided a big impetus to the growth of the maquila industry in all of the Central American countries. It is important to note, however, that the benefits of the CBTPA would be available during a transition period from October 1, 2000 until either September 30, 2008, or the date that the Free Trade Area of the Americas, which is still under negotiation, enters into force. As explained below, the Central American countries decided not to wait until the expiration of the CBTPA but rather make most of its provisions permanent under CAFTA.

4.2. ATC

Although this section looks primarily at the impact of CAFTA on the maquila industry in CAM, the ending on January 1 2005 of the Uruguay Round Agreement on Textiles and Clothing (ATC) may well have the largest long-term impact on textile and apparel production in the region in the near to medium-term future. As of January 1 1995, the MFA was succeeded by the ATC which provided for the gradual elimination of textile quotas. While in theory developing countries were supposed to favour implementation of the ATC, many countries (including the ones in CAM) were concerned that the removal of quotas would actually harm their maquila industry since the ATC makes the outsourcing of apparel assembly to less-developed countries less attractive. More importantly, however, once quotas are removed completely, CAM countries would lose their relative advantage obtained through the CBTPA and market share will go to countries with the highest comparative advantage in textile production. Comparative advantage is determined not only by the cost of labour and the price of raw materials, but also by factors such as timeliness and percentage rejects. Since China has the cheapest labour and arguably one of the strongest textile industries in the world, its accession into the WTO in December 2001 and the end of the ATC have left CAFTA countries wondering what will happen to their maquila industries in the future. Between 2001 and 2003, Chinese market share of the total value of U.S. apparel imports jumped from 13.7% to 40.2%. Between 2001 and 2002 alone, China's total exports of textile products to the U.S. jumped by 125%. Nearly all of this growth was in products that had been freed from quota restrictions. Chinese textile exports to the U.S. increased by 50% in the first half of 2005. This caused the U.S. government to invoke a textile safeguard provision intended to protect domestic industries that China had accepted upon joining the WTO in 2001. Under this provision, a temporary 7.5% limit was immediately imposed and in November of 2005, the U.S. and China reached a longer-term bilateral agreement to be implemented that will hinder China's full access to the U.S. market until December 31, 2008. For other countries that export to the U.S. (including CAM), the 'safeguard' tariffs placed by the U.S. on Chinese textile goods as a mitigating measure for market disruptions provide no more than a three-year window (until 2008) to take advantage of the quota-free trading environment and win precious market share in the U.S. and EU before China is able to fully compete.

The net effect of the expiration of the ATC has been positive for Asian exporters (particularly China and India), but negative for CAM and Mexico (Table 2). The latter have lost market share in the U.S., which has especially hurt apparel producers who face difficulties competing with Asian countries. Textile producers in CAM have suffered less since raw material

Table 2. Percentage shares of different countries in U.S. textile and clothing imports.

Country	2003	2004	2005[2]
China	14.9	17.2	25.1
India	4.3	4.6	5.3
Mexico	10.8	9.9	8.7
CAFTA countries[1]	11.6	11.0	10.0

[1] Includes the Dominican Republic.
[2] January-September.
Source: based on data of the U.S. International Trade Commission (USITC).

sourcing is shifting somewhat from the U.S. to CAM (and Mexico), but the net effect is still negative.

4.3. CAFTA

The bottom line of the (complicated) provisions in the CAFTA agreement for maquila production is that they consolidate the (unilateral and temporary, and therefore revocable) preferential plans provided by the USA through the CBTPA which, just like the CBI/ CBERA, is scheduled to expire in 2008. In particular, CAFTA makes permanent the liberalized rules of origin for inputs to the maquila industry granted temporarily under the CBTPA, in this way providing CAFTA countries a continuation of the preferential treatment for apparel assembled in CAM from regional or third-country (non-CAFTA) cloth. This preferential treatment included in the CBTPA entitles some apparel produced in the CAM to be imported into the U.S. at reduced or zero tariff rates even when produced with non-American cloth, as long as American yarn had been used in cloth production. In addition CAFTA expands on CBTPA's limited benefit for regionally made fabrics which applied only to knit apparel. These are very important achievements for Central American maquila producers because during the early stages of CAFTA negotiations, many U.S. apparel companies and retailers asserted that the decision to continue, eliminate, or expand apparel assembly in Central American maquilas depended largely on the details of CAFTA, given the largely negative impact of the 2005 quota elimination. The way in which CAFTA allows for a number of regional and third-country fabrics can be summarized as follows:

First, CAFTA contains a 'yarn forward' rule of origin which first appeared in NAFTA. In practical terms, this rule stipulates that apparel enters the U.S. tariff-free if the fiber, yarn, cloth, and sewing thread are from the U.S. or CAFTA countries, and the cut and assembly takes place in CAFTA countries. In this way, CAFTA changes the original U.S. yarn requirement of the CBTPA into a U.S. or regional yarn requirement. It also lifts the prohibition on dyeing and finishing of fabrics. This rule applies to about 90% of Central American apparel that is currently exported to the U.S.

Second, garments made of silk, linen and other yarns and fabrics which are generally not produced in the U.S. or any other CAFTA country, are subject to a more lenient 'single transformation rule.' This rule allows a garment to maintain its status as an originating product, provided that the cutting and sewing take place in a CAFTA country. Fabric and yarn from any origin does not affect the item's ability to qualify.

Third, CAFTA contains a set of new rules concerning the component that determines a garment's classification. Basically the new ruling only requires the main component and pockets to be originating, thus eliminating the need for a trimmings rule, as is found in the CBTPA. In addition, while the CBTPA only allows apparel to include limited foreign articles - up to 25% of the value of all components – CAFTA generally allows unlimited foreign accessories such as snaps and buttons.

Fourth, Nicaragua and Costa Rica obtained temporary preferential access quotas (called Tariff Preference Levels (TPLs)) meaning that if apparel made of cotton or MMF are cut or knit to shape and assembled, they qualify as originating, even if the fabric or yarn is not from the CAFTA region. Up to 100 million square meter equivalents (SMEs) of Nicaraguan apparel annually can enter the U.S. for the first ten years. This is equivalent of about 75% of Nicaragua's current third-party input use. Costa Rica obtained a two-year (with a possibility of extension) TPL for 500,000 SME for tailored wool apparel that would enter the U.S. free from rules of origin restrictions at a tariff level equivalent to 50% of the tariff applied for most-favoured nations. Since Costa Rica has not yet ratified CAFTA, it is not clear when this two-year period will start.

Finally, a so-called 'accumulation' rule with Mexico and Canada for woven garments allows, within certain limits, some Canadian and Mexican inputs (especially yarn) to be used in regionally-produced apparel without disqualifying the apparel from duty-free benefits.

Trade preferences in general, and particularly the liberalization of the rules of origin in through the CBTPA and now CAFTA, have turned out to be a big stimulus for a transformation of the Central American maquila industry from pure assembly of clothing from imported inputs to 'full-package production'. Production of intermediate textile products such as yarn, thread and fabrics is increasing throughout the region and some countries (Honduras, El Salvador, Nicaragua) are also reporting renewed cotton plantings. Table 3 shows the rapid decline after 2000 of the proportion of intermediate inputs coming from the U.S.: domestic suppliers apparently were able to successfully move back up the supply chain, thus dramatically increasing the domestic content of maquila exports.

It can be expected that the increasing concentration of CAM maquila exports towards the U.S. seen after the CBTPA will be further strengthened by CAFTA. For example, in El Salvador and Honduras, maquila exports in 2003 accounted for about 80% of their total exports to the U.S., even though the share is only 18% in the case of Costa Rica. Even though with the expiration of the ATV the U.S. apparel market is no longer protected by quotas (at least in theory), the completion of the ATV does not mean the end of tariffs. Therefore the continued preferential access of Central American countries to the U.S. market through the CAFTA agreement is essential for maintaining their comparative advantage in this market. Given that the end of the ATV may eventually result in average export price reductions of the order of 20%, it is imperative that countries that enjoy such (nearly always temporary) preferential access use this opportunity to improve their competitiveness.

Table 3. Intermediate imports to maquila from U.S. as % of maquila exports to U.S., and exports of maquila to U.S. (10⁶ US$).

Year	Costa Rica Intermediate imports/ exports	Costa Rica Exports	Guatemala Intermediate imports/ exports	Guatemala Exports	Honduras Intermediate imports/ exports	Honduras Exports	Nicaragua Intermediate imports/ exports	Nicaragua Exports	El Salvador Intermediate imports/ exports	El Salvador Exports
1989	62.2	337.9	81.6	138.2	74.7	88.8			97.6	43.8
1990	58.0	397.4	73.0	203.2	75.3	116.0			81.1	56.2
1991	62.1	453.9	63.7	350.9	64.7	201.7		1.3	71.4	93.7
1992	60.1	607.1	72.9	477.9	55.5	377.2		3.6	57.5	171.7
1993	61.2	671.9	78.7	573.9	50.1	522.1	55.0	11.5	50.9	259.7
1994	63.5	704.7	78.0	623.6	52.5	666.3	34.0	29.7	46.6	411.8
1995	62.0	777.9	78.9	714.1	52.6	956.0	23.0	76.5	51.4	600.3
1996	72.0	722.0	77.9	831.6	59.7	1267.6	19.0	146.1	48.0	740.1
1997	70.7	869.1	71.4	1001.8	56.3	1725.4	21.0	186.6	49.1	1078.0
1998	76.3	839.7	61.5	1183.7	62.4	1946.1	20.0	237.4	53.4	1198.9
1999	60.9	846.6	40.1	1280.2	55.0	2243.6	21.0	284.1	42.9	1360.7
2000	41.3	846.8	7.0	1545.2	0.9	2463.3	4.0	345.8	5.3	1640.9
2001	39.7	791.0	6.0	1677.5	0.8	2485.7	3.0	390.6	6.1	1671.2
2002	36.7	747.2	14.9	1727.0	0.8	2556.3	3.0	445.8	5.0	1712.7

Source: based on Table 10 in Morley, 2006.

5. Current situation and trends in the maquila industry in the Central American CAFTA countries

Although CAFTA opens new opportunities for the Central American region, the experience of Mexico suggests that a free trade agreement and proximity to the U.S. alone are not sufficient to remain competitive. With or without quotas, the average U.S. tariff rate on imported clothes is 17%. Although the U.S. has an incentive to source from countries with tariff preferences, a lack of tariffs not always compensates for the great advantages Asian producers have in costs. Quota removal could eventually lead to an oversupply and significant drop in world prices. Thus far, however, Central American countries have been fairly successful in maintaining their market share in the U.S. (see Table 4) but this may be to a significant extent due to the bi-lateral agreement accorded between the U.S. and China that will restrain Chinese exports until 2008. What will happen after 2008 is anybody's guess. What does remain clear, however, is that given the substantial wage differences between CAM and China[34], CAM would be hard pressed to compete on production costs[35] alone and therefore must develop those characteristics that make them competitive: efficiency, quality, dependability, speediness of supply and flexibility.

The following sub-sections describe the main challenges faced by the maquila sector in each of the five Central American CAFTA countries.

5.1. El Salvador

El Salvador's maquila industry currently consists of about 250 firms of which 180 are apparel assembly firms and only 32 engage in full-package production (the remaining are accessories firms). Even though during the past 5 years El Salvador has more or less maintained its position and market share in the U.S. import market, the biggest negative factor affecting El Salvador' future competitiveness in the U.S. apparel market is its heavy dependence on U.S. preferential treatment. Thus, there is a clear need to diversify maquila production and move further towards full-package production. Another negative factor for the apparel industry is labour costs which are considerably higher in El Salvador than in most Asian countries. Also, El Salvador's location advantage in speed-to-market is diminishing as Asian

---

[34] According to IADB (2005), the average hourly wage in maquila operations (including benefits) in China is US$0.88 (but only $0.48 according to INCAE 2004), as opposed to $0.92 in Nicaragua, $1.48 in Honduras, $1.49 in Guatemala, $1.58 in El Salvador, and $2.70 in Costa Rica, respectively. These figures compare with hourly labour costs of about $18 and $16 in textile operations in the USA and Canada, respectively (Parada Gómez 2004).

[35] A 2001 study by the Apparel and Footwear Association compared total cost up to shipment, of a shirt that takes 20 minutes to cut, sew, and finish and is destined for the U.S. retail market. The least to most expensive countries in which to produce were: China ($1.12), Nicaragua ($1.50), Dominican Republic and Honduras ($1.70), Guatemala ($1.80), El Salvador ($1.85), Costa Rica ($2.00), Mexico ($2.20) and the United States ($5.00).

Table 4. Market participation of the twenty largest exporters of textile and clothing in the U.S. market.

Rank	Country	1996	Country	2000	Country	2004	Country	2005
1	China	13.3	China	14.6	China	16.0	China	25.1
2	Hong Kong	10.4	Mexico	10.5	Mexico	10.3	Mexico	9.1
3	Mexico	9.9	Hong Kong	7.6	Hong Kong	5.8	India	4.5
4	Taiwan	4.8	**Honduras**	**4.1**	**Honduras**	**4.1**	Indonesia	4.0
5	Dominican Republic	4.5	Dominican Republic	4.0	Vietnam	3.8	Hong Kong	4.0
6	Philippines	3.9	South Korea	3.8	Indonesia	3.6	**Honduras**	**3.8**
7	South Korea	3.8	Indonesia	3.5	India	3.4	Vietnam	3.6
8	Indonesia	3.5	Taiwan	3.3	Dominican Republic	3.1	Bangladesh	3.1
9	Italy	3.3	Bangladesh	3.3	**Guatemala**	**2.9**	**Guatemala**	**2.7**
10	India	3.3	Philippines	3.2	Bangladesh	2.8	Dominican Republic	2.6
11	**Honduras**	**3.3**	India	3.1	Thailand	2.7	Thailand	2.6
12	Thailand	2.8	Thailand	3.1	South Korea	2.7	Philippines	2.5
13	Bangladesh	2.7	Canada	3.0	Philippines	2.7	Sri Lanka	2.3
14	Sri Lanka	2.7	**El Salvador**	**2.7**	**El Salvador**	**2.6**	**El Salvador**	**2.3**
15	Canada	2.5	Italy	2.6	Italy	2.4	Cambodia	2.2
16	**Guatemala**	**2.1**	**Guatemala**	**2.5**	Sri Lanka	2.3	Italy	2.2
17	Macao	2.0	Sri Lanka	2.5	Canada	2.2	Canada	2.2
18	**El Salvador**	**1.9**	Macao	1.9	Taiwan	2.2	Pakistan	1.8
19	**Costa Rica**	**1.9**	Turkey	1.8	Macao	2.1	South Korea	1.7
20	Malaysia	1.7	Pakistan	1.6	Cambodia	2.1	Jordan	1.5

Source: based on Table 9 in Hernández et al., 2006.
Central American countries in bold.

producers improve shipping times. The serious security situation in El Salvador is also reason for concern.

On the more positive side, most apparel buyers are well aware of production risk and therefore never source from one country only. This means that El Salvador may try to compete with Asian suppliers as a second-best source, through credibly reassuring buyers that Salvadoran-made goods have minimal reputation risk. There is an obvious and essential role for the public sector in this respect in terms of developing and implementing a workplace standards monitoring program through which El Salvadoran apparel can earn a distinctive label in the world market.

5.2. Honduras

Honduras remains the main exporter of apparel volume in CAM. Because the country imports the majority of its supplies for cloth production from the U.S., 89% of U.S. apparel imports from Honduras qualify as duty-free (compare with an average of 52% for the entire CBERA region). Just like El Salvador, Honduras is trying to take advantage of CAFTA by moving away from assembly-only towards full-package production. However, the latter is not yet widespread: of the 200 textile maquilas in Honduras, just over 50 engage in full-package production. The current scarcity of domestically produced cloth acts as a serious limitation on further expansion of full-package production, even though the cloth industry has received a significant push as a result of CAFTA. Domestic cotton production is currently very limited and faces strong opposition from shrimp producers in the Southwest, who are concerned that the pesticides used in cotton cultivation would damage their livelihood through water pollution.

Factors positively affecting the competitiveness of the Honduran maquila industry include relatively short end-to-end logistics time, competitive wages compared to Costa Rica, Mexico and the U.S., good facilities in free export zones and an efficient seaport (Puerto Cortés). As a result of these positive factors, Honduras has been able to considerably strengthen its position in the U.S. import market during the past 5 years (Table 4). The completion of the ATC is expected to stimulate co-investment between U.S. and Honduran investors, and at least in the medium-term larger firms are expected to survive competition with China due to closer, more direct relationships with U.S. retailers. In the long term however, it remains an open question if Honduras' U.S. preferential treatment is enough to compete with lower wage costs and cheaper inputs of Asian suppliers. The worsening security situation in Honduras is also reason for concern, particularly in Tegucigalpa and San Pedro Sula where many maquilas are located.

5.3. Guatemala

Guatemala has the most developed and vertically integrated maquila industry and is the leader in full-package production in the Central American region, despite the lack of a domestic cotton industry. Guatemala imports cotton not only from the U.S. but also from Asia (most firms are funded by Asian capital), despite that that disqualifies apparel for tariff preferences in the U.S. export market. The Regional Directory lists 50 cloth-producing firms, 231 apparel firms, and 260 firms offering accessories and services such as thread, packaging, embroidery, labels, buttons, and zippers.

Factors negatively affecting the competitiveness of the Guatemalan maquila industry include relatively high electricity costs, slowness in custom procedures, and increased competition from China and other Asian countries. On the positive side, Guatemala has the most developed cluster and full-package production system in CAM, the highest degree of product diversification, and less reliance on more expensive U.S. cloth in order to qualify for preferential treatment. Guatemalan maquila labour is relatively well-trained. Just as in the case of El Salvador, Guatemala may try to compete with Asian suppliers as a second-best source, through credibly reassuring buyers that Guatemalan-made goods comply with international standards regarding labour rights and environmental stewardship.

5.4. Nicaragua

The maquila sector in Nicaragua is one of the fastest growing maquila industries in the world. Nicaraguan apparel exports to the U.S. registered a 20% increase in 2005 over 2004, and another 12% increase in 2006 (until June 30), at a time when the CAFTA region as a whole is losing market share in the U.S. export market. As in the case of Guatemala, Asian investors in Nicaragua have chosen to import most supplies and fabrics from Taiwan, Korea, and Hong Kong, even if this disqualifies the apparel from preferential tariff treatment in the U.S.

The Regional Directory lists 35 apparel assembly firms (but this number is said to be growing) of which only 12 undertake full-package production. The weak vertical integration and insufficient cluster development in the maquila industry in Nicaragua can partially be attributed to the relatively young age of the industry. Cluster development could be further delayed due to the TPLs granted to Nicaragua under CAFTA, which allow third-country yarn and cloth if equal amounts of U.S. cloth are imported. While preferential access granted through the TPL provisions in CAFTA make the country less vulnerable to the expiration of the ATC, the late addition of a 'one-for-one' requirement forcing every unit of non-originating cloth in Nicaraguan garments to be matched by a unit of originating cloth makes production using non-originating cloth less attractive.[36] Nevertheless many maquila firms still expressed interest in using TPLs. Many of those currently sourcing at least some textile from the U.S. expressed plans to increase usage of American textile to 50% in order to take advantage of TPLs. Managing TPLs in such a way that one-for-one is met will be a challenge for the industry – and fines for non-compliance are under negotiation.

The TPLs, low dependency on preferential access and lower costs of Asian fabric, extremely low labour costs (only China has lower labour costs), proximity to the major U.S. markets (resulting in relatively fast response times) all are important factors determining Nicaragua's competitiveness in the U.S. apparel market relative to Asian producers. Finally, the security situation in Nicaragua is such that it does not act as an impediment to attract qualified foreign middle management.

Factors that negatively affect the competitiveness of the Nicaraguan maquila industry include high electricity costs, lack of a large-size vessel port on the Atlantic coast and the need for improvements at Puerto Corinto on the Pacific Coast (currently most of Nicaraguan apparel enters the U.S. through Miami via the Honduran port of Puerto Cortés), limited telecommunications services, limited professional services and insufficient access to local financing, and political risks and inconsistent government policies over the years that affect the business climate. Compared to its Asian competitors, the Nicaraguan maquila industry also suffers from relatively high defect levels and problems with on-time delivery and filling quick-response orders.

In order to further develop the Nicaraguan maquila industry in the short term, further development of apparel assembly plants and increased focus on value-added operations should have high priority. But in the longer term, also Nicaragua will need to move towards increased full-package production.

5.5. Costa Rica

The situation regarding the maquila industry in Costa Rica exhibits a number of characteristics that makes it somewhat different from its sister industries in the other Central American CAFTA countries.

First, the relative importance of the maquila industry in the Costa Rican economy is much less than in the other Central American CAFTA countries (see Table 1). Second, the size of the industry has been gradually decreasing since the implementation of NAFTA in 1994

---

[36] One industry expert estimates that as soon as U.S. cloth is 40% more expensive than Asian cloth, TPLs cease to be beneficial. Since many types of U.S. cloth are 40% - 50% more expensive than their Asian counterparts, the 'one-for-one' requirement may well significantly reduce the benefits of the TPLs for the Nicaraguan apparel sector.

– whereas before 1994 the maquila industry in Costa Rica generated employment for about 40,000 workers, currently some 13,000 people find jobs in the textile maquilas. Thus Costa Rica already experienced a substantial restructuring and adjustment process of its maquila industry in the 1990s, with only 40 firms remaining today, and this makes its remaining maquila industry less vulnerable to the effects of the expiration of the ATC.

Third, Costa Rican firms see their geographical position towards the U.S. (where over 90% of its maquila exports go) as one of their main elements of its comparative advantage. Most Costa Rica firms have specialized in so-called 'short run production'[37] which requires rapid response in which it does not have to compete with China. Whereas the total value of maquila exports from the CAFTA region as a whole to the U.S. decreased by 13 percent for the year ending June 30 2006 compared to the previous 12-month period, Costa Rica's exports decreased by less than 7 percent (and compare to the decrease in Guatemala's exports by nearly 15%).[38] Fourth, the maquila industry in Costa Rica is nearly entirely full-package, despite the fact that the country does not have a cloth industry and sources most textile from the U.S.

Finally, the fact that CAFTA is not yet implemented in Costa Rica has a clear negative effect on the maquila industry. The delay in approval of CAFTA not only causes substantial business insecurity but the fact that Costa Rica does not yet qualify as a CAFTA country already has had a number of repercussions for the industry which depends for 89% of its total export value on the US market. For example, the 'single transformation rule' although it applies to boxer shorts (among other products), as long as CAFTA is not implemented boxer shorts from Costa Rica do not qualify which is costing Costa Rica about US$ 10 million per year[39] in import duties payable in the USA. This is because even though Costa Rica is the largest provider of boxer shorts in the US market, without CAFTA these remain subject to a 16.5% import tariff.[40] The same tariff applies to cotton underwear imports for which Costa Rica is the fourth largest supplier in the US market. Another example relates to the 'accumulation of origin rule' in CAFTA which without CAFTA does not apply to intermediate products[41] produced in Costa Rica which in practice often prevents textile plants in other countries from using Costa Rican produced inputs in their manufacturing process.

---

[37] As opposed to 'long runs' which involve production of large quantities of relatively simple apparel with a much longer lead time.

[38] But exports in the first two months of 2007 were 14% lower than in the same period of 2006 ($ 66 million versus $ 77 million), largely due to the fact that Costa Rica has not yet ratified CAFTA.

[39] Source: Miguel Schyfter, president of the Costa Rica Association of Exporters of the Textile Industry.

[40] Other textile imports from Costa Rica are subject to even higher import duties without CAFTA: for example, blue jeans 28.6%, woolen clothing 28.2%, brassieres 23.5%.

[41] For example, without CAFTA textile exports from other Central American countries to the US that use intermediate inputs made in Costa Rica of cotton, wool or synthetic material are subject to import duties of respectively 13.5%, 18.8% and 18.8%.

6. Macro-economic and poverty impacts of the maquila provisions in CAFTA

Mathematical models for Honduras, El Salvador, Nicaragua and Costa Rica[42] were used to simulate the impact of the maquila provisions in CAFTA on national income, employment, factor incomes and poverty. In order to understand the macro-economic effects it is useful to remember that as far as the maquila industry is concerned, CAFTA does not just liberalize. Rather, for maquila the focus of CAFTA is on consolidating the gains for the CAM countries achieved under the CBTPA.[43] Therefore, the analysis compared the situation *with* and *without* CAFTA, rather than *before* and *after* CAFTA.

The mathematical models used for the simulations are of the recursive dynamic general equilibrium type (commonly referred to as Computable General Equilibrium or CGE models). Analysis of economy-wide effects of trade liberalization measures are especially important for small and relatively open economies such as those found in CAM. The models are solved in two stages of which the first is to generate a solution for a one-year equilibrium using a static CGE model according to which the (short-term) state of the economy is entirely based on the underlying Social Accounting Matrix (SAM). The CGE models are based on the standard model format developed by IFPRI (Lofgren *et al.*, 2001). Subsequently a model between periods is used to handle the dynamic linkages that update the variables that drive growth in the longer term. The model is solved one period at a time. The variables and parameters used as linkages between periods are the aggregate capital stock (updated endogenously), population, domestic labour force, factor productivity, export and import prices, export demand, tariff rates, and transfers to and from the rest of the world (all of which are modified exogenously). Whereas the individual country CGE models differ somewhat in structure due to differences in the underlying SAMs and other reasons, they all use similar model closures with the exception of El Salvador.[44]

Finally, in order to assess the implications of the changes in macro-economic indicators for poverty, a micro-simulation methodology is used to translate the labour market outcomes of the CGE models into a distribution of income across households. This methodology makes use of a household survey as close as possible to the base year of the CGE to get a base period distribution of the labour force across the households represented in the survey. Then in the first step the labour force is divided among the various skill groups represented in the CGE model, and rates of unemployment for each are calculated. Then random numbers are assigned to the group which will shrink in size and that group is ranked according

---

[42] Unfortunately no such model is available for Guatemala.

[43] Besides consolidating existing but temporary privileges, CAFTA also further liberalizes the rules of origin for a number of apparel products by extending tariff-free access to products wholly produced in CAFTA countries or produced with non-CAFTA fibers.

[44] Due to a different exchange rate regime, the external closure in the El Salvador model differs from the one used in the models for the other countries.

to the random numbers. Subsequently the procedure moves down the ranked list of the unemployed until a sufficient number have been found to reach the amount of employment given by the CGE solution. Then, working with the new simulated labour force by type, one repeats the procedure to change the skill or sectoral composition of that labour force. At a final stage, the wage of the new labour force with the composition determined by the CGE solution is changed in accordance with the CGE solution. At this point the new labour force with the new wage structure is reassembled into the households from the base period survey and new levels of household income per capita as well as poverty and income distribution statistics are calculated. For more details see Ganuza *et al.* (2002).

6.1. Honduras

The simulation for Honduras is based on a model developed by Morley and Piñeiro (2007a). This model is solved for 1997 (the base year for the data in the underlying SAM) and then solved recursively until the year 2020, using the following assumptions:
1. Model closures:
   a. External balance: since Honduras has a flexible exchange rate, foreign savings is fixed.
   b. Fiscal balance: government consumption and income taxes are fixed (making government savings endogenous).
   c. Savings-investment balance: the saving rates of households and government are fixed which makes total saving and investment positively related to the level of income (and therefore endogenous in the model with total investment determined by total savings).
2. Excess supply of unskilled and semi skilled labour at a fixed real wage rate. Within each period labour is mobile across sectors, equating real wages across sectors for these two types of labour. There is an upward-sloping supply curve for skilled labour. The simulations determine the amount of employment of unskilled or semiskilled labour that is consistent with the supply of skilled labour (and capital as well as the other macro constraints).
3. Capital is fully employed and sector specific, i.e. profit rates are free to vary across sectors and capital moves freely between sectors.

The maquila simulation run with the Honduran model gives us the growth path for the Honduran economy for the period 1997-2020 under the changes in the rules of origin for maquila introduced by CAFTA. This growth path is then compared to the one obtained with the base simulation (under the assumption of no maquila provisions from neither CAFTA nor the CBTPA) in order to assess the impacts of implementing the changes in the maquila scheme introduced by CAFTA. To model the impact of the liberalization of the rules of origin for maquila of the CAFTA agreement, the level of intermediate imports to the textile industry is kept constant at the observed level of 1997 prior to the passage of the CBTPA in 2000. Then starting in 2005 these intermediate imports are reduced to the low levels observed after 2000. This simulation then shows the positive effect of domestically producing a greater share of the intermediate inputs to the maquila industry.

The base simulation projects that the Honduran economy will grow at a relatively slow rate of 3.06% per year from its 1997 base up to 2020. The relaxation of the liberalized rules of origin for maquila increases the annual growth rates of the economy by 1.4% (Table 5). Import substitution in the maquila sector and the resulting increase in employment increase demand and output in all other sectors as well. The effect on employment and the profitability of capital and its growth rate are all also substantial. While the annual growth rate of employment for both sexes jumps to around 4.5% (an improvement of around 1.4% over the baseline), the rate of return to capital jumps from 12% to 16%. From there to 2020 the rate of growth of capital increases to 4.5% per year (as opposed to only 3.5% in the baseline scenario). That is enough to bring the down rate of return, but even in 2020 the return is still 30% higher than under the baseline scenario. There is also a much higher level of employment for the unskilled and higher wages for the skilled. In terms of the distributional impact of the maquila previsions in CAFTA between factors of production, it turns out that relative to the 1997 baseline, in 2005 the capital share rises relative to that of labour. However, the rate of capital formation rises too and therefore drives down the rate of return to capital. As a result, even in 2020 the share of capital is still higher than in 1997. Thus, the maquila previsions in CAFTA favour capital at the expense of unskilled labour. In other words, the rate of growth of unskilled employment, even though quite large, is not as rapid as the growth rate of the economy.

In conclusion, without the maquila provisions of CAFTA, Honduras' annual rate of economic growth is likely to be 1.4% lower than with these provisions, while employment for unskilled and semi-skilled labour would be nearly 27% less. Of course these simulations do not take into account the impact of the end of the ATC which may well dampen the positive impact of CAFTA.

6.2. El Salvador

Just as in the case of Honduras, the simulation for El Salvador is based on a recursive dynamic CGE model developed by Morley and Piñeiro (2007b), but in this case the base year for the data in the SAM is 2000. The static solution is generated using the same closure assumptions as employed in the case of Honduras, with one important exception: since El Salvador dollarised its economy in January 2001, the exchange rate is fixed which means that foreign savings is flexible (i.e. endogenous to the model). For the recursive model, the same assumptions apply as discussed earlier for the Honduras model. To simulate the growth path for the economy for the period 2000-2020 under the changes in the rules of origin for maquila introduced by CAFTA, the level of intermediate imports to the textile industry are kept constant at the observed level of 2000 prior to the passage of the CBTPA. Then starting in 2005 these intermediate imports are reduced to the much lower levels observed after the implementation of the CBTPA.

The base simulation (reflecting the El Salvadoran economy without CAFTA) projects a rather optimistic growth rate of 4.8% per year from its 2000 base up to 2020 (Table 5). The relaxation of the liberalized rules of origin for maquila increases the annual growth rate of the economy by 0.4%. Nevertheless, the effect on employment of unskilled labour in general, and that of females in particular, is more significant. Since the rate of growth of both maquila and non-maquila commodities increases, more of production is shifted over to unskilled, labour-intensive commodities, resulting in a jump in employment of urban and rural unskilled female labour by respectively 0.8% and 0.5% per year. Clearly, unskilled urban females are the big winners from the maquila provisions in CAFTA. Due to a contraction of the government sector (decreased consumption of the public sector caused by the savings-investment closure of the model and the sharp reduction in foreign saving given El Salvador's fixed exchange rate after the dollarisation) employment growth of skilled labour is sharply reduced but still positive.

Earnings inequality is expected to rise in El Salvador due to excess demand for skilled labour and excess supply of unskilled labour, and this is the case with or without CAFTA. Between 2000 and 2020, the earnings position of unskilled labour relative to that of their skilled counterparts deteriorates significantly, by about 14%. This is entirely due to higher wages of skilled labour: unskilled labour benefits from the maquila provisions in CAFTA only through higher employment rates. Unlike in the case of Honduras, the data in the SAM for El Salvador used in the model simulations permit conclusions regarding the relative earnings position of rural and urban workers. The maquila provisions in CAFTA do not seem to affect rural-urban differences in wage rates for skilled labour which is perhaps not surprising given that maquila uses mostly unskilled labour. A similar conclusion holds for the relative earnings position of males and females.

The maquila provisions of CAFTA have a rather limited, but positive impact on the rate of growth of capital formation which leads to a capital stock in the year 2020 that is 8% higher than in the absence of these provisions. However, and in line with the rising share of unskilled labour in total GDP, the factor share of capital decreases over time, even though the decrease is much smaller in the maquila scenario compared to the baseline.

In conclusion, whereas the impact of the CAFTA maquila provisions in El Salvador is small, the impact on the growth of the textile sector itself is substantial. However, the biggest impetus of the maquila provisions in CAFTA is on labour, in particular on the creation of unskilled jobs for urban females where the additional 1% annual growth rate in employment creation results in 19% more jobs by the year 2020.

6.3. Nicaragua

Our discussion of the results for Nicaragua is based on a paper by Sánchez and Vos (2005). Their CGE model follows the IFPRI standard model described in Lofgren *et al.* (2001) which facilitates comparison with the results obtained by IFPRI for Honduras and El Salvador. The static solution of the model is generated for the base year 2000 and the dynamic part of the model is solved recursively until the year 2012. The static solution is obtained under similar assumptions as in the Honduras model with one exception: in the Nicaragua model, labour is subdivided according to occupational type (salaried and own-account), skill level (skilled vs. unskilled) and gender. Real wages are fixed for all types of labour (so all labour markets, including the one for skilled labour, adjust via quantities). The discussion below refers to the version of the Nicaragua model that uses the same external, fiscal and savings-investment closures as the Honduras model.[45] The recursive model is solved in the same manner as in the case of the Honduras and El Salvador models.

Unlike Honduras and El Salvador, Nicaragua enjoys preferential access to the US market via TPL provisions. The discussion below is based on the assumption that Nicaragua will be able to completely fill these preferential quotas.

The net effect of the maquila provisions in CAFTA on economic growth is an additional 0.65% per year during the period 2006-12. This means that by the year 2012 the maquila provisions in CAFTA add about 4% to Nicaragua's GDP. The size of the maquila sector itself is simulated to about double during the period 2006-12, an annual growth rate of 12%. Textile exports will also more than double, increasing Nicaragua's dependency on maquila exports without attaining a more diversified economy. Exports of non-traditional agricultural and agro-industrial exports will become less competitive due to appreciation of the real exchange rate.

The maquila provisions in CAFTA have a positive effect on employment which is expected to be nearly 4.5% higher by 2012 compared to the scenario that only involves tariff reduction and agricultural quotas. But unskilled labour is again the clear winner: by 2012 employment of unskilled female wage earners will be over 16% larger than without the maquila provisions, and their wages will be 4% higher, compared to only 1.5% for male unskilled wage earners. So unlike in the case of El Salvador, unskilled labour will benefit not only through higher employment rates but also through higher wage rates, even if the size of the increase remains relatively limited. But generally, wages of unskilled labour relative to those of skilled labour will improve somewhat.

---

[45] Despite the fact that the exchange rate regime in Nicaragua is of the 'crawling peg' type (mini-devaluations) which basically implies a fixed nominal exchange rate, due to unsatisfactory model results obtained under this assumption the simulations discussed in this section were generated with a nominal exchange rate that is allowed to vary, keeping foreign savings constant.

In summary, the impact of the maquila provisions in CAFTA on the Nicaraguan economy will be positive but again not dramatic. But just as in Honduras and El Salvador, the biggest impetus will be on employment of unskilled female wage labour: whereas by the year 2012 the Nicaraguan economy is expected to be 4% larger compared to a situation without CAFTA, employment of unskilled female wage labour will be 16% higher. Again however, these calculations do not take account of the potentially negative impact of the completion of the ATC.

6.4. Costa Rica

Our discussion regarding the impact of the maquila provisions in CAFTA in Costa Rica is based on a CGE model developed by Sánchez (2007) which again follows the IFPRI standard model described in Lofgren *et al.* (2001). The static solution of the model is generated for 2002 (the base year of the data in the SAM) and the dynamic part of the model is solved recursively until 2026, under the assumption that CAFTA will take effect in 2007.

As in the Nicaragua model, labour is subdivided according to occupational type (salaried and own-account), skill level (skilled vs. unskilled) and gender. Whereas for non-salaried labour nominal salaries are assumed fixed, for salaried labour it is assumed that real salaries are 'downward sticky' (i.e. they can go up but not down). The same model closures are used as in the Honduras and Nicaragua models, i.e., the nominal exchange rate is endogenous and therefore foreign savings remain fixed[46]; fixed household and government saving rates make total investments endogenous in the model, while fixed government consumption and income taxes endogenise government savings.

Like Nicaragua, Costa Rica has negotiated a number of TPL provisions even though these allow less liberal access to the US market than is the case for Nicaragua. Again we assume that the preferential quotas of the TPL arrangements will be completely filled.

Unlike in Honduras and El Salvador where maquila plays a dominant role, the positive economic effects of CAFTA in Costa Rica are nearly entirely due to tariff reductions which permit cheaper imports of industrial inputs. The impact of the maquila provisions in CAFTA on the Costa Rican economy is very modest, which is not surprising given the much lower weight of the textile maquila industry in Costa Rica (Table 5). By the year 2026, the maquila provisions in CAFTA result in a GDP that is a paltry 0.2% higher than without these provisions, and their effect on total employment and wages is negligible (even though employment in maquila is 0.6% higher).

---

[46] The endogenous nominal exchange rate in the model is in line with the recent (October 2006) policy changes that made the Costa Rican exchange rate regime more flexible, by doing away with the 'crawling peg' and introduce a free movement within a certain band width.

Table 5. Impact of CAFTA textile maquila provisions on income, employment and poverty.

	Honduras[1]	El Salvador[2]	Nicaragua[3]	Costa Rica[4]
Average annual % change over baseline scenario				
GDP	1.4	0.4	0.6	0.01
Total employment	1.4	0.5	0.7	0.005
Percentage point difference with baseline scenario over entire simulation period				
Total poverty	-7.3	-3.7	-0.7	0.0
- urban	-6.7	-3.3	-0.8	0.0
- rural	-7.8	-4.5	-0.7	-0.02
Extreme poverty	-8.3	-1.5	-0.4	-0.01
- urban	-6.3	-1.3	-0.6	-0.04
- rural	-10.2	-1.9	-0.4	+0.02
Income distribution	positive	positive	positive	neutral

[1] Based on simulations for 2004-2020
[2] Based on simulations for 2005-2020
[3] Based on simulations for 2006-12 (GDP and employment) and 2006-2010 (poverty)
[4] Based on simulations for 2007-2026

6.5. Poverty and distributional impacts

In general, and as expected given the positive impact on economic growth, the maquila provisions of CAFTA are poverty reducing (Table 5). With the exception of Costa Rica, these provisions unambiguously help both the rural and urban poor. And again with the exception of Costa Rica, the poverty effects are of a non-trivial magnitude. They are expected to be largest in Honduras which is not surprising given the enormous importance of the maquila industry in that country. By the year 2020 and compared to the baseline scenario, the maquila provisions in CAFTA will have cut the national poverty rate by more than 7 percentage points (to 59% down from 66%). Extreme rural poverty in Honduras is 54% in the baseline scenario and is expected to decline to 44% by 2020 under the maquila scenario which represents a 1.3% annual percentage improvement over the baseline scenario. While the overall income distribution becomes less skewed as a result of the maquila provisions in CAFTA, labour income inequality is expected to increase in Honduras with or without CAFTA, due to a projected increase in the supply of skilled labour (2%) that is smaller than the projected increase in demand for skilled labourers (wages for the unskilled and semiskilled are fixed by the assumption of an excess supply of those types of labour). The maquila provisions in CAFTA are projected to strengthen the tendency towards greater wage inequality in urban areas because the higher rate of GDP growth increases wages of the skilled relative to the unskilled. However, it is important to note that, relative to the baseline scenario, CAFTA increases the earnings of both the skilled and the unskilled in

Hans G.P. Jansen and Samuel Morley

Honduras. While for the latter the improvements come in the form of more jobs at the same wage, the improvement for skilled labour comes in the form of higher wages only. Moreover, many poor urban households have unemployed skilled labour resources and putting these to work would actually be good for poverty, even if it increases labour earnings inequality. In conclusion, the benefits of the maquila provisions in CAFTA for the poor stem primarily from increased employment opportunities.

In El Salvador relative to the baseline scenario, the maquila provisions in CAFTA increase per capita income by an additional 0.4% per year while cutting the overall poverty rate by nearly 4 percentage points (from 23.6 to 19.9%) over a period of 15 years, equivalent to a 1.1% annual percentage decrease in the overall poverty rate. Just as in Honduras, the benefits of the maquila provisions in CAFTA for the poor in El Salvador stem primarily from increased employment opportunities: the very big increase in the demand of unskilled (mostly female) urban labour leads to a decrease in both urban and rural poverty, with the latter decreasing even more than the former due to urban-rural demand linkages. Thus employment growth is critical to reducing poverty. Furthermore, if the economy creates urban employment that pulls unemployed or inactive workers out of the countryside at the same time that the increase in urban employment and income increases the demand for agricultural production by urban households, then all of this has a favourable impact on rural poverty. Relative to the baseline scenario which predicts a significant decrease in income inequality over time, the maquila scenario narrows the average income differential between urban and rural households. While skilled urban labour gains more (in terms of wage increases) than unskilled urban labour (for which wages are assumed fixed) which leads to an increase in urban labour income inequality, the linkages between the urban and rural labour markets for the unskilled ensure that the poverty effects in the rural sector dominate those in the urban sector. In this way the positive effect of job creation on the distribution of income is greater than the associated rise in the skill differential.

In Nicaragua, the impact of maquila on poverty is much smaller than in Honduras and El Salvador. There are two basic reasons for this. First, CAFTA impacts the Nicaraguan economy primarily through increased preferential access to the U.S. market via quotas. Second, the simulation results are for a much shorter period (2006-10) than in the case of Honduras and therefore somewhat less insightful. Nevertheless, the poverty effects are certainly of non-trivial magnitude. The maquila provisions in CAFTA impact nearly equally on rural and urban poverty, both of which will be about 0.8 percentage points lower than without these provisions. It is interesting to mention that while reducing tariffs alone increases rural poverty in Nicaragua somewhat, the maquila provisions more than compensate this negative distributional effect. The effect of maquila on the distribution of household income is positive (decrease in the Gini) but very small. That the poverty effects are less than the effect on GDP is explained by the fact that the benefits for the poor mainly are through higher employment and not so much via higher wages.

Given the marginal effects of the maquila provisions on income and employment in Costa Rica stemming from the small share of the sector in the total economy, the poverty impacts of these provisions can be expected to be extremely small. Indeed they turn out that way, with no measurable effect on aggregate poverty. However, by the year 2026 rural poverty and extreme urban poverty are a marginal 0.02 and 0.04 percentage points less compared to the baseline scenario. As indicated earlier, the largest economic effects of CAFTA in Costa Rica stem from tariff reductions which are responsible for virtually all poverty reduction (estimated at 0.6 percentage points by 2026 in Sanchez (2007)). Even though these reductions as well as the maquila provisions particularly benefit the urban poor, also the rural poor gain insofar as those who lost their job in agriculture benefit from increased employment opportunities in the non-agricultural sectors such as maquila. Given the very small effects on poverty, the overall income distribution is not affected.

7. Concluding remarks

The economic importance of the textile maquila industry in CAM is undeniable and CAFTA has changed the playing ground to a considerable extent. Historically, U.S. trade agreements have included incentives that limited the development of forward and backward linkages in the textile and apparel sectors in exporting countries: most inputs needed to be U.S.-made and most value-adding processes needed to be undertaken in the U.S. to qualify for duty benefits. Compared to previous agreements such as the CBI and CBTPA, CAFTA has considerably widened the scope of the trade relations between CAM and the U.S. as far as the apparel trade is concerned, to such an extent that it may provide additional opportunities for growth in the sector, additional jobs, and increased vertical integration.

However, CAFTA alone will not guarantee these results and does not replace an integrated approach to improve competitiveness at the enterprise level in the Central American region. Past successes of Central American maquila exports to the U.S are to a substantial extent due to preferential access. While Central American exports of textile and apparel are heavily focused on the U.S. market, 72% of all exports to the U.S. in 2005 consisted of only six major products[47], virtually all of which were heavily protected by the quota system under the ATC. Given that the end of the ATC may result in average export price reductions of the order of 20%, it is imperative that CAM increase their competitiveness since the increased Asian competition could easily erase benefits provided by CAFTA. Maquila in CAM is unlikely to compete with Asian suppliers on the basis of cost and therefore needs to develop and exploit other (potential) strengths that will give it a comparative advantage in the U.S. market.

---

[47] These are cotton and MMF underwear (33% of total exports to the U.S.), knit shirts and blouses (24%), trousers and slacks (12%) and nightwear (3%).

Factors that will help the survival and further development of the maquila industry in CAM include the following:

- *Increased emphasis on quality*. Central American firms need to overcome bottlenecks that prevent the sector from shifting away from simple, low-price, high-volume products in which Asian countries excel (China in particular), towards higher value-added items and those that are needed quickly and frequently (fast-response items).[48] Quality issues are widely reported to be the result of insufficiently skilled labour, pointing to the importance of improving quality through worker training programs.

- *Improved responsiveness*. This involves exploiting the region's geographical location advantage which allows shipments to reach clients much faster than from Asia. Increased regional integration and simplification of customs processes will help improve delivery time. Employing personnel at customs who deal specifically with businesses in the Zonas Francas would speed the process, since specific regulations apply to exports produced in these zones. Firms can improve the speed of shipments by training a member of staff to serve as a broker that can approve shipments. Incorporating the newest manufacturing methods, moving from line to module production, for example, can increase the speed of delivery. Increasing communication between the shelf of the retailer and the factory through Point of Sale (POS) technology can also improve speed to market.

- *Full-package production*. Clients increasingly prefer to source from suppliers capable of full-package production which also provides obvious economic benefits through forward and backward linkages. In this respect it is important to draw the attention to the risk issue. Compared with pure apparel assembly, the risk factor is much higher in full-package production because even a partially cancelled order incurs a great loss: the loss of the wages paid to workers for their assembly and overhead costs, but also the loss of the inputs, particularly cloth that had already been purchased by the maquila. This can be overcome by increasing efforts to enter into contracts with risk-sharing firms.

- *Creation/improvement of the general conditions required to attract foreign investors*. These include efficient transport systems, reliable and competitively priced electricity, adequate supply of (semi-)skilled workers, enforcement of labour laws, security etc. In this respect it is important to note that the U.S. Department of Homeland Security's Bureau of Customs and Border Protection (Customs) has designated the Port of Cortés in Honduras a Container Security Initiative (CSI) port. The partnership Customs has formed with the Port of Cortés, the largest port in CAM will enable expedited entry for cargo entering the U.S. from that port. This designation is a good example of the sort of measures that Central American countries need to aim at in order to further enhance the competitiveness of their maquila industry.

---

[48] An example in this respect is Honduras where expertise mainly lies in sub-segments of the apparel sector – simple and cheap garments such as t-shirts - that will face fierce competition with China due to the end of apparel quotas. Building on the country's long tradition of T-shirt production, a shift to T-shirts with embroidery, screen printing etc. may be warranted.

Finally, there may be something to be learned from the Mexican experience. Mexico's textile and apparel sector boomed after NAFTA, but has been continually weakening since 2001. In 1999, 15% of all U.S. apparel imports came from Mexico, but as of May 2005, Mexico had slipped to only 8.6% of U.S. market share in terms of units and 9.5% in value terms. Mexico traditionally has concentrated its principal maquila exports in categories that receive preferential access which held back the implementation of some of the measurements needed to be able to compete with Asian exporters (China in particular) in the U.S. market after the implementation of the various stages of the ATC during the period 1995-2005. In addition, preferential access provided by NAFTA has had the unintended effect of creating distortions in sub-sectors of the maquila industry in which Mexico does not enjoy a comparative advantage. As a result, China has increasingly taken over Mexican apparel exports in the U.S. and this trend is only expected to get stronger once transitory U.S. restrictions on Chinese textiles and apparel imports will be revised (and probably reduced significantly) in 2008. The lesson to be learned from Mexico is that preferential treatment for Central American maquila sector offered by CAFTA, in combination with the temporary quota on Chinese textile imports into the U.S., should not delay the implementation of a strategic approach needed to confront the post 2005 challenges. Rather it should be carefully used by the CAFTA countries to upgrade their maquila industry and re-allocate resources to those parts of the industry where a solid comparative advantage can be developed and exploited in the U.S. apparel market. Given the small sizes of the individual Central American countries, a possible road may consist of developing a regional virtual vertical integration so as to compete against Asian competitors as a region, allowing for the development of intra-regional trade of intermediary products (e.g. fabric and trims) with the final product exported duty-free to the U.S.

References

Ganuza, E., R. Paes de Barros and R. Vos, 2002. Labour market adjustment, poverty and inequality during liberalisation. In: R. Vos, L. Taylor and R. Paes de Barros (eds.) Economic Liberalisation, Distribution and Poverty: Latin America in the 1990s. Cheltehham, Edward Elgar, pp. 54-88.

Gereffi, G., and O. Memedovic, 2003. The global apparel value chain: What prospects for upgrading by developing countries. Sectoral Studies Series, UNIDO, Vienna, Austria.

Hernández, R., I. Romero and M. Cordero, 2006. Se erosiona la competitividad de los países del DR-CAFTA con el fin del acuerdo de textiles y vestuario? International Trade and Industry Unit, Serie Estudios y Perspectivas No. 50, ECLAC, Mexico.

IADB, 2005. The emergence of China: Opportunities and challenges for Latin America and the Caribbean. Draft paper for discussion, Integration and Regional Programs Department, IADB, Washington, DC.

IDS, 2006. Monthly reports on textiles and apparel. International Development Systems Inc., Washington, DC. www.ids-quota.com.

INCAE, 2004. El Sector Textil Exportador Latinoamericano Ante la Liberalización del Comercio. Costa Rica.

ITC, 2006. Improving competitiveness of the clothing industry in Central America. Project proposal. International Trade Centre (ITC), UNCTAD/WTO, Geneva Switzerland.

Lofgren, H., R. Harris and S. Robinson, 2001. A Standard Computable General Equilibrium (CGE) Model in GAMS. Trade and Macro Division Discussion Paper No. 75, (IFPRI, Washington, DC.

Morley, S., 2006. Trade Liberalization under CAFTA: An Analysis of the Agreement with special reference to agriculture and smallholders in Central America. DSGD Discussion Paper No. 33, Development Strategy and Governance Division, IFPRI, Washington DC.

Morley, S. and V. Piñeiro, 2007a. The Impact of CAFTA on Employment, Production and Poverty in Honduras. IFPRI, Washington, DC.

Morley, S. and V. Piñeiro, 2007b. The Impact of CAFTA on Poverty, Distribution and Growth in El Salvador. IFPRI, Washington, DC.

Parada, Gómez and Alvaro Martín, 2004. La cadena global de prendas y vestir. Manuscript, National University, Heredia, Costa Rica.

Sánchez, M., 2007. Liberalización Comercial en el Marco del DR-CAFTA: Efectos en la Producción, la Pobreza y la Desigualdad en Costa Rica. Series Estudios y Perspectivas No. 80, Unidad de Desarrollo Social, ECLAC, Mexico City.

Sánchez, M. and R. Vos, 2005. Impacto del CAFTA en el Crecimiento, la Pobreza y Desigualdad en Nicaragua: Una Evaluación Ex-Ante usando un Modelo de Equilibrio General Computable Dinámico. Paper prepared for the Nicaraguan Ministry of Industry and Trade (MFIC) and UNDP Nicaragua. The Hague, The Netherlands and Mexico City, Mexico.

Smallholder procurement in supply chain development: a transaction costs framework

Ruerd Ruben and Froukje Kruijssen

Abstract

Transaction costs play an important role in linking the smallholder to markets. There is a lot of information and negotiation in the early phases of market development and monitoring costs tend to be high, but these can be better controlled once retailers establish contractual relations with selected producers. Such preferred supplier arrangements are helpful in reducing risks related to delivery frequency and product quality, but require substantial investments. Therefore, they only become feasible when opportunistic behaviour can be adequately controlled. The shift from wholesale markets towards preferred supplier regimes therefore requires that savings in governance costs offset increasing capital costs. This chapter uses insights derived from transaction cost theory to analyse the development and evolution of fresh produce procurement channels by supermarkets in developing countries. Special attention is given to the particular conditions that permit smallholders to shift from wholesale selling towards a preferred supplier system. We assess the role of economic motives (fixed investments, variable production costs, economies of scale) compared to the expected behavioural changes (in governance and opportunistic behaviour) for entering into contractual delivery. We conclude that the nature of supplier-buyer relationships alters in a number of subsequent phases from chain optimisation to integral chain care. Smallholders in developing countries need special assistance to become engaged in such procurement regimes.

1. Introduction

Supply chain integration is rapidly becoming a major strategy for guaranteeing adequate sourcing of perishable products to urban market outlets in many developing countries. Local farmers use, however, different mechanisms and procedures for deliveries to markets, giving rise to a wide variety of supply chain sourcing and management arrangements. Some retailers are developing preferred-supplier arrangements (PSA) with local growers, while others still purchase their products from dedicated wholesalers at the local or regional market (LWM). The choice between these different procurement strategies critically depends on differences in competitive relations and consumer demands which lead to a particular structure of transaction costs perceived in the local market (Dorward, 2001; Pitelis, 1993).

The shift in procurement regimes from (general or dedicated) wholesalers to preferred supplier contracts is related to the balance between transaction and real investment costs. Closer ties between smallholders and buyers could lead to considerable savings in governance and control costs, but usually require substantial investment to reduce the risks of quality underperformance and/or delayed deliveries to satisfy the supply conditions imposed by modern retailers. Overcoming these trade-offs between governance and investment costs is a fundamental condition for enabling smallholder producers to become part of preferred supplier regimes.

This chapter focuses attention on the development pathways and evolutionary regimes of fresh produce procurement channels by supermarkets in developing countries and the potential options and strategies for including smallholder producers in preferred supplier arrangements. Several studies indicate that the increased emphasis by (inter)national supermarkets on reliable sourcing and higher (private) quality standards tends to reinforce tendencies towards vertical coordination (Berdegué *et al.*, 2005; Gulati *et al.*, 2007; Dolan and Humphrey, 2000). This is easily accompanied by the exclusion of smallholder producers who face major difficulties in meeting the stringent scale, consistency and quality requirements. Their potential engagement in preferred supplier regimes is critical for overcoming these constraints, but requires major support in terms of pre-investment facilities and organisational development for reaching scale economies.

In the remainder of the chapter we discuss the conditions for smallholder supply chain integration and the implications for transaction costs and welfare. We use a stylised optimisation approach with an explicit transaction cost module in order to identify the trade-offs between governance costs and fixed investments (entry costs) for quality upgrading under two markedly distinct procurement regimes. We simulate the conditions which invoke supermarkets to switch from wholesale procurement to preferred-supplier relations. Subsequently, these results are used to discuss the current evolution of supply chain integration regimes in different types of developing countries and to identify business support programs that might enable smallholder participation.

2. Procurement regimes in supply chains

Procurement regimes for fruits and vegetable are usually characterised by high-frequency, constant delivery and stable quality. Delivery arrangements between producers and processors or retailers are primarily based on observable output characteristics (like volume, size and colour), but increasingly also include more detailed controls on input requirements (e.g. type of seed, fertiliser use, pesticides applications, packaging, etc). In the latter case, the buyer tries to enforce particular resource-use decisions made by growers in an effort to reduce the uncertainty regarding the desired product attributes (taste, quality, food safety and freshness).

Supermarkets' procurement strategies - including functional aspects such as warehouse operations, transport management and packaging - are shaped according to the characteristics of specific product categories. For perishable products, these characteristics concern technical attributes like seasonality (reliable supply), storability (shelf life), transportability and processing options at the supply side, related to consumer preferences and their willingness to pay for these attributes. Compliance with quality criteria is of particular importance in the case of fresh fruits and vegetables. Buyers face problems in monitoring the quality, safety and shelf-life of products. Pesticide residues and phytosanitary aspects of production and trade are difficult to detect, but affect the business reputation of sellers vis-à-vis buyers. In order to guarantee adequate traceability, retailers search for partnership relationships with producers that reduce such information and screening costs and reinforce trust amongst chain agents.

Transaction costs play an important role in the design and organisation of procurement regimes. While most attention is usually given to transport costs, other aspects related to the relationships between producers and buyers in fact deserve far more attention (Ruben *et al.*, 2003). Following North (1990), we distinguish three main categories of transaction costs: (a) information and search of partners, (b) screening and monitoring, (c) bargaining and enforcement of contracts. Sales to traders or wholesalers, deliveries to dedicated wholesalers and preferred-supplier arrangements can be characterised according to their transaction cost characteristics and requirements in terms of scale and investments (see Table 1). Wholesale is the least demanding in terms of investment, but usually involves high transaction costs. Specialised or dedicated wholesalers establish stable relationships with producers who are therefore better able to intensify their production systems. Preferred supplier arrangements tend to involve higher (fixed and variable) production and handling costs but save on the afore-mentioned governance costs and thus reduce the buyers' exposure to risks resulting from substandard quality and out-of-time delivery.

The shift from wholesale purchase towards preferred supplier arrangements is strongly influenced by changes in consumer preferences and adjustments in supermarket formats. Once urban consumers begin to appreciate quality, freshness and safety as important attributes for the selection of vegetables, supermarkets start looking for a selective group of producers that are able to guarantee the delivery of these products. This usually requires input

Table 1. Characteristics of procurement regimes.

	Trader & wholesale purchase	Specialised wholesale	Preferred supplier
Information & search	High	Low	Low
Screening & monitoring	High	Medium	Low
Bargaining & enforcement	High	Medium	Low
Variable costs (quality)	Low	High	High
Fixed investments (scale)	Low	Low	High

delivery (seed and chemicals) and technical assistance services to be in place, permitting farmers to upgrade their product quality. Eventually, the produce may be labelled or even certified in order to safeguard consumer confidence. At this stage, dedicated wholesalers are usually excluded from the delivery process.

In a dynamic market, retailers frequently adjust their formats in order to position the fresh department as the centrepiece of their operations. The permanent supply of fresh fruits and vegetables is considered to be a major strategy for attracting clients on a daily base. Therefore, supply chain coordination requires suppliers to be able to guarantee constant and reliable deliveries of a consistent quality. Moreover, more attention is given to product innovation as part of the strategy for responding to the consumer's variety-seeking behaviour. The assortment can then be expanded to include, for example, pesticide-free, fair trade or organic products. In these circumstances, close partnerships become a vital condition for realising specific investments, and occasional sourcing from smallholders may yield a competitive advantage.

The economic and institutional conditions for such shifts in supply chain regimes are still not very well understood. Regarding the position of smallholders, both exclusive and inclusive tendencies are observed in empirical case studies, depending on the level and degree of vertical coordination. In the following, we explore the basic requirements for supply chain development in order to identify to which areas public or private support should be directed to enable smallholder participation in more integrated procurement systems. We therefore present a stylised model approach that enables the disentanglement of critical institutional and behavioural constraints, and compare these findings with a number of empirical studies that address governance and trade perspectives in fresh produce chains in several developing countries.

3. Transaction costs approach for supply chain integration

We use the basic model framework for contractual arrangements as developed by Dorward (2000) for a detailed assessment of the interactions between production, investment and delivery conditions and the implications for optimal channel choice. The approach is based on a non-linear optimisation model with a target MOTAD formulation (Tauer, 1983). The objective function includes the maximisation of a function of buyers' and sellers' utility and bargaining power. Expected incomes of buyer and seller are optimised, which implies that the differences between the acceptable and actual sums of deviations of incomes below target incomes become zero. The model accounts for differences in governance costs, changes in trust relationships, economies of scale for buyers and sellers, and differences in opportunism. A major adjustment compared to the original model is the incorporation of fixed investments (accounting for asset specificity) in order to assess the trade-offs between reduction in governance costs and the associated costs for establishing higher quality market outlets.

This modelling framework enables us to analyse the critical constraints for providing smallholders with access to higher rewarding preferred supplier relations with retail supply chains. The key variables in the model reflect the conditions that are usually considered to be important for smallholder inclusion: (1) the production technology, (2) the institutional organisation, and (3) the governance regimes. At the production side, both fixed investment costs (like irrigation, warehouses or greenhouses) and intensification in variable costs (high-quality seed material, chemical or organic fertilizers, etc) are considered to guarantee improved quality management and more reliable supply. At the institutional level, the impact of economies of scale on marketing operations is explored, enabling savings in transaction costs. Governance aspects are addressed by including differences in (ex-ante) bargaining costs for buyers and sellers, and the (ex-post) implications of contracts for the degree of opportunistic behaviour. The comparison of these three aspects could yield important insights into key areas where interventions are required to enable successful smallholder participation in integrated supply chains.

Stylised data were used to run the model in GAMS. We are interested mainly in the optimal contractual choice, i.e. purchase at the Local Wholesale Market (LWM) or delivery through Preferred Supplier Arrangements (PSA) under different market conditions (i.e. low and high supply) and various behavioural assumptions (low, moderate or high opportunism). Since the model does not consider diminishing rate of returns, the total commodity volume will be fully transacted either on the LWM or under PSA. Only differences in variable production costs made by sellers and fixed investments made by buyers (e.g. warehouses, transport equipment) between LWM and PSA are considered. A baseline was established in which capital costs are assumed to be equal for LWM and PSA, governance costs under PSA are one third of those on LWM, and the effects of opportunistic behaviour are assumed to be zero. In the base run, the total produced volume (1000 units) is transacted on the LWM.

We specify nine different simulation settings and assess the impact of changes in the production and delivery conditions for the market outlet choice. In all optimisations, asset specificity and risk preferences for sellers are held constant, and differences between producers are initially disregarded.[49] Figure 1 presents a graphic representation of the scenario results. The diagram shows the total cost structure for a single unit of the commodity traded under nine different types of procurement arrangements. The simulation results indicate the relative weight of real investment and governance costs, and the efforts made to control opportunistic behaviour, while the resulting margin reveals which market outlet provides the highest net returns (and thus will be the preferred channel for the delivery and purchase of the produce).

---

[49] The parameter definition and the simulation results are presented in an appendix available to readers on request.

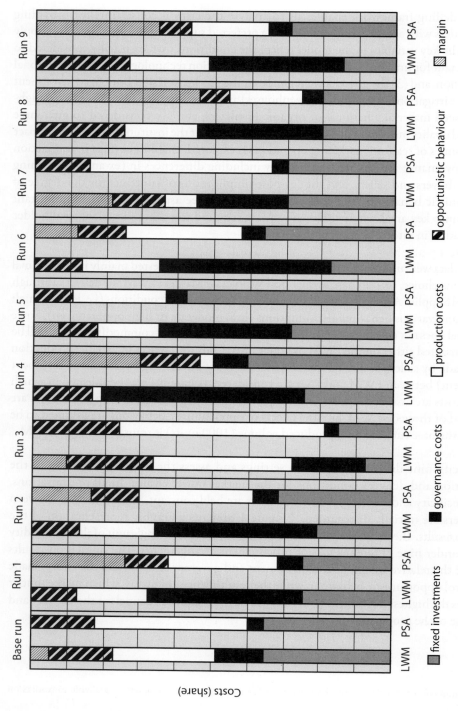

Figure 1. Simulations of effects of transaction costs on procurement regimes.

The first simulation (run 1) assesses the effect of changes in governance costs. Governance costs under PSA procurement are reduced by 50 percent, while fixed investments remain the same under both contractual forms and only variable production costs increase. Under these (rather unlikely) circumstances a switch from wholesale purchase to preferred suppliers takes place, since retailers can take advantage of lower control costs that outweigh the higher production costs required for the latter channel. The next two simulations (run 2 and 3) maintain the difference in governance costs as specified in the first analysis, but simulate the effects of an increase in buyers' fixed investments under PSA of respectively 50 percent and 65 percent. The results show that an increase in fixed costs could easily outweigh the expected gains in governance costs. A switch towards PSA is thus only feasible at relatively low investment levels (run 2). When fixed investments rise further (run 3) the break-even point is reached, which means that purchase at the LWM involves lower total transaction costs compared to establishing PSA's.

The fourth simulation shows that economies of scale can be favourable for transacting under PSA. Compared to the third run, the commodity volume has been increased ten-fold. Since fixed investments can now be spread out over a larger number of transactions, PSA provides an optimal solution (run 4). The effect of a further lowering of governance costs under PSA, while increasing fixed investments by a hundred percent, are shown in the fifth and sixth simulation. If gains in governance costs under PSA are low due to small differences in trust amongst agents, preferred supplier regimes cannot compensate for the higher fixed costs (run 5). Only when substantial reductions in governance costs become feasible, is switching towards PSA perceived as the optimum solution (run 6).

Whereas fixed investments refer to the lumpy investments made by the buyer, producers might also face increasing variable production costs under PSA. These costs increase due to higher input demands for guaranteeing quality requirements and more demanding handling procedures. We analyse the effect of an increase in variable production costs between LWM and PSA (run 7). This results in an optimum channel choice in favour of the wholesale market, indicating that quality requirements cannot be too demanding for producers to ascertain the feasibility of PSA. Finally, changes in opportunistic behaviour are taken into account. A major advantage of PSA is expected when reduced opportunistic behaviour can be enforced. This is indeed confirmed even under relatively low levels of fixed investment (run 8) and is reinforced when investment costs further increase (run 9).

The results of these simulations illustrate that only under certain specific conditions can savings in governance costs offset capital requirements related to supply chain integration. PSA can become a feasible sourcing strategy for smallholders, especially when fixed investment costs remain at a reasonable level and/or sufficient scale economies in trade can be realised. The same holds for the incentives for increasing variable production costs to comply with more stringent quality requirements, which are usually far from sufficient in enhancing smallholder involvement in vertically coordinated transactions. Otherwise, possibilities for reducing opportunistic behaviour and benefits arising from economies of scale seem to offer sound incentives for switching to preferred-supplier arrangements. Institutional reforms and behavioural change are thus usually required to support smallholder adjustments of production systems as a pre-condition for engaging in coordinated supply chain transactions.

4. Empirical evidence

Against the background of the presented simulation results, we review a number of recent empirical studies regarding the development of fresh produce supply chains in different parts of the world (see Table 2). Special attention is given to the relative importance attached to efforts for improving production practices, institutional supply chain organisation and innovations in governance regimes that influence smallholder participation.

Table 2. Critical factors for fresh produce supply chain coordination.

Country	Products	Key issues	Authors
East Africa (Kenya & Zimbabwe)	Vegetables	Monitoring & auditing regimes	Dolan and Humphrey (2000)
Ghana	Pineapple	Institutional organisation (coop of outgrowers)	Yeboah (2005)
Senegal	Vegetables (french beans)	Institutional organisation (standards-driven)	Maertens and Swinnen (2006)
India (Andhra Pradesh)	Vegetables	Production (crop mix) & governance	Deshinkar *et al*. (2003)
Southeast Asia	High-value crops	Contract farming arrangements	Gulati *et al*. (2007)
Thailand	Horticulture	Upgrading of production practices and reliable deliveries	Ruben *et al*. (2007); Boselie *et al*. (2003)
China (Beijing region)	Horticulture	Smallholders linked to local traders	Dong *et al*. (2007)
Central America	Horticulture	Production (technology), quality standards & institutional innovation	Berdegué *et al*. (2005)
Latin America	Fruits & vegetables	Consolidation of procurement through standards and grades.	Reardon and Berdegué (2002)

4.1. Africa

The production of fresh fruit and vegetables for export is growing rapidly in several sub-Saharan African countries (Dolan and Humphrey, 2000). Procurement under contract for European supermarkets is usually based on strict specification of product varieties, production practices and quality surveillance regimes. Outgrower arrangements are frequently used as a device to guarantee the quality of deliveries at reduced investment costs, while most value-added activities are controlled by a small group of downstream processors and exporters. EUREPGap rules are now generally imposed by supermarkets to safeguard safety and quality standards. Most attention is given to monitoring and surveillance, increasingly based on full tracking and tracing programs. In some cases, exporters also provide technical assistance to support compliance with standards.

Controlling governance costs is a key issue in many African sourcing arrangements, but at relatively low investment levels for improving primary production practices. Preferred supplier relationships are therefore likely to remain weak and limited in scope to better-off categories of producers (with a large segment of expatriates). Co-investment arrangements between local exporters and producer associations are sometimes used to improve coordination. A further step is taken by small-scale pineapple outgrowers in Ghana that created a marketing cooperative Farmapine, including two former exporters as shareholders (Yeboah, 2005). The cooperatives provides credit to its members and guarantees exports at stable (dollar indexed) prices. Such arrangements prove highly significant in reducing uncertainties and provide the necessary incentives for improving quality performance and reliable deliveries.

Important parts of fruit and vegetables production in sub-Saharan Africa are located on plantations or middle/large-scale farms. Maertens and Swinnen (2006) indicate for Senegal that the tightening of European food standards induced structural changes in the supply chain, including a shift from smallholder contract farming to large-scale integrated estate production. Interestingly enough, poor households could benefit from this as a result of the increasing demand for wage labour and the higher wage rate. Whereas at the production side higher investment requirements tend to wipe out smallholder producers, the preferred supplier regimes do provide opportunities for pro-poor development, taking advantage of the labour market mechanism.[50]

---

[50] Barrientos *et al.* (2003) notice, however, that African fruit and vegetables exports provide much employment for temporary, mainly female workers often in poor working conditions. Codes of conduct remain limited to formal (male) workers, while contract workers have few legal employment entitlements (see also: Kritzinger *et al.*, 2004).

4.2. Asia

In many Asian countries, smallholders are important producers of high-value crops. Ruben *et al.* (2007) and Boselie *et al.* (2003) analyse retailers' procurement strategies in Bangkok (Thailand) and find considerable advances towards integrated supply chain management, mainly based on institutional innovation and in-depth investments in improved production systems. The preferred-supplier system of the Thai supermarkets permits substantial savings in governance costs by improving delivery reliability, reducing out-of-stock losses, and gradually incorporating new stock keeping units, thus partly offsetting the fixed investments made by the farmers and the company. Producers are increasingly involved in farm-level value-added activities (like quality control, cutting, trimming, packing, etc.) while receiving technical assistance from the company. Concentration of procurement can thus be accompanied by decentralisation of investments.

On the other hand, supermarkets in Nanjing (China) select their suppliers mainly on quality and price criteria rather than short lead-time, and most fixed investments are located at the company level. Economies of scale are still difficult to reach and governance costs for controlling defaults remain quite high. Similar results concerning Chinese fresh produce procurement regimes in the greater Beijing region are reported by Dong *et al.* (2007). The expansion of horticulture production is largely the result of smallholder initiatives that mainly deliver their produce to local and regional traders This is partly explained by the fragmented land tenancy structure and the absence of farmers cooperatives, but more importantly the small trader-dominated system prevails because quality monitoring and safety standards are not yet fully in place (Hu *et al.*, 2003).

In India, the transformation of procurement regimes is occurring somewhat slower than elsewhere in Asia, since most trade in fresh vegetables is still at a local and sub-regional level. Given the employment potential of horticultural crops, public investment schemes have been launched to promote production, mainly favouring medium-size producers that have irrigation facilities. Deshingkar *et al.* (2003) therefore argue that changes in institutional regimes (for access to land, capital and water) and adjustments in governance arrangements are required to guarantee better prospects for smallholders. Emerging partnerships between cooperatives and associations of lower cast farmers with private agents can be successful in breaking the power of the traditional elite. Apart from technological innovations, changes in organisation and supply chain governance are thus of critical importance in establishing inclusive preferred supplier regimes.

4.3. Latin America

Horticulture trade in Latin America is mainly driven by rising incomes and strong urbanisation, occasioning important adjustments in procurement systems of supermarket chains. Berdegué *et al.* (2005) point to the rapid diffusion of preferred supplier regimes linked to centralised procurement through regional distribution centres. In addition to cost-saving arguments, this institutional change is especially motivated by the requirements for functional upgrading. Advances in (mainly private) quality and safety standards lead to the selection of most qualified farmers that are capable of delivering consistent volumes of a consistent quality.

In a similar vein, Reardon and Berdegué (2002) discuss the dominance of supermarket chains as a key outlet for national fruit and vegetables production, which already represents 2-3 times the value of exports. Local traditional markets are rapidly loosing ground as an outlet for producers. Demands for scale and quality imply large investments in transport, packaging and cold chains that can only be met by medium to large producers with substantial capital reserves, or by associations of smallholders if access to credit and management skills can be guaranteed.

5. Policy options

Supply chains for fresh products to local and international supermarkets are increasingly structured around selective preferred supplier arrangements. Major arguments that motivate retailers to shift from wholesale purchase to vertical coordination are related to the requirements for consistent and reliable sourcing and the proliferation of more demanding private food quality and safety standards. Smallholders are easily excluded, since they face major constraints on the intensification of their production systems and meeting the required economies of scale.

Empirical evidence from field studies regarding the restructuring of market-oriented fruit and vegetables supply chains in developing countries suggests four major stages in the evolution of procurement regimes:
1. In the early stage of linking producers to market outlets, most transactions take place on the spot market with traders and intermediaries that purchase small quantities of the produce and deliver at the wholesale market. Subcontracting and sharecropping arrangements are major pathways for reducing governance costs and risk, providing some opportunities for resource intensification and quality upgrading (as illustrated in Scenario Runs 1-3).
2. In subsequent stages of supply chain development, retailers start putting the emphasis on *Chain Optimisation*, aiming at the further reduction of transaction costs related to handling practices and transport by focusing on a smaller group of dedicated wholesalers. This generates lead-time reduction by changing from fragmented and uncoordinated store deliveries to a system of centralised ordering and finally the establishment of multi-functional distribution centres capable of handling larger volumes and maintaining stable relations with the selected producers (see Scenario Runs 4 - 6). This strategy reaches its limit when associated suppliers receive too few tangible benefits to compensate for the initial investment costs.

3. Increased investments for quality and safety assurance mark the shift towards *Integral Chain Care* through a coalition with preferred suppliers. The introduction of good agricultural practices (GAP) and good manufacturing practices (GMP) reduces the amount of rejection and contributes to a stronger image. Information costs can be controlled through tightly integrated supplier-buyer networks. Producers are also encouraged to start quality controls at the farm level and become locked into the network through specific investments, thus increasing the gap between certified and non-certified suppliers (Scenario Runs 7-8).

4. Finally, *Joint Product Development and Innovation*; based on the introduction of new varieties and new production technologies (e.g. hydroponics vegetables, MD22 pineapples) or the development of organic products make production more demand-oriented. This phase could also include the branding of fresh produce by established large scale producer/wholesalers and is characterised by a full partnership between suppliers and retailers (see Scenario Run 9).

Smallholder producers face four major limitations preventing them from participating in this full sequence of supply chain transformations: (1) limited fixed capital investments, (2) low input intensity, (3) small volume (limited scale), and (4) high opportunism. Public interventions and private arrangements should reinforce each other to overcome these constraints. While in the early stages input intensification and scale receive most attention, in subsequent stages financial regimes and legal standards become increasingly important to guarantee further development of supply chain linkages.

Scale and investment problems are particularly critical in export-oriented supply chains originating from sub-Saharan African countries. Therefore, emphasis is usually given to efforts for improving production practices and governance regimes to enhance consistent deliveries, economies of scale and farmers' organisation. Chain optimisation programs involve specific supplier assistance packages with pay-back in kind after the harvest, together with production protocols to enhance compliance with input applications. Savings in transport and transaction costs are considered a key condition for stimulating farmers to make more in-depth investments. This is frequently accompanied by further concentration of deliveries in the hands of medium and large-scale producers, but may otherwise create considerable positive welfare effects for resource-poor households through the creation of local employment.

In Southeast Asia, supply chain integration is first and foremost oriented towards local and regional market outlets. Major efforts are directed towards improved institutional arrangements to reduce chain governance costs, thus enabling producers to make substantial investments for quality upgrading. This may eventually lead to a win-win situation for both farmers and retailers, since higher prices can be paid and deliveries are guaranteed. However, in many Asian countries, traditional practices based on personal (trust) relationships still prevail in the purchasing business. Preferred supplier programs that cut through established (personal) interests may therefore meet with a high degree of opportunism that can only be controlled through asset-specific fixed investments.

In Latin America, where urbanisation has progressed most, it is to be expected that the fresh fruit and vegetable sourcing systems will rapidly adapt to the fast-changing retail environment. Fruit and vegetables are the cash-cows and traffic builders for the (inter)national supermarket chains that require efficient and effective procurement. More demanding quality and safety standards require close partnerships and alliances to support traceability of products from farm to fork. This implies that the focus of supermarket management moves from chain optimisation towards a deepening of the supplier-buyer relationship, with far-reaching repercussions for the timeframe of contractual arrangements and the number and type of outgrower relationships. Preferred supplier systems bring stability and efficiency to the fresh produce value chain, but at the same time are only accessible to a limited segment of producers with sufficient access to capital and information resources.

The switch from wholesale purchase towards preferred-supplier can be conceptualised as a dynamic process that requires a critical balance between technological upgrading, innovations in governance and progressive engagement in supply chain coordination. The exclusion of smallholders can be occasioned by each of these processes, but takes a different shape and sequence depending on the particular conditions of market integration. Dovetailing the technical, institutional and behavioural changes remains the main challenge for public-private partnerships in equitable market integration.

References

Barrientos, S., C. Dolan and A. Tallontire, 2003. A Gendered Value Chain Approach to Codes of Conduct in African Horticulture. World Development 31: 1511-26.

Berdegué, J.A., F. Balsevich, L. Flores and T. Reardon, 2005. Central American supermarkets' private standards and quality and safety in procurement of fresh fruits and vegetables. Food Policy 30: 254-269.

Boselie, D.M., S. Wertheim and M. Overboom (eds.), 2003. Best in Fresh. Building Fresh Produce Supply Chains. KLICT, Den Bosch.

Deshingkar, D., U. Kulkarni, L. Rao and S. Rao, 2003. Changing Food Systems in India: Resource-sharing and Marketing Arrangements for Vegetable Production in Andhra Pradesh. Development Policy Review 21: 627-639.

Dolan, C. and J. Humphrey, 2000. Governance and Trade in Fresh Vegetables: the Impact of UK Supermarkets on the African Horticulture Industry. Journal of Development Studies 37: 147-176.

Dong, X., H. Wang, S. Rozelle, J. Huang and T. Reardon, 2007. Small Traders and Small Farmers: the Small Engines driving China's Giant Boom in horticulture. In: J.F.M. Swinnen (ed.), Global Supply Chains, Standards and the Poor. Wallingford: CAB International, pp. 109-121.

Dorward, A., 2001. The Effects of Transaction Costs, Power and Risk on Contractual Arrangements: A Conceptual Framework for Quantitative Analysis. Journal of Agricultural Economics. 52:59-73.

Gulati, A., N. Minot, C. Delgado and S. Bora, 2007. Growth in High-value Agriculture in Asia and the Emergence of Vertical Links with Farmers. In: J.F.M. Swinnen (ed.), Global Supply Chains, Standards and the Poor. Wallingford: CAB International, pp. 91-108.

Hu, Dinghuan, Yu Haifeng and T. Reardon, 2003. The operation of fresh and live non-staple foodstuff food in Chinese supermarkets and the consumer buying behaviour. Chinese Rural Economy 8: 12-17.

Kritzinger, A., S. Barrientos and H. Rossouw, 2004. Global Production and Flexible Employment in South African Horticulture: Experiences of Contract Workers in Fruit Exports. Sociologia Ruralis 44: 17-39.

Maertens.M. and J.F.M. Swinnen, 2006. Trade, Standards and Poverty: Evidence from Senegal. LICOS Discussion Paper 177. Leuven: Katholieke Universiteit Leuven.

North, D., 1990. Institutions, Institutional Change and Economic Performance. New York: Cambridge University Press.

Pitelis, C., 1993. Transaction Costs, Markets and Hierarchies. Cambridge Mass: Blackwell.

Reardon, T. and J. Berdegué, 2002. The Rapid Rise of Supermarkets in Latin America: Challenges and Opportunities for Development. Development Policy Review 20: 371-388.

Ruben, R., D. Boselie and Hualiang Lu, 2007. Vegetables Procurement by Asian Supermarkets: A Transaction Costs Approach..Supply Chain Management: An International Journal 12: 60-68.

Ruben, R., Hualiang Lu and E. Kuiper, 2003. Marketing Chains, Transaction Costs and Quality Performance: Efficiency and Trust within Vegetable Supply Chains in Nanjing City. Paper presented at SERENA Seminar, Nanjing. October 2003.

Tauer, L.W., 1983. Target MOTAD. American Journal of Agricultural Economics. 64: 606-610.

Yeboah, G., 2005. The Farmapine Model: a Cooperative Marketing Strategy and a Market-based Development Approach in sub-Saharan Africa. Choices 20: 81-85.

Mobilizing innovation: sugar protocol countries adapting to new market realities

Johannes Roseboom

Abstract

The EU Sugar Reform has negative consequences for the sugar industries in Sugar Protocol (SP) countries as their export revenues will decline sharply. In order to adapt to new market realities the EU offers the SP countries a development assistance package (€ 1.2 billion) to restructure their industries, as well as various other forms of assistance. At the same time, the Everything-But-Arms (EBA) agreement has created new opportunities for sugar-producing EBA countries to export to the EU sugar market. Another major driver for innovation in the sugar industry is the use of sugarcane or sugarcane waste as a source of renewable energy. The latter option (using sugarcane waste for energy production) is the better bet for SP countries in the short run, but in the long run energy production may become the core business of the sugarcane industry and sugar just a by-product. It will depend, among other things, on the oil price, the efficiency with which sugarcane can be produced and transformed into energy, continued access to preferential markets for sugar, and subsidies for renewable energy.

1. Introduction

In recent years, analysis of the impact of the EU Sugar Reform on third countries (and in particular the Sugar Protocol countries) has been dominated by trade models. However, now that the key parameters of the EU Sugar Reform have been set and are being implemented, the policy focus has shifted to how these countries can adapt to new market realities. The *EU Action Plan on Accompanying Measures for Sugar Protocol Countries* (EC, 2005a) with a total budget of approximately € 1.2 billion for eight years is playing a lead role in this process. But various other instruments are also being mobilized to facilitate the adaptation process, including an ACP Sugarcane Research and Innovation Programme to be funded by the European Development Fund (EDF). A budget of € 13 million has been set aside for this (still-to-be-approved) five-year programme.[51]

This chapter aims to provide an overview of the strategic policy choices that the Sugar Protocol countries are confronted with and focus on the question of how innovation, both technically and institutionally, could facilitate their adaptation strategies. First, however, an overview of the broader context will be sketched in order to better understand the drivers for innovation.

---

[51] This chapter is based on the feasibility study for this programme (Roseboom *et al.*, 2007). I would like to thank my colleagues Taco Kooistra and Claudia Pabon for their input into this study.

Johannes Roseboom

2. The global sugar market and the EU sugar reform

Sugar is an important ingredient in people's diet the world over and sugar production (both from sugarcane and sugar beet) is widely distributed. During the 2005/2006 season, total world sugar production reached nearly 150 million ton – of which 76% originated from sugarcane and 24% from sugar beet. Since the 1980s, all growth in sugar production (on average 2% per year) comes from sugarcane, while sugar production based on sugar beet has been stagnant or contracting. Currently, 69% of the world's sugar is consumed in the country of origin whilst the balance is traded on world markets. This makes it one of the more intensively traded agricultural products in the world.

What makes sugar in particular a 'difficult' commodity is that sugar markets have a long history (often going back to colonial times) of being heavily regulated by governments the world over. Two such regulations are:

1. The *EU Sugar Regime*, which regulates the internal production, import and export of sugar of the European Union by means of fixed, national production quota, import restrictions, and an internal intervention price that is substantially higher than the world market price;
2. The *ACP/EU Sugar Protocol*, signed by the EU and some 18 ACP countries,[52] regulates that these countries have the right to export a certain quota of sugar (i.e. approximately 1.3 million ton in total) to the EU at a guaranteed price (related to the price paid to European farmers) and on a duty-free basis. While at the time of signing the treaty (1975) the internal EU price was close to (or even below) the world market price, in most of the years after 1975 the EU sugar price exceeded the world market price quite significantly. In recent years, for example, the EU sugar price has been roughly three times the world market price. For countries that hold large Sugar Protocol (SP) quotas (such as Mauritius, Fiji, Jamaica, and Swaziland, which jointly hold 80% of the SP quota) this Treaty has been particularly profitable. Some SP countries, for example, import sugar for their own consumption in order to fill their SP quota. In addition, further market access is given by the Agreement on Special Preferential Sugar (SPS), granting temporary import quotas for some 17 ACP countries (some 200,000 ton in 2002/03).

Despite the fact that at the pre-reform price level the EU produced more than enough sugar (19-20 million ton) to cover internal consumption (16-17 million ton), the SP meant that it had to import the agreed quota and at the same time dump a surplus of sugar (4.7 million ton) on the world market at high cost. This practice has been under criticism as unfair to

[52] Barbados, Belize, Congo (Republic of), Cote d'Ivoire, Fiji, Guyana, Jamaica, Kenya, Madagascar, Malawi, Mauritius, Mozambique, St. Kitts & Nevis, Swaziland, Tanzania, Trinidad, Zambia, and Zimbabwe. Surinam and Uganda were originally also Sugar Protocol countries, but stopped exporting to the EU many years ago. Their quotas have been redistributed.

developing countries for a long time. It is one of the classic examples of the No Aid but Trade Campaign. However, the two major reasons that triggered the EU Sugar Reform were:

1. A successful complaint by Australia, Brazil, and Thailand (three major sugar exporters) at the WTO that some of the EU sugar export practices are in conflict with international trade agreements. These agreements have placed explicit limits on the volume of subsidized sugar export by the EU;

2. The Everything-But-Arms (EBA) agreement, which gives the poorest countries in the world free access to the EU market, including the EU sugar market for which there is a special transitional regime. Sugar import quota and levies for EBA countries exporting to the EU are to be phased out by 2009. In contrast to the SP countries, no import quota restrictions will apply to the EBA countries in the future. Given the high EU sugar price, the EU sugar market fears being swamped by sugar from EBA countries after 2009. Studies differ in terms of the magnitude of the immediate supply response by the EBA countries, but a major supply response in the long run in the case of no price adjustment is the most likely scenario upon which to base policy decisions.

In response to these emerging problems, the EU decided to reform its sugar policy drastically and reduce the internal EU price for sugar by 36% over the period 2006 to 2009. The expected outcome of this reform is that the European sugar production will contract sharply as high-cost producers will stop producing and hand in their quota. Their quota will not be redistributed to low-cost producers in other European countries. All-in-all, it is hoped that this reduction in production (in combination with an increase in imports due to the EBA agreement) will be sufficient to reduce subsidized EU sugar exports to acceptable levels under WTO agreements (EC, 2005b).

The reduction of the internal EU price for white sugar by 36% will also affect the EU sugar imports from SP countries. The quotas as such remain untouched (and unused quotas can be redistributed), but the price paid for sugar will decline significantly. Despite the sharp price reduction, the EU internal sugar price will still be double the world market price, and hence will remain an interesting export market for most SP countries. Nevertheless, the reduction in revenues requires that they adjust and streamline their sugar sectors significantly.

Oxfam, one of the more vocal players in the EU Sugar Reform debate, proposed a reduction of the EU sugar production quota by 25% rather than a reduction in the EU sugar price (Oxfam, 2002). In that way the SP countries (as well as the EBA countries) could continue to benefit from a high EU sugar price. However, the EU opted to induce internal quota reductions indirectly through a major price reduction. In that way high-cost sugar producers in Europe are expected to give up their quota first and new sugar imports from sugar-exporting EBA countries will be dampened.

While most parties involved seem to accept the price reduction as inevitable in order to rationalize the global sugar market, the EU is facing criticism because of the short time period within which the price reduction will take place (i.e., four years) and the limited amount of adjustment funding made available for the SP countries. At the same time, sugar-exporting developing countries that fall under the EBA agreement may increase their export to the EU market in the coming years as import restrictions (quota and import tariff) will be gradually lifted. At the same time, however, they are also confronted with the lower EU sugar price due to the EU Sugar Reform. Hence the supply-response by these countries will be more limited than in a scenario without an EU sugar price reduction. In the latter instance, some studies predict an additional 3 million ton of sugar entering the EU by 2010, while with the EU price reduction the EU sugar import from EBA countries is estimated in the range of 0.2-0.9 million ton of sugar, depending on the assumed substitutability between European sugar and sugar from EBA countries (Van Berkum *et al.*, 2005).

While the trade models show that the countries that formally complained about the EU sugar export practices (Australia, Brazil and Thailand) hardly gain from the EU Sugar Reform, it is clear that the SP countries will be the big net losers of the EU Sugar Reform. They will see their export revenues from sugar decline by some 36% (assuming there is no reduction in the volume of export). On an annual basis this will be a loss of some € 245 million in export revenues for the SP quota (1.3 million ton) and some € 36 million for the SPS quota (0.2 million ton). At the same time, the EBA countries will gain, but substantially less than under a scenario of no price reduction.

In the following sections, we will have a closer look at the SP countries and their options for responding to this drastic reduction in income.

3. Some key characteristics of the sugar protocol countries

In terms of sugarcane production, the 18 SP countries represent only a very small share (3%) of the worldwide sugarcane production (Table 1). The real big producers in the world are Brazil and India, which together are good for half of the world's sugarcane production. Other important producers are China, Thailand, Pakistan, Mexico, Colombia, Australia, USA, Indonesia, Cuba, South Africa and the Philippines (all producing more than 25 million ton of sugarcane per year). Most sugarcane is transformed into sugar, but some is also used to produce alcohol and (increasingly) ethanol as well as many other products.

The cost price of sugar varies greatly across the various sugar exporting countries (Figure 1), indicating that the reduction of the EU sugar price (from roughly three times to two times the world market price) will affect some SP countries a lot harder than others. Even at the high pre-reform price level, the sugar industries in several Caribbean countries were already making losses (Barbados, St Kitts & Nevis, and Trinidad & Tobago). For these countries adjustment to the lower price level will be difficult, if not impossible. St Kitts & Nevis and Trinidad & Tobago, for example, have decided to stop producing sugar. Barbados is in the same league, but wants to transform its industry from producing sugar to producing energy. Jamaica stands out as the country for which the dice could go either way – either

Table 1. Worldwide sugarcane production.

	Sugarcane production (2001-2005 average)	
	(million ton)	(percentage)
Sugar Protocol countries (18)	39.4	3.0
Brazil	388.9	29.7
India	270.0	20.6
Other countries (98)	610.7	46.6
Total (118)	1,309.0	100.0

Source: FAO production statistics, downloaded April 2006.

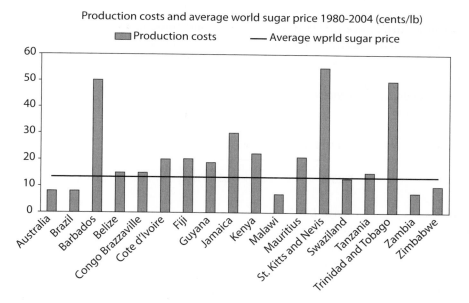

Figure 1. Sugar cost price comparison (Garside et al., 2005).

the industry is rationalized and significantly modernized and will survive, or it will have to close down. Explanations for the high cost of sugar production in the Caribbean are: (a) a lack of economies of scale; (b) a high-cost environment (relatively expensive labour and land); (c) poor management due to government ownership; and (d) lack of capital to renew plants and improve sugar fields. The sugar industries in these countries have already been in decline for some time (as reflected by declining trends in production and yields) and the EU Sugar Reform just gives the final blow to an already weak industry.

Apart from the sugar cost price, factors that are important in terms of understanding the vulnerability of SP countries as a consequence of the current EU Sugar Reform are:

1. The relative dependence of the sugar industry in SP countries on the EU market. Table 2 characterizes the SP countries on the basis of two variables, namely the sugar export as a percentage of sugar production plus import and the share of the EU in the country's sugar export. Countries with sugar industries located in the lower, right-hand corner are the ones most affected by the EU price reduction. Interestingly enough, some exporters are big importers themselves. We also identified the SP countries that fall under the EBA agreement and hence will benefit from no longer being bound by EU import quota after 2009.

2. The relative size of the sugar industry in the overall economy (in terms of share in GDP and employment) and in terms of export earnings. In four SP countries (Fiji, Guyana, Mauritius, and Swaziland), the sugar industry represents more than 5% of GDP and in six SP countries (as above plus Belize and St Kitts & Nevis) sugar export earnings exceed more than 10% of total export earnings. Hence, sharp reductions in sugar earnings in these countries can easily lead to macro-economic instability. In addition, social and environmental concerns play a major role in these countries.

Table 2. Dependency of the sugar industries in sugar protocol countries on export to the European Union.

Share EU in country's sugar export	(Sugar Export) / (Sugar Production + Import)		
	0-25%	25-75%	75-100%
0-25%	*Congo, DR* *Mozambique* *Zambia*		
25-75%	Cote d'Ivoire Zimbabwe	Belize *Malawi* St. Kitts and Nevis Swaziland	Fiji Guyana
75-100%	Kenya* *Madagascar* *Tanzania*	Jamaica* Trinidad and Tobago*	Barbados Mauritius

Countries in italics also fall under the EBA agreement and hence after 2009 their sugar exports to the EU will no longer be constrained by EU import quota.
* Major sugar importers.

While the sugar price reduction by the EU is looming as a big threat to at least some of the SP countries, at the same time new opportunities are emerging for the sugarcane industry in the form of creating additional value by producing bio-ethanol and electricity from sugarcane waste. Many of these energy-generating activities have become profitable in recent years due to a high oil price as well as technological developments. This opportunity is particularly attractive for SP countries that are struggling with high energy import bills. In addition, there is a rapidly growing market for bio-ethanol due to the renewable energy policies of the EU and the USA.

4. EU policy measures to accommodate the EU sugar reform

In recognition of the socio-economic consequences of the EU Sugar Reform on SP countries, the European Commission has promised a package of accompanying measures to facilitate the adaptation of these countries to a new market situation. This package was first presented in January 2005 under the title 'Action Plan on Accompanying Measures for Sugar Protocol Countries Affected by the Reform of the EU Sugar Regime' (EC, 2005a).

Considering the differences between the SP countries, in terms of the intensity of the impact of the reform as well as possible responses, the Action Plan offers a broad range of support options, to be tailored to each situation. It includes both trade measures and development assistance to help the SP countries to adapt. Favourable trade measures are expected to emerge from the current round negotiations within the WTO as well as from the bilateral negotiations of the EU with the ACP countries on Economic Partnership Agreements. Although these trade measures are important, in this chapter we will focus on the second component of the Action Plan, namely the package of development assistance that the EU has offered to the SP countries for the period 2006-2013 in order to adapt to the new market situation. The total budget envelope for this assistance runs to approximately € 1.2 billion.

The Action Plan argues that the development assistance can be provided mainly along three different axes: (1) enhancing the competitiveness of the sugar sector; (2) promoting the diversification of sugar-dependent areas; and (3) addressing broader adaptation needs, such as employment and social services, environmental impact, and macro-economic stability. A positive assessment of the viability of at least part of the sugar industry is an essential prerequisite for directing EC support towards enhancing competitiveness. If not, priority should be given to measures along the other two axes. The Action Plan discusses various possible measures along each of the three axes, but leaves the actual selection of the required measures to each of the SP countries.

To get access to the Action Plan funding, the 18 SP countries have each been requested to develop a multi-annual, comprehensive adaptation strategy in close consultation with all the stakeholders in the sugar industry. Although an important first step in shaping up the development assistance package under the Action Plan, the process took place under a cloud of political friction between the EU and the SP countries. Accepting the EU Action Plan basically meant accepting the EU Sugar Reform. By mid-2006, however, 13 of the 18 SP countries had developed and submitted to the EC a national adaptation strategy for their sugar sector. However, the quality of these national adaptation strategies differs greatly and has often been negatively affected by an atmosphere of animosity and the idea of trying to squeeze as much assistance out of the EU as possible. One problem is that many of the submitted strategies do not really prioritize the required actions (they want to do everything under the sun), nor indicate how they will be financed and implemented. The total estimated costs of the national adaptation strategies are many times higher than the available EU budget. In many instances it is unclear how the full strategy will be financed, taking into account the different possible funding sources. This basically disqualifies many of the strategies as effective planning tools. Nevertheless, the EC has accepted the national adaptation strategies as a starting point and has selected in each country key areas on which it will focus its assistance. What will happen with those parts of the strategy that do not receive EU funding is often unclear or highly insecure.

Despite the overwhelming emphasis on improving the competitiveness of the sugar sector, investment in sugar research features only in a few of the strategies in a significant way. Moreover, in selecting priority areas for EC support this opportunity of supporting competitiveness further disappears into the background and is left to the industry to take care of. This is rather unfortunate, particularly in upcoming sugar-producing countries that still lack any significant sugar research capacity of their own.

The three most common themes across the national adaptation strategies put forward by the SP countries are: (1) increased productivity at both field and factory level; (2) diversification into ethanol and electricity production (which will require major capital investment); and (3) social measures to support people that will lose their job due to the restructuring of the sugar industry. Interestingly enough, many SP countries (in particular EBA countries) are optimistic about their future opportunities in the sugar industry and aim to expand production. The strategies of those countries usually also include further investment in infrastructure (irrigation, roads, railways, etc.) and the opening up of new land for sugar production.

Enhancing the overall productivity of the sugar industry stands out as an essential requirement for the SP countries to stay in business and continue to benefit from a substantially less attractive, but still significant preferential trade agreement with the European Union (the new EU sugar price will still be roughly double the world market sugar price).

5. Sugar research and innovation capacity

Sugarcane is a plantation crop, organized and managed predominantly in large production units. Due to their size, sugar estates can afford to employ agricultural specialists to bring advanced technical expertise to the enterprise. The sugar mills also employ the necessary technical specialists for chemical analysis and controlling the various factory processes. All-in-all, sugar companies with stakes in both sugarcane production and processing usually employ a substantial cadre of trained technical specialists. However, the research capacity of these in-company technical services (other than some adaptive trials and testing) is usually rather limited.

A common phenomenon in many sugar-producing countries (often inherited from colonial times) is for local sugar companies to organize and finance their technical services jointly. Depending on the size of the sugar industry, this joining of forces allows them to move into more advanced research activities for which they individually lack the capacity. Classic examples of industry-based and funded sugar research institutes are the South Africa Sugar Research Institute (SASRI), the Mauritius Sugar Industry Research Institute (MSIRI), the Sugar Industry Research Institute of Jamaica (SIRI), and the West Indies Central Sugarcane Breeding Station (WICSCBS). The latter is a regional entity.

Table 3 provides an overview of the sugar industry in the different SP countries, including their sugar research capacity. In about half of the SP countries the sugar industry is monopolized by just one public or private company that owns all the sugar mills. Such companies often also own major sugar plantations, but do not necessarily control all sugarcane production. Some of the production may come from independent smaller sugar plantations, but increasingly also from smallholders who operate as outgrowers for the larger plantations. Prices paid to independent sugarcane producers under such monopolistic situations are a permanent cause of conflict. In many countries institutions have been created to deal with this problem (like Sugar Boards and Authorities) and have been given the authority to set a fair sugarcane price for both producers and processors. Nevertheless, conflicts are still quite common, in particular about incomplete and delayed payment and corruption.

Sugar research and innovation is first and foremost an in-company activity and particularly so in countries with just one sugar company. Specialized, stand-alone sugar research institutes and technical services only occur when there are multiple local sugar companies.[53] However, the balance between intramural and extramural technical capacity can differ quite a bit from country to country. In some countries, like Mauritius, there is a long and strong tradition of collaboration and so the industry depends heavily on MSIRI for technical input, while in other countries, like Mozambique, the technical collaboration between the sugar companies has been relatively weak (or non-existent) and far less research and technical services are undertaken jointly.

---

[53] The exception, Fiji, has only recently created a sugar research institute after complaints by sugar growers that the Fiji Sugar Corporation was not doing enough.

Table 3. Overview of the sugar industry and sugar research capacity in the sugar protocol countries.

Country	Sugar companies	Sugar research
Barbados	The sugar industry is owned by the government, which has contracted Barbados Agricultural Management Company (BAMC) to manage it.	BAMC has an Agronomic Research and Variety Testing Unit as well as a Sugar Technology Research Unit. Barbados also houses the West Indies Central Sugarcane Breeding Station (WICSCBS), which has a regional mandate.
Belize	Belize Sugar Industries Ltd (BSI) and Petrojam Ltd. are both privately owned, but have been granted monopolies by the government.	BSI: R&D Unit
Congo, Republic of	Société Agricole de Raffinage Industriel du Sucre (SARIS-Congo) is owned by SOMDIAA, a group of food-processing industries in various French-speaking African countries.	No specific information available.
Côte d'Ivoire	Industry dominated by two companies: Sucrivoire and Sucre Africain (SUCAF).	Centre National de Recherche Agronomique (CNRA): conducts contract research for the two sugar companies
Fiji	Fiji Sugar Corporation (FCS) holds a sugar monopoly and is state owned.	Sugar Research Institute of Fiji (SRIF), established in 2005 as a tripartite partnership between FCS, the Fiji Sugarcane Growers Council and the Government of Fiji. Previously the responsibility for sugar research rested with FSC.
Guyana	Guyana Sugar Corporation Inc. (GUYSUCO) holds a sugar monopoly and is state-owned.	GUYSUCO: Agricultural Research Centre
Jamaica	Seven mills (Frome, Monymusk, Bernard Lodge, Appleton, Worthy Park, St. Thomas, and Trelawny) are still in operation of which at least three are expected to be closed down. All mills are state-owned. Proposed restructuring of the industry includes privatization of the mills.	Jamaica Sugar Industry Authority: Sugar Industry Research Institute (SIRI)
Kenya	Some 8 sugar factories in operation: Busia Sugar Company, Mumias Sugar Company, Muhuroni Sugar Company, Nzoia Sugar Company, Chemelil Sugar Company, South Nyanza Sugar Company, West Kenya Sugar Company, and Miwani Sugar Company. Most companies are partly owned by the	Kenyan Sugar Research Foundation (KESREF), established in 2000, took over all sugar research previously conducted by the Kenyan Agricultural Research Institute (KARI). This restructuring was the result of an attempt to shift the full responsibility (including financing) for sugar research back

Madagascar	Industry dominated by two companies: Siramamy Malagasy (SIRAMA) and Sucrerie Complant de Madagascar (SUCOMA)	Centre Malgache de la Canne et du Sucre (CMCS)
Malawi	Illovo Sugar (Malawi) Ltd. (previously a government monopoly).	Illovo Sugar (Malawi) Ltd: R&D unit
Mauritius	As part of the adaptation strategy, the milling capacity has been rationalized significantly. Only four mills will stay in operation, namely: Savannah, Rose Belle, Mon Loisir, and Mon Desert Alma. The first two mills are owned by the 'Societe Usiniere de Sud' in which various shareholders participate.	Mauritius Sugar Industry Research Institute (MSIRI) and the University of Mauritius (several sugar-related departments)
Mozambique	Industry comprises four privately owned companies/mills: Maragra Mill (Illovo), Mafambisse Mill (Tongaat-Hulett), Marromue Mill (Sena Holdings Ltd), and Xinavane Mill (Tongaat-Hulett). All four mills are in foreign hands.	Centro de Promoção da Agricultura (CEPAGRI) (formerly Instituto Nacional de Açucar)
St. Kitts and Nevis	St. Kitts Sugar Manufacturing Corporation (state owned): in the process of being closed down.	St. Kitts Sugar Manufacturing Corporation: Agronomy and Research Department
Swaziland	Industry dominated by four companies: Mhlume and Simunye (Royal Swaziland Sugar Corporation), Tambankulu Estates (Tongaat-Hulett), and Ubombo (Illovo).	Swaziland Sugar Association: Technical Services.
Tanzania	Industry dominated by three companies: (i) Kagera Sugar Company Ltd. (Sugar Industries Ltd.); (ii) Killombero (Illovo); and (iii) Mtibwa Sugar Estates Ltd (Tanzania Sugar Industries).	Kibaha Sugarcane Research Institute (Ministry of Agriculture) and the National Sugar Institute. The latter focuses primarily on training, but conducts some research as well.
Trinidad and Tobago	Caroni Ltd. (state owned) has been dismantled. Caroni's sugar refining business will continue as the Sugar Manufacturing Company Ltd, processing imported raw sugar.	
Zambia	Nkambala, owned by Zambia Sugar PLC / Illovo, covers some 90% of the market. Kafue (Consolidated Farming Ltd) covers the remaining 10%.	Nkambala depends, through its mother company Illovo, on sugar research capacity in South Africa.
Zimbabwe	Industry dominated by two plants: Triangle Mill and Hippo Valley Estates. Tongaat-Hulett, which already owned Triangle Mill, has recently also taken over Hippo Valley Estates.	Depend, through Tongaat-Hulett, on sugar research capacity in South Africa.

A complicating factor in the case of Mozambique (and also in other African countries, like Malawi, Swaziland, Zambia and Zimbabwe) is that the ownership of the sugar companies is no longer exclusively national. In particular, South African sugar companies nowadays own quite a number of subsidiaries throughout southern Africa and rely heavily on the technology base at home (i.e., SASRI and Sugar Milling Research Institute [SMRI]). Rather than investing in building local sugar research capacity, these companies prefer to contract out research to SASRI and SMRI. In the short run this gives them the best research results money can buy. In the long run, however, this will keep the host countries dependent on imported sugar technology. A national sugar research and innovation strategy may counterbalance such dependence and prioritize those areas where building local capacity is most needed.

Characteristic of sugar research (in contrast with most agricultural research) is that it is primarily organized and financed by the sugar industry itself. The monopolistic / oligopolistic character of the industry means that the commodity chain is usually relatively well-organized (i.e. a few powerful players that can take the lead) and can be taxed easily to finance a public good like sugar research (i.e. no free riders and low collection costs). This model seems to work well as long as: (a) ownership of the industry is predominantly national (foreign companies have different loyalties -- see above); and (b) state ownership in the sugar industry is not undermined by political interference.

One handicap of sugar research being financed within the industry is that when revenues are low research will be affected as well. This is one of the big threats that sugar research in SP countries is facing at the moment.

6. Enhancing sugar productivity: an innovation agenda

The EU Sugar Reform is forcing the SP countries to push through major rationalizations within a short period of time. The most dramatic ones that have been proposed are the complete shutting down of the sugar industry in St Kitts & Nevis and a partial shutting down in Trinidad & Tobago. In other countries, consolidation of milling capacity and elimination of marginal sugarcane fields are being proposed as well. For those parts of the industry that intend to stay in business, however, increased productivity (both at field and factory level) will be crucial.

In order to enhance the overall productivity of the sugarcane industry, three major innovation clusters within the industry can be identified, namely: (1) sugarcane breeding; (2) agricultural practices in sugarcane production; and (3) sugarcane processing and products. We will discuss each of these clusters in detail in the following three sections.

6.1. Sugarcane breeding

Sugarcane breeding is a well-established practice in the sugarcane industry and has a long and successful history. Leading sugarcane breeding centres among the SP countries are the West Indies Sugarcane Central Breeding Station (based in Barbados, but servicing the whole Caribbean), MSIRI (based in Mauritius), and indirectly SASRI (based in South Africa, but servicing many neighbouring SP countries). Most of the funding for this breeding work is coming from the local sugar industry. However, for breeding work done for third parties these centres usually charge a fee or royalties. In particular, many African countries lack local sugarcane breeding programmes and hence their sugar industries rely on imported sugarcane varieties. Their own involvement is usually limited to variety testing only.

For a long time, sugarcane breeding has focused primarily on high yields and high sucrose. With the rapidly emerging interest in producing electricity out of bagasse (the waste left after the sugarcane has been milled), high fibre content has suddenly become a desirable characteristic. While in the past low fibre was preferred, now the selection has started to move in the opposite direction.

Electricity companies are only interested in a steady, year-round supply of electricity. In order to get around this bottleneck, sugar companies are: (a) installing generators that can work on both bagasse and other sources of energy (i.e., oil, coal, or gas); and (b) trying to lengthen the sugarcane harvesting season. This has resulted in a demand for early-maturing, high-sucrose sugarcane varieties. In this business model, sugar production is still the lead activity. By adopting a business model in which energy production is leading and sugar a by-product, as in the case of Barbados, breeders are looking for sugarcane varieties that can be harvested year-round and are less concerned about the sucrose content.

In addition to these characteristics required for electricity production, breeding programmes continue to emphasize disease resistance (such as to ratoon stunting disease, yellow spot and yellow leaf syndrome) and improved agronomic characteristics such as rapid covering of the inter-row, erectness, tolerance to drought and frost, and optimal nutrient uptake (Glaz, 2003).

Genetically modified (GM) sugarcane varieties are currently under development in various countries, but most importantly in Australia and Brazil. They have both announced the commercial introduction of GM sugarcane by 2011. However, for SP countries exporting to the EU market, sugar from GM sugarcane may encounter problems of acceptance by European and other consumers. Hence some caution in introducing GM sugarcane (and investing in the development of them) in these countries is warranted. GM sugarcane can expect less resistance when it is used exclusively for non-food applications like the production of bio-ethanol. Among the SP countries, only Mauritius has invested in a sugarcane biotechnology programme to date.

A common problem in many sugar-producing countries is the relatively slow uptake of new sugarcane varieties by sugarcane growers. While the standard recommendation is to replant sugarcane fields every 6-8 years, many sugarcane growers (and in particular the smaller ones) ratoon their sugarcane for a far longer period, sometimes for up to 20-30 years. Because replanting is costly, a slowdown in the spread of new varieties is usually a sign that growers are pessimistic about the sugar market prospects.

Greater investment in sugarcane breeding will not lead to immediate successes in sugarcane fields. Developing a new variety takes time (13-15 years) and the uptake of improved varieties tends to be slow due to high replanting costs. An intervention in the latter may help sugarcane planters to increase their yields per hectare and reduce their costs per ton sugarcane produced in the short run. In other words, they should reduce the backlog there is in adopting improved sugarcane varieties. This is a one-time, short-term advance that can be made. Speeding up the sugarcane breeding programmes in general is the longer-term solution – this requires more funding as well as the adoption of better breeding techniques (e.g., molecular markers).

6.2. Agricultural practices in sugarcane production

Principle areas of attention with regard to agricultural practices in sugarcane production are:
- *Crop rotation*. In most countries sugarcane is grown as a mono-crop without any crop rotation. Research, however, has shown that long-term cultivation of sugarcane can lead to changes in soil pH, loss of organic matter, and adverse changes in soil biota. Crop rotation can reverse these developments and help increase productivity levels (Glaz, 2003). Moreover, crop rotation and multi-cropping in sugarcane areas may lead to a more ecologically sound use of the land and diminish the dependency of farmers on a single crop for their income.
- *Crop protection*. The use of chemical pesticides and insecticides in sugarcane production is quite common, although application levels seem to be relatively moderate compared to some other crops. Nevertheless, the total costs of this chemical protection are quite considerable in monetary terms as well as in terms of health risks and environmental damage. Hence there is a permanent drive to develop cheaper alternatives that have less negative externalities. In countries like Australia, India and South Africa the use of bio-pesticides and biological control is a major research topic.
- *Water management*. Sugarcane is a relatively hardy tropical or sub-tropical crop, which has been adapted to grow both in high rainfall areas and in desert conditions. In the latter situation it is entirely dependent on irrigation. Often the volume of water available determines the area that can be planted with sugarcane. Hence improving the irrigation efficiency is high on the research and innovation agenda in many sugarcane growing areas. Too much water can also constitute a problem (both in irrigated and rain-fed production), hence the importance of adequate drainage.

- *Soil management.* Important aspects of soil management in sugarcane production are maintaining soil fertility, avoiding soil compaction and reducing the incidence of soil erosion. The latter is a particular problem in hilly areas and not only affects the soil quality of the sugarcane fields, but also creates huge negative externalities downstream as rivers and lakes get filled up with sediment.
- *Nutrient management.* Nitrogen and phosphorus are crucial nutrients for adequate development of sugarcane, while at the same time they can cause environmental pollution when not properly managed. Reported improved management practices are the re-use of trash, the application of micro-organisms, and the optimization of nutrient application by using spectroscopy tools for assessing nutrient status in the cane in real time and adapting the management practices accordingly (Glaz, 2003).
- *Mechanization.* The introduction of mechanized harvesting is usually steered by cost-benefit considerations. In countries where labour is relatively cheap and capital expensive, harvesting is still done predominantly by hand. Topographic characteristics also influence the choice for a human cutter instead of machinery. Most of the machines used by sugarcane producers have been developed by the private agricultural machinery industry. The role of sugarcane research has usually been limited to looking at how best machinery can be used in the field and what type of adjustments are needed (e.g., the optimal width between sugarcane rows). Innovation in this domain is taking place especially in Brazil and Australia (Ridge, 2003). New developments focus on refinement of cane transport equipment, harvesting machinery, trash management to optimize nutrient application, and cutting for replanting.
- *Burning of sugarcane.* Burning of sugarcane prior to harvesting is still a common practice in many countries. Increasingly, however, this practice is under attack (and some countries have introduced legislation that forbid this practice) because the smoke it causes is a health hazard and causes environmental pollution. In addition, there is evidence that burning often negatively affects the quality of the sugarcane. Currently research is taking place to further optimize the use of the trash either for energy cogeneration or in the field as a natural fertilizer.
- *Optimization models and geographic information systems.* Better understanding of sugarcane growth has been brought about by the use of models. APSIM-Sugarcane and CANEGROW are models successfully used for estimating yields and making irrigation decisions not only in Australia and South Africa, where they were developed but also in other countries like Mauritius. Successful application of GIS tools has been reported in Argentina, Cuba and Thailand (Glaz, 2003).

So far, we have discussed agricultural practices in sugarcane production without considering the characteristics of the farming households involved. Traditionally, sugarcane has been very much a plantation crop, grown on large estates. Increasingly, however, smallholder sugarcane growers (between 0.5 and 10 ha) are entering the scene. In particular, in new sugarcane growing areas, smallholder settlements are quite popular. However, the yields per hectare of smallholder sugar growers tend to be substantially lower (10-20%) than that of neighbouring sugar estates. The exact reasons for this difference are not clear, nor the interventions needed to eliminate this gap. Lack of adequate technology transfer mechanisms is one of the factors that may come into play.

6.3. Sugarcane processing and by-products

Sugar cost price differences are not only determined by the efficiency of sugarcane production, but also by that of sugarcane processing. Factory efficiency is determined by both the quality of the plant infrastructure as well as by its management. Although modern plants are usually a lot more efficient than older ones, the quality of the management of the plant (process and quality control, logistics, administration, etc.) can still make a major difference in the efficiency of the plant. For example, the quality of raw sugar from many SP countries tends to be relatively poor, resulting in price penalties in the EU market. This may not be such a problem when you receive a high, protected sugar price, but at a substantially lower price such penalties are felt a lot more. Better quality control (including paying farmers for the quality of their sugarcane rather than sheer volume) may reduce such penalties considerably.

Energy saving and co-generation of electricity at the plant may also result in major cost savings. However, such interventions often require important modifications to the existing facilities. Sugar mills are also coming under increasing scrutiny for the way in which they manage their water use and waste streams due to tightening environmental standards (Blackwell, 2002).

The most important recent development regarding sugarcane processing efficiency has been the better utilization of waste products. Sugarcane residues are produced either as post-harvest residues or as the result of its processing into final products. Harvesting the sugarcane will leave as by-product the trash, i.e. tops, dry and green leaves, which are usually burned in the fields directly after harvesting. Processing into sugar will yield residues, such as bagasse (solid resulting after juice extraction), cachaza (material remaining after cleaning the juice, which is a mix of juice, coagulated proteins and minerals), molasses (thick syrup obtained in the preparation sugar by repeated crystallization) and water (Pabon Pereira *et al.*, 2006).

Bagasse, for example, is rich in energy and produced in large quantities. Rather than burning it very inefficiently in order to get rid of the waste, the aim now is to recover as much energy as possible and sell the surplus in the form of electricity to the national grid. This requires investment in better boilers as well as in generators. National electricity companies are usually only interested in a steady, year-round supply of electricity. The answer to this challenge has been the introduction of generators that can work on both bagasse and other sources of energy (gas, coal, biomass, etc.). Still, in quite a number of countries national electricity companies seem to be hesitant to adopt the idea of buying electricity from sugar companies or only offer a very low price (e.g., South Africa). The technology is there, but many institutional and managerial hurdles still need to be mounted. Nevertheless, in most national adaptation strategies co-generation of electricity is included as one of the more important measures to be pursued.

Another major waste product of sugar production is molasses, which is rich in sugar and can be used in many different ways, such as the production of ethanol, glycerol, fructose syrups, solvents, organic acids, amino-acids, and vitamins. It is nowadays often also directly used as animal feed and fertilizer. Bio-ethanol can be produced from molasses or straight out of sugar juice, but the latter is only an interesting economic proposition when sugarcane can be produced very cheaply and there is no alternative but to sell it at the low world market price (e.g., Brazil and Australia). Bagasse can also be used to produce bio-ethanol, but the conversion process is more demanding and still in development.

For sugar industries that have access to markets that pay sugar prices substantially higher than the world market price, producing sugar as the primary product rather than bio-ethanol is the more attractive proposition at present. However, this may change as the oil price continues to rise, hence pushing up the price of bio-ethanol. Most national adaptation strategies, however, propose entering the bio-ethanol market by producing bio-ethanol on the basis of molasses. This requires investment in bio-ethanol plants. Whether or not this is an attractive economic proposition all depends on the oil price and local government policies regarding renewable energy and taxation, as well as on the production costs associated with the sugarcane production itself, since biomass costs play a large role in the economics of the bio-ethanol industry.

While in most national adaptation strategies the production of electricity and bio-ethanol are conceptualized as important by-products of the sugar industry, the national adaptation strategy of Barbados clearly adopts a different business model for the sugar industry. In this model, the production of electricity constitutes the primary source of income, while ethanol and sugar production come in second.[54] Although there are doubts regarding the feasibility of this business model at present (in particular because of the high-cost environment of Barbados), it may give us a hint as to how the sugarcane industry will transform itself into a renewable energy industry in the long run. It all depends on what oil prices do and what type of alternative (and thus competing) renewable sources of energy may emerge.

---

[54] The projected revenues after the reforms of the sugar industry in Mauritius are 75% sugar, 15% electricity, 7% ethanol, and 3% carbon credits. In the case of Barbados, the projected, post-reform revenues are 50% electricity, 25% ethanol and 25% sugar.

7. Conclusions

Competition is the main driver for innovation to take place in any industry, including the sugar industry. In addition, two other major factors are driving innovation in the sugar industry in SP countries at the moment, namely:

- Policy-driven changes in market opportunities: the *EU sugar reform* and the *EBA agreement*, both being phased in at the moment, are affecting the opportunities of the SP countries on the EU sugar market. The EU price reduction will expose SP countries to stiffer competition (although still relatively protected). However, not all SP countries will be affected equally. The impact depends on the SP quota held (which is based on historical rights), the overall competitiveness of the local sugar industry, and on whether the country falls under the EBA agreement or not. While in some SP countries sugar production will contract, in others (in particular those with EBA status and a competitive advantage) it will expand. Mobilizing the necessary capital for such expansion will constitute an important bottleneck.
- High oil prices and a greatly increased interest in renewable energy have turned the spotlights on sugarcane as the most efficient crop for producing bio-energy at the moment. Economically, the most interesting opportunity at the moment for SP countries is to use sugarcane waste to co-generate electricity and bio-ethanol. It is a business model in which energy production is a by-product of sugar production. The pace at which this business model will be adopted depends on the availability of capital and technical know-how in the sugar industry in SP countries. A possible next step is to make energy production the core business of the sugarcane industry and sugar a mere by-product. However, this is still a hotly debated scenario in terms of environmental soundness (some argue that more energy goes into the production of sugarcane than we get out of it and we are therefore substituting one type of pollution for another) and moral acceptability (energy production competing with food production).

Although on-going and planned sugarcane research activities promise interesting improvements in the future efficiency of the sugar industry, most of them will come on board too late to save all the high-cost sugar producers in SP countries from bankruptcy. Short-run solutions have to be found more in the sphere of catching up on adopting existing technologies. Subsidies to improve sugarcane fields, to speed up the adoption of new sugarcane varieties, and to promote better agricultural practices will most likely result in more immediate improvements in sugarcane yields.

Also at the factory level immediate gains can be made by catching up on existing technologies and managerial practices. As experience in South Africa has shown, benchmarking of sugar mills and sugar industry operations is a useful tool for identifying those areas that can be improved easily and quickly. The International Sugar Organization (ISO) as well as private consultancy firms offer international benchmarking services.

References

Blackwell, J., 2002. Recent developments in sugarcane processing. International Sugar Journal 104: 28-42.

European Commission (EC), 2005a. Action Plan on Accompanying Measures for Sugar Protocol Countries Affected by the Reform of the EU Sugar Regime. Commission Staff Working Paper. Brussels: CEC.

European Commission (EC), 2005b. The European Sugar Sector: Its importance and its future. Brussels: CEC.

Garside, B., T. Hills, J.C. Marques, C. Seeger and V. Thiel, 2005. Who Gains from Sugar Quotas? London: Overseas Development Institute (ODI) and London School of Economics and Political Science.

Glaz, B., 2003. Integrated crop management for sustainable crop production: Recent advances. International Sugar Journal 105: 175-186.

Oxfam, 2002. The Great EU Sugar Scam: How Europe's sugar regime is devastating livelihoods in the developing world. Oxfam Briefing Paper No. 27. London: Oxfam.

Pabon Pereira, C.P., J.B. van Lier, W.T.M. Sanders, M.A. Slingerland and R. Rabbinge, 2006. The Role of Anaerobic Digestion in Sugarcane Chains in Colombia. In Proceedings VIIIth Latin American Workshop and Symposium on Anaerobic Digestion, Punta del Este, Uruguay, 2-5 October 2005.

Ridge, R., 2003. Trends in sugarcane mechanization. International Sugar Journal 105: 150-154.

Roseboom,, J. , T. Kooistra and C. Pabon, 2007. Support to Research in the ACP Sugarcane Sector. Brussels: Transtec.

Van Berkum, S., P. Roza and F. van Tongeren, 2005. Impacts of the EU Sugar Reforms on Developing Countries. The Hague: LEI.

Institutions and governance

Importance of institutions and governance structures for market access and protection of property rights of small farmers in developing countries

Guido Van Huylenbroeck and Ramon L. Espinel

Abstract

Institutions and governance are important for the proper functioning of markets. In this chapter we illustrate the importance of adequate institutions and governance structures for protection of property rights and market participation of small farmers. To this end we consider three cases: cattle trading in Uganda, water management in the Peninsula of Santa Elena, and biodiversity conservation in relation to indigenous knowledge. The results indicate that the asymmetry of information embedded in present markets and governance structures hinders smallholders to fully benefit from the values incorporated in their property rights.

1. Introduction

Development economics has focused on the role of the agricultural sector in the development process since early writings on the field. For developing countries this is a complex matter with many factors influencing the process. Main theories explaining agricultural contribution to development focus on factors such as accumulation of capital and surplus labour from agriculture (Lewis, 1954), access to inputs and technology (Schultz, 1964), induced technological innovation (Hayami and Ruttan, 1971), trade, dependency, and unequal exchange (Prebisch, 1981; Dos Santos, 1970). Only from the early 1990s the importance of institutions and governance structures for development was emphasized in the seminal contribution made by Hoff *et al.* (1993). Since then also the World Bank devoted more attention to the role of institutions and adequate governance structures in development processes (see e.g. World Bank, 2002 and 2003). Also Arie Kuyvenhoven was interested in the topic of institutions as proven *inter alias* by Ruben *et al.* (2003) but also by the many projects and PhD research in this area he supervised.

In this chapter we try to look into the importance of institutions for development and market participation of small farmers. We hereby mainly focus on the results of some research in Africa and Latin America to show the importance of governance structures for market participation and protection of property rights. The chapter is organised as follows. First we shortly mobilize the theoretical insights on the importance of institutions and governance structures for small holders. Next we describe three cases in which the role of governance structures is important and evident. The first case refers to commodity output markets and shows that market participation of small holder beef farmers in Uganda is low because of high transaction costs; the second example shows that also in input markets and more in particular for the functioning of water markets proper governance structures are important. This is illustrated using the case of poor performance of the irrigation system in the Peninsula of Santa Elena in Ecuador. In the third case we emphasize the role of institutions in protecting biodiversity by developing the case of property rights on indigenous knowledge. We end with some conclusions.

2. The neo-institutional economics framework to analyse governance structures

The neo-classic paradigm starts from the assumption that markets regulated by demand and supply result in optimal allocation of resources and provision of commodity and non commodity outputs. However this paradigm is only true in a frictionless economic environment. In practice such frictionless economic environment does not occur and organising market transactions requires assignment of property rights, bringing together transaction partners, searching for information, negotiation and protection of interests. This means that constructing a market involves costs which we call transaction costs. In his seminal paper of 1937, Coase was the first to point out that it is precisely the presence of these transaction costs that produces market imperfections and market failures. The neo-institutional economic theory developed this further and is at the basis of the present interest in institutions in economic research. This theory states indeed that in the presence of transaction costs other governance structures may be superior to market governance.

Transactions involve costs because they require agents to search for a partner with whom to exchange, to screen potential trading partners to ascertain their trustworthiness, to bargain with potential trading partners (and officials) to reach an agreement, to transfer the product, to monitor the agreement to see that its conditions are fulfilled and to enforce the exchange agreement (Holloway *et al.*, 2000). These costs increase with the frequency of the transactions, the specificity of the assets involved, and the uncertainty of the context (Williamson, 1985). In such a context people will induce in higher costs to protect their interests. This leads North (1990) to define transaction cost as the costs of defining and protecting the property rights of goods. New Institutional Economics (NIE) emphasizes that the success of a market system will be dependant upon the institutions that facilitate the protection of property rights and thus facilitate efficient transactions. Institutions are thereby defined as any combination of formal and informal rules. This may be a combination of informal rules such as customary laws, trusts and norms which in a dense social network lead to the development of fairly stable informal structures for organizing activities and more formal rules such as political and judicial rules as well as economic rules and contractual rules, the function of which is to facilitate political or economic exchange. Within this framework of institutions market players may then develop specific governance structures or institutional arrangements in an attempt to limit their own part of the transaction costs and facilitate their successful participation in the market in order to take value of their property rights.

These arrangements may on their turn lead to a need to adapt the formal rules. Therefore some authors (inter alias Mittenzwei, 2004) advocate a more game theoretical approach to institutions in which these are regarded as equilibria of a repeatedly played game in which players are capable of shaping institutions through their strategic interaction. Note that the two perspectives of institutions are closely interrelated as both study the relationship between human-made constraints and the players' strategic interaction. The only difference indeed is the causal connection between both the constraints and the interactions. According to Mittenzwei (2004), advocates of the institutions-as-equilibria view study how equilibrium behaviour leads to the establishment of human made-constraints, while proponents of the institutions-as-rules view seem to focus on the impact of human made constraints on the (equilibrium) behaviour of individuals.

Neo-institutional economics has developed a consistent body of principles and concepts to identify aspects of the physical, cultural and institutional settings that are likely to affect the determination of who is to be involved in a situation, the actions they can take, the cost of those actions, the outcomes that can be achieved, what information is to be available, how much control individuals can exercise, and what payoffs are to be assigned to particular combinations of actions and outcomes (see for more explanation on these principles among others North, 1990; Ostrom, 1992; Eggertson, 1990).

This consistent framework may in particular be useful for development problems in developing countries, where the lack of economic institutions to produce correct incentives has put many of them in a perverse path of underdevelopment. Indeed, the high occurrence of market failure and incomplete markets (because of higher transaction costs and information asymmetries) can often be brought back to the non-existence or poor functioning of governance structures and institutions. This is so because many of the institutions or formal rules of behaviour that are taken for granted in developed countries and that facilitate market exchange there are absent in low-income countries. Thus, NIE is a robust framework that could help us to determine the sort of institutions needed (either formal or informal) to improve economic performance in developing countries. The NIE theory also helps to explain why in many cases primary producers are vulnerable to exploitation by other more powerful organisations (middlemen, retailers, large companies, foreign countries and so on). The reason is that they often lack the capacity (or finances) to set up governance structures or arrangements which favour their interests. Lack of adequate organisation power of primary producers results in intermediate governance structures enabling the end market to function but at the expense of the primary producers' margin. This result is a logic expectation, as in a competitive environment the end price is often determined by market forces. If the functioning of markets in which small farmers operate is not well organised this leads to pressures on the interests of primary producers, or directly affect the prices received by them. Besides total transaction costs decompression, we should therefore also look to the division of transaction costs among transacting parties involved and in particular to the transaction cost of small producers to participate in the market or to protect their interests. Our basic hypothesis is indeed that in the absence of tailored-institutions protecting smallholders' interests, small producers face so high transactions costs that their market participation is hindered or that the benefits of market participation are captured by others. This situation is common in agriculture (see Hoff *et al.*, 1993).

In the remaining of the chapter we illustrate this for three cases. The first case shows that too high transaction costs may impede farmers from full participation in the market and to sell in lower value markets. The second case shows that underdeveloped governance institutions in input markets may hinder market participation, while the third case elaborates on the role of institutions in protecting property rights on nature and biodiversity.

3. Market participation affected by transaction costs: the case of the beef market in Uganda

Our first example is taken to illustrate the importance of market organisation and intermediate structures for market participation of small farmers. The evidence is based on a study of beef market participation of cattle farmers in Uganda (Kyeyamwa *et al.*, 2007). Reason for the study was the observation of low off take of cattle (defined as the percentage of cattle sold to the market) which remained stable around 10 to 12% in the past 25 years and which apparently does not follow the market signals. Therefore the study adopted a theoretical model presented by Renos *et al.* (2003) to test the hypothesis that some households may not participate in the market due to high transaction costs. Based on a survey of more than 650 individual transactions of cattle the model tries to provide a measure of both proportional and fixed transaction costs. A distinct character of this method is that it recognizes that households face a three-stage decision in accessing output markets. The first decision is whether or not to transact, the second is how much to transact and the third is where to transact. (for details on the model see Kyeyamwa *et al.*, 2007.)

The fieldwork for this study was undertaken in the districts of Soroti, Kiruhura and Kamuli located within what is commonly referred to as the 'cattle corridor' of Uganda during the period of September - October 2005. Through face-to-face interviews with livestock owners in front of their herds, the study reconstructed the demographic and transaction histories of a total of 696 transactions completed between August 2004 and August 2005. A transaction in this context means an agreement between a farmer and a buyer to exchange livestock (regardless of the number) for payment. In almost all cases the respondents were male household heads, mainly husbands who took the final decision to sell, where to sell, and to whom to sell.

Farmers have four common channels through which they market their livestock. Informal (on-farm or farm-gate) sales accounts for approximately 38% of total sales, while the formal sales account for 62%. Livestock keepers exchange formally at either the primary/collection market (58%), the larger secondary market (3%) or the much larger tertiary/terminal market

(1%). These formal market places are classified into the above three categories according to their throughput and associated infrastructure. The prices received by the sellers are highly variable for any given type of animal. The unweighed mean coefficient of variation (regardless of gender, age and grade) is 23.7%, which is quite high in comparison to livestock markets in high income countries. Cattle sold in the primary, secondary and tertiary markets brought in 14.9%, 37.5% and 43.7% respectively higher prices than those sold at the farm gate price during the same period. These differences suggest poor spatial integration in the cattle markets as also found in livestock markets elsewhere in Africa (Bailey *et al.*, 1999; Fafchamps and Gavin, 1996).

As expected, the results of the model show that the quantity transacted in the local market is dependent on proportional and fixed transaction costs. Proportional transaction costs, represented by time spend on a transaction, have a negative impact on selling in the market. The results indicate that if time spent travelling to the local market increases above three hours, the probability of participating in the market decreases by 22%. Similarly, fixed transaction cost related variables also affect the market choice decision. If farmers have access to market information, their search costs are reduced. Having market information increases the probability of participation in the local market by 9%. This shows the importance of lowering the search and information costs. Experience and level of education increase the farmer's ability to search for market information. The two factors positively influence quantity sold. Bargaining and negotiation costs also impact on market choice. Existence of a grading system or knowledge of the animal quality reduces information and negotiation costs as sellers and buyers are expected to agree more quickly on the animal attributes. The model shows that efforts to improve the grading of animals could increase the participation in the local market by 70%.

The influence of transaction costs is also shown by the sign of the coefficient of the selectivity parameter in the equation for predicting prices in the local market. This selectivity parameter controls for the fact that prices can only be observed when farmers actually participate in the market. The fact that this coefficient is negative and highly significant suggests that the unobserved costs related to the transaction do lower the real price obtained by farmers and thus hinder participation in the market. Figure 1 shows a simulation of market choice due to changes in net prices as a result of changes in proportional transaction costs and fixed transaction costs (represented by market price information cost). Compared to the base scenario being the farm gate sales, a 10% increase in net prices received at the local market as a result of reduction in transportation costs leads to approximately 10% switch from base scenario to market scenario. Interestingly, full knowledge of market prices has potential to facilitate this switch by 15%. A combination of full knowledge of market price and a 10% increase in net prices received has a simulated effect of increasing local market participation by 20% as opposed to the base scenario. These findings further underscore the relevance of transaction costs in market choice and market participation.

 Guido Van Huylenbroeck and Ramon L. Espinel

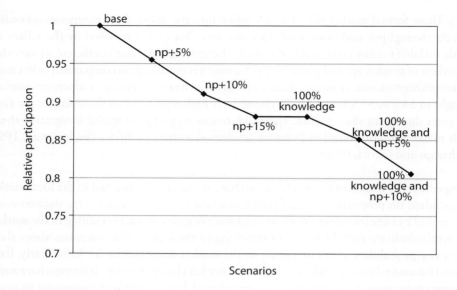

Figure 1. Relative participation in farm gate sales under different scenarios (Kyeyamwa et al., 2007).

4. Governance of irrigation systems affected by deficiency of institutional arrangements: the case of the Santa Elena Peninsula in Ecuador

The second case illustrates that the lack of adequate governance structures do not only hinder participation in output markets but may also hinder the functioning of input markets in particular in the case of common pool resources such as water. For this kind of goods the governance structures need to be carefully designed to articulate between the collective and private property rights on water. The reason is that when irrigation systems are implemented at the regional level, usually the infrastructure is developed by public intervention. As a result, when there are a large number of potential beneficiaries of such works, a situation of public good arises with respect to the management of water. A structure of governance for such systems is required. Theoretically a pricing system for water should be able to allocate the water to the most efficient user. However, the establishment of such market is not easy. The lack of sufficient institutional arrangements is conducive to the mal functioning of water markets with the consequent frictions amongst the participants and the inefficient use of resources, particularly land as a result.

We illustrate the importance of good governance institutions for the irrigation management for the case of the Peninsula of Santa Elena in Ecuador. In this region a large and expensive irrigation system has been build but a large underutilisation of this infrastructure has been observed. The antecedent dates to the early years of 1970. A major flood control and irrigation system was built in the central coast of Ecuador with the largest dam of the country affecting more than 200 thousand hectares of agricultural areas. Once the complete structure was installed, the problem of disposing of excess water arose. The simplest solution was to drive such excess channelled towards the sea, at about 150 km. distance and through the semi-desert area of the Santa Elena Peninsula.

Since ancient times the Peninsula played an important role in prehispanic agriculture (Pedro Cieza de León, 1984). The the confluence of large uncultivated tracts of land and the possibility of large volumes of water, gave rise to the project of a parallel irrigation system to be developed to incorporate some 40 thousand hectares of good soils for cash income crops.

The land ownership structure in the Peninsula obeyed an ancestral form of communal property instituted since the times of the colonial regime under Spanish rule. As the whole area was affected by desertification due to climate change speeded by deforestation, the system of property rights was largely untouched by the time the projected irrigation system was conceived. As it often happens, the last people to be informed of a major infrastructure work that affects production conditions are the inhabitants of the region to be affected (Ostrom, 1992). This was the case in Santa Elena.

Because of asymmetry of information land was purchased at very low prices by speculators as it was perceived that irrigation and further infrastructure, like roads and port facilities, were to be developed. As a result of these transferred lands, the communal structure of the Peninsula was altered and the land near the canals was concentrated in the hands of a few large owners (Espinel, 2002).

As Herrera (2005) points out, the institutional arrangements governing the irrigation system resulting from the canals layout were poorly designed and lacked enough structure to hold the new relations between economic agents, governmental officers, and local institutions. As a result a complex system dominating the economic and social environment was established. The salient features of the system can be summarized in three aspects.

In the first place, the Cabildos, that is, the local governance institution for the communal organization of the Peninsula, of which there are more than 50, lost ground as a form of authority directing the use of natural resources including land. These was a result arising from the fact that Cabildos were the negotiating instance to transfer land from the common structure of property to the private property of the newcomers that purchased it (Castillo, 2004).

Second, the outsiders that became the new proprietors (most of them purchasing land from speculators that bought it from Cabildos) were, in some cases, entrepreneurs intending to develop agriculture made possible by the new irrigation infrastructure, but the majority were rent seekers acquiring large tracts of land with the intention of reaping off the benefits of new investments to take place in the surrounding area, such as freeways, and other public facilities. When the major works were completed, it became clear that further development was to be postponed because of the restrictions imposed to the Ecuadorean economy with the world financial crisis at the end of the 1970s (Espinel, 2002).

Finally, the public institution building regional infrastructure also became the ruler of the irrigation system implanted for the usage of the excess water coming from the far away dam in an adjacent region. The system of irrigation at the Peninsula, termed the TRASVASE, was brought to life and administered by a government based institution. Irrigation infrastructure and water, then, were largely seen as a public good (Herrera *et al.*, 2005).

As a general outcome of this process, the intended area to benefit from irrigation, conceived originally of about 40 thousand hectares, was reduced to less than 25 thousand, and actually developed to agriculture was just about 4 to 5 thousand hectares (Herrera, 2005; Cornejo *et al.*, 2006).

In addition to the failure in utilization of the infrastructure in place, conflicts made an early appearance between the stakeholders involved in the irrigation system. The three major participants, i.e., communal farmers, landowners, and government officials were soon entangled in contradictions with respect to the use of the canals and the irrigation water. Communal farmers were claiming their right to access to water, demanding additional infrastructure to conduct water to communal lands separated of the canals by the lands of the new proprietors. Landowners realized that usage of water for irrigation demanded costly private infrastructure in the form of secondary and tertiary canals to conduct the water from the main canals to the agricultural fields, for which credit was demanded. The governmental institution soon started to face budgetary constraints that impeded further actions, and demanded return payments from users of the infrastructure. But this requires also an adapted governance structure. Theoretical work of among others North (1990), Ostrom (1990 and 1992), Challen (2000) and Vermillion (1999 and 2004) suggests that the best way for making water users responsible for the maintenance of irrigation infrastructure is to shift property rights to so called Water User Organisations (WUO). This devolution of governance structure (from State governance to self-governance) is also promoted by the World Bank (1993).

In order to analyse whether such shift would increase the use of the irrigation infrastructure, Herrera (2005) performed a contingent valuation survey to analyse whether farmers in the irrigated area of the Peninsula would increase the use of water if the governance was improved. This was done in two steps. In a first study it was analysed which factors determine the engagement of farmers in more water demanding activities (going from livestock activities demanding no irrigation over yearly crops with low water requirements to perennial crops with high water requirements). It was observed that mainly TC related factors influence the choice. Contact with extension officers (lowering the information and search costs), increased possibilities of access to irrigation equipment (because more non farm income available) and a larger farm area in the irrigation perimeter were factors positively contributing to perennial cropping while factors such as being a small farmer, lower education level and no other sources of income lead to more engagement in less water depending crops or livestock farming. This indicates that in order to increase participation in irrigation the WUO should not only maintain the infrastructure but also rent some services to water users (such as access to extension, renting of irrigation equipment, even access to output markets). This hypothesis was further tested by investigating the willingness-to-pay a higher price for the water if the services provided by the WUO would be increased. As the results in Figure 2 show the mean willingness to pay (WTP) for water under this improved situation is 0.05 US$ per cubic metre. This is higher than the current price of 0.035 US$ per cubic metre and the WTP increases with higher use of the land and dependence of the income on farming. It means that in particular the small holders are willing to pay a higher price for better services which proves again that their market participation can be improved if the right institutions can be found to provide them with better information, reduced risks of participation and higher access to specific capital (in this case irrigation equipment).

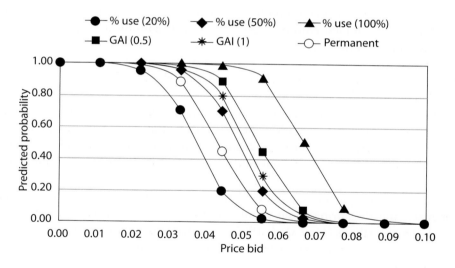

Figure 2. Predicted probabilities of accepting the price bid under different situations of land use and income (Herrera et al., 2005).

5. The preservation of biodiversity as a problem of property rights on indigenous knowledge

The third case looks into small-holder agriculture in the framework of globalisation (Espinel, 2006). Competitiveness for these farmers in local agriculture is lost very fast as developed countries' agriculture increasingly specializes on the same crops cultivated under capital intensive but land extensive methods.

On the other hand, peasants in the humid tropics of less developed countries are farming in the middle of abundant biodiversity which is not only important from the environmental point of view but increasingly from the genetic resource point of view. Its knowledge is presently of no market value to these farmers, while it is highly valuable to 'state of the arts' farming in developed countries and in some less developed economies (Mazoyer, 2004).

The capital-intensive, land-extensive agriculture of large modern farms sustains largely on productivity jumps based on biotechnology which rest on renewed genetic material collected in the humid tropics.

In developing countries legislation protecting biodiversity and indigenous knowledge is weak, little known, and less enforced. Thus, there exists a problem of specification and assignment of property rights. A loss of biodiversity is observed because of non discretionary extraction. Sensitive to over-exploitation are micro-organisms, and species that carry genes used in genetic engineering for crops (Powers and McSorley, 2001).

Correction of this situation needs arrangements in which participate: 1) small- and medium-size farmers, governments and NGOs, 2) institutions related to agriculture, trade, and environmental issues, and 3) corporate firms and large-scale farmers requiring genetic material that can only be acquired from biodiversity-rich areas in developing countries.

However, these multiple and complex relations give rise to problems of information asymmetries, and thus bring about problems of moral hazard, adverse selection, and principal-agent situations (Hoff *et al.*, 1993). The challenge for the immediate future of biodiverse developing countries is to develop suitable models that incorporate biodiversity richness into peasant and small-holders' productive activities.

To illustrate the loss of biodiversity we look at the case of Ecuador where small farmers in rice production account for more than 80 percent of the supply of the grain, a significant staple in the country's diet (Espinel, 2006). According to the Ecuadorian Agricultural Census of 2000 peasants holding farm units ranging in size between 1 and 5 hectares accounted for 68 percent of all the farm area dedicated to the cultivation of rice.

Research by Hildebrand *et al.* (2005) shows that small and medium farms, up to holdings of 10 hectares, are multi-cropping units where as many as a seven crops in average is found. This study took place in the provinces of Los Rios, Bolívar and Cotopaxi. The main crop is rice; the other agricultural products include fruits, vegetables, medicinal plants, and timber, plus small animals and fowl. The smaller farms (less than 3 ha) reported to market about two thirds of their rice production, while the larger farms (5 to 10 ha) sell about 80 percent. In all cases, farmers report a constant reduction of the area dedicated to crops other than rice.

The reason for the increase in rice is the increased requirement for cash income. In effect, rural families experience enlarged expenditures for the purchase of consumables and agricultural inputs, while at the same time the prices of staple crops are going down. This situation is also reported elsewhere with respect to other countries in Latin America (see, for instance, Giarraca (2004) on Argentina's small farmers).

The combination of increased cultivation of one crop and the reduction of multi-cropping results in a loss of agricultural genetic diversity. This brings a sharp impact on the habitat of small-holders agriculture and homogenizes the life-world of crop areas, and by doing so reduces its capacity to adapt and reproduce in changing biological conditions (Risler and Mellor, 1996).

There is evidence that this type of outcome is taking place in different parts of the developing world. Of about 75 types of vegetables existing at the beginning of the 20th century, about 97 percent of the varieties of each type are now extinct (The New Internationationalist, March 1991, p. 17). At the same time, thousands of local varieties of rice, wheat, maize, and potatoes have been eliminated.

In Ecuador, most of the rice produced by peasant farmers for own consumption comes from varieties that are unknown at the market level. Local names of varieties like Chileno, Patucho, and at least half a dozen more, are cultivated in different locations, while the so called improved breeds are seeds patented by research centres and international corporations. These patented seeds cannot be reproduced by farmers, as property rights are already established over them.

Also in response to international market patterns, the commercialisation of agricultural products is more and more a matter of standardization through supermarkets (Reardon *et al.*, 2003). Consumers in all countries are now facing a homogenized bundle of food commodities that narrows production to a handful of varieties for each crop.

The consequence of such pattern is the reduction of varieties at the rural level. These 'global' varieties have been selected through years of research from the local (indigenous) varieties of developing countries, in a process that started at the international level with CIMMYT at Mexico and IRRI in the Philippines (Eicher and Staatz, 1998). Gradually this research was displaced to private research units controlled by multinational firms. The direction of research moved then towards a model of extensive mono-cropping, changing the pattern of the world system of food supply (McMurtry, 1998).

The combination of increased productivity and reduced cost per unit of land achieved in modern agriculture of developed countries explains the continuous trend of declining prices for staples. It is unquestionable that the efficiency gains obtained in developed countries in terms of productivity per unit of land, or per unit of labour input reflect the great advance of modern agriculture which relies more and more in the application of science to agricultural production. Biotechnological breakthroughs achieved through genetic engineering are inseparable of today's agricultural accomplishments (Enriquez, 2001). But it also represents limits to agricultural development.

The impact of intensive agriculture on the environment is inevitable and, thus, continuous action has to be taken to manipulate soil nutrients and soil structure in order to maintain levels of productivity, but above all actions for genetic renewal become necessary after a given time (Parker, 2000). The variation in habitat that agriculture continuously imposes to the environment occasions the inadequacy plants to current conditions and makes change necessary (Altieri and Nicholls, 2001). This change is possible only if new genetic material is available, and thus modern agriculture requires returning to original pools of genetic reserve (Vera, 2001). These are in the tropics, in peasant farmland in less developed countries.

The following cycle expresses the paradox for modern agriculture. Increased productivity in developed countries relies more and more on the genetic improvements that alter plants. Efficiency gains obtained in production of grains such as rice translate into declining prices. Price reduction has an impact on the incomes of small farmers in less developed countries which drive them to increase the area cultivated of those same grains instead of local varieties in order to increase cash receipts. Increasing area of cash crops reduces multi-cropping and consequently affects biodiversity. The impact on the habitat of small agriculture in less developed countries results in a reduction of genes of selected species and limits the germoplasm at disposition for new genetic improvement. This becomes a vicious cycle that at large will result in a bottleneck to general improvement on the efficiency of crops. How can we change this pattern and turn it into a virtuous cycle of sustainable agriculture and poverty reduction?

At the base of the market failure associated to the loss of biodiversity there is a problem of asymmetry of information. Corporations and researchers do give a market value to specimens extracted from the freely accessible genetic pool in diversity rich areas. As soon as basic genetic material is incorporated to a new technology, it is appropriated and adequately priced for commercialisation. A property right is sought for and enforced as soon as lawfully permitted. But at the beginning of the genetic improvement change, the owners of the natural pool of resources and of the knowledge of the properties of such material remain uninformed of the potential to be obtained from them. No definite form of property is found at that level, the only right present being the ownership of the habitat from which the knowledge and the material are extracted.

If adequate information were available to peasant farmers in bio-diverse areas concerning the potential properties of the plants and specimens found in their agricultural habitat, there would develop a consciousness of market value that would allow the specification and assignment of property rights in favour of farmers. It would help to internalise the externality affecting biodiversity and a market for biodiversity would be established.

But according to Furubotn and Richter (2000) specification and assignment of property rights is not a sufficient condition to ensure a successful market operation. Economic incentives are affected by the ownership structure of resources and behaviour changes correspondingly. The incentives of private property rights help to economize on transaction costs and thus contribute to the economic welfare of society. But the situation is not entirely favourable. As a consequence of positive transaction costs, property rights cannot be fully assigned, perfectly enforced, or priced. In the case of environmental wealth it is the impossibility of making complete assignment of property rights in resources to individuals, and not private ownership as such, that explains environmental problems.

In our case transaction costs are present in the form of inadequate transmission of knowledge and lack of research to identify species that result in the correct genetic material to be developed and incorporated into seed and plant production (Burton, 2000). Thus large costs exist at the corporate level for the trial and error process of specifying the right material to be developed. On the side of farmers, asymmetric knowledge misguides the biodiversity conservation and dampens the development of correct exploitation and preservation techniques.

Transaction costs occur mainly because of bounded rationality that affects decision making. These give rise in general to opportunistic behaviour, which in turn results in rent seeking actions. For example, the uncertainty in the results of research on new species and genetic material drives corporations to look for a 'finders' keepers' attitude precluding the spreading of scientific acquired knowledge. As a consequence the 'public good' characteristic of natural resources owned by peasant farmers is appropriated, creating by it an externality. The appropriation of the genetic resource in a monopolistic fashion induces a rent seeking behaviour as resources are made scarce by property rights in the form of patents, thus capturing rents in the form of economic transfers.

Also bounded rationality is exacerbated by the lack of information on the side of farmers. Consequently biodiversity is exploited in excess. The common pool of natural resources is affected by depletion, degradation, and destruction. The advancement of monoculture thus takes place with the impulse coming from two sides. On the one hand, rent-seeking exploitation of biodiversity requires exclusion. As ownership of the bio-diverse resources is not well specified and property rights have not been adequately assigned, there is incentive to eliminate the fountain of used-up genetic material, i.e. there is incentive to limit biodiversity. The advance of monoculture in peasant farmer's areas also contributes to this result, as the bio-diverse resources are not correctly valued. As a result of this process, in the medium-to-long run farmers at both sides are the losers. As genetic pools deplete, the cost of genetic engineering and biotechnology applications to agriculture will rise.

On the other side, small holder agriculture in less developed economies will also suffer by the limitations imposed by reduced incomes in the face of increased costs. The increase of a monoculture system of agriculture will further reduce diversity. The environmental change will impose a greater toll on underdeveloped agriculture as a consequence of the impact on the habitat. One inevitable outcome of such trend would be the increase in poverty and reduction of living conditions.

This calls for state action (see, for example, Buchanan and Tullock, 1962; Alchian, 1977; Williamson, 1979; Barzel, 1989; Ostrom, 1990). There is a clear role for government intervention in making information available to farmers through research and diffusion, as well as in developing and enforcing a system of property rights to correct the actual situation.

The role of the state and non-government organizations (NGOs) is further important for the creation and promotion of incentives in the use of biodiversity by small-holders and peasant farmers. The development of banks of germoplasm in the rural areas of the humid tropics is required to preserve the existence of ecological diverse material that constitutes the pool of natural resources needed for the sustainability of agriculture, both in developed and underdeveloped countries. Specificity in the use of biodiversity should also be investigated and promoted. This in turn will increase the knowledge of rural peoples on the importance and uses of biodiversity.

There is thus a need of effective institutions adapted to the conditions in rural areas in developing countries. This requires the participation of civil society in the determination of the political, economic, and social framework in which the markets for biodiversity will have to function.

6. Conclusion

The three cases described above show the importance of finding and creating adequate institutions and governance structures for protection of property rights and market participation of small farmers. As showed, this is true for marketable and non-marketable (or less tradable) goods. Adequate market arrangements and governance structures must enforce the protection of property rights al lower transaction costs than is presently the case.

The asymmetry of information embedded in present markets and governance structures make it impossible for smallholders to fully benefit from the values incorporated in their property rights. This is the case in the beef market in Uganda, where because of badly operating market information structures farmers sell their animals at much lower prices than they could get in the final market. It is also the case in the water example in the Peninsula of Santa Elena, where smallholders have sold their land a much lower price than they would be able to get if they were fully informed about the irrigation possibilities and development plans. In this case also high transaction costs to bring into operation the high-investment infrastructure explain the small fraction of irrigation-available land incorporated to production. If adequate institutional arrangements were made to promote access to the resource and adequately price the water much better performance would be achieved.

It is also certainly the case in the example of genetic diversity where farmers do allow people on their land for searching for genes because they are uninformed about the value of these genes for large multinational companies. Clear property rights assignment is necessary to make biodiversity exploitation sustainable and, most important, properly globalise agricultural development.

Therefore, we make a pledge for continuous research and development to find adequate institutions and governance structures so that market failures precluding farmers to participate in markets under better conditions are eliminated or importantly reduced.

References

Alchian, A.A., 1977. Economic Forces at Work. Indianapolis: Liberty Press.

Altieri, M. and C.I. Nicholls, 2001. Ecological Impacts of Modern Agriculture in the United States and Latin America. In: O.T. Solbrig, R. Paarlsberg and F. di Castri (eds.) Globalization and the Rural Environment. Massachusetts: Harvard University Press, David Rockefeller Center for Latin American Studies.

Bailey, D., C.B. Barret, P.D. Little and F. Chabari, 1999. Livestock markets and risk management among East African Pastoralists: A review and Research Agenda. GL-CRSP Pastoral Risk Management Project, Technical Report No.03/99. Utah State University, Logan. 46 pp.

Barzel, Y., 1989. Economic Analysis of Property Rights. Cambridge: Cambridge University Press.

Buchanan, J.M. and G. Tullock, 1962. The Calculus of Consent. Ann Harbor: University of Michigan Press.

Burton, L., 2000. Agrociencia y Tecnología. Madrid: Editorial Paraninfo.

Castillo, M.J., 2004. Land Privatization and Titling as a Strategy to Diminish Land Loss and Facilitate Access to Credit: the case of communal landowners in the Peninsula of Santa Elena, Ecuador. Master of Science Thesis: University of Florida, Department of Food and Resource Economics.

Challen, R., 2000. Institutions, Transaction Costs and Environmental Policy: institutional reform for water resources. Cheltenham: Edward Elgar.

Cieza de León, P., 1984. *La Crónica del Perú* (Introducción de Franklin Pease). Lima: Pontificia Universidad Católica del Perú.

Coase, R.H., 1937. The Nature of the Firm. Economica 4: 386-405.

Cornejo, C., D. Haman, R. Espinel and J. Jordan, 2006. Irrigation Potential of the Trasvase System, Santa Elena Peninsula, Guayas, Ecuador. Journal of Irrigation and Drainage Engineering. ASCE. September-October: 453-462.

Dos Santos, T., 1970. The Structure of Dependence. American Economic Review 60.

Eggerston, T., 1990. Economic Behavior and Institutions. Cambridge: Cambridge University Press.

Eicher, C.K. and J.M. Staatz, 1998. International Agricultural Development (3rd. Edition). Baltimore: Johns Hopkins University Press.

Enríquez, J., 2001. As the Future Catches You: how Genomics and Other Forces are Changing your Life, Work, Health, and Wealth. New York: Crown Publishing Corporation.

Espinel, R., 2002. Contribución al Desarrollo de la Peninsula de Santa Elena. ESPOL: Propuestas. Año V, No. 7, Noviembre: pp. 5-17.

Espinel, R., 2006. La Globalización y sus Efectos en la Agricultura: los pequeños y medianos productores y sus alternativas. Asociación Latinoamericana de Desarrollo Rural ALASRU. Mexico. No. 4: 265-280.

Fafchamps, M. and S. Gavin, 1996. The Spatial Integration of Livestock Markets in Niger, Journal of African Economics 5: 366-405.

Furubotn, E.G. and R. Richter, 2000. Institutions and Economic Theory: the contribution of New Institutional Economics. Michigan: The University of Michigan Press.

Giarraca, N., 2004. Introducción in Norma Giarraca and Bettina Levy (eds.) Ruralidades Latinoamericanas. Buenos Aires: CLACSO.

Hayami, Y. and V. Ruttan, 1971. Agricultural Development: An International Perspective. Baltimore: Johns Hopkins University Press.

Herrera, P., 2005. Institutional Economic Assessment of the Governance of Irrigated Agriculture: the case of the Peninsula of Santa Elena, Ecuador. Doctoral Dissertation. Belgium: Faculty of Bioscience Engineering, Ghent University.

Herrera, P., G. Van Huylenbroeck and R. Espinel, 2005. An application of the contingent valuation method to asses the efficiency of the institutional structure of irrigation property rights: the case of the Peninsula of Santa Elena. International Journal of Water Resource Development 20: 537-551.

Hildebrand, P., E. Bastidas, A. Anzules, J. Castro, E. Chica, C. Diaz, R. Espinel, M. Hartman, E. Jiménez, J. Peña, M. Quilambaqui, F. Royce, E. Stonerook and C. Zambrano, 2005. Sondeo de los Pequeños Productores en la Zona de Influencia de la Estación Experimental Tropical Pichilingue del INIAP. INIAP-UF-ESPOL. Gainesville: University of Florida.

Hoff, K., A. Braverman and J.E. Stiglitz, 1993. The Economics of Rural Organization: Theory, Practice and Policy. A World Bank Book. Washington, D.C.: Oxford University Press Inc.

Holloway, G., C. Nicholson, C. Delgado, S. Staal and S. Ehui, 2000. Agro-Industrialization Through Institutional Innovation: Transaction Costs, Cooperatives and Milk-Market Development in the East-African Highlands. Agricultural Economics 23: 279-288.

Ruben, R., A. Kuyvenhoven, and P. Hazell, 2003. Institutions, technologies and policies for sustainable intensification in less-favoured areas, In Staying Poor: Chronic Poverty and development Policy. Contributed Paper for International Conference on Staying Poor: Chronic Poverty and Development Policy IDPM Manchester, 7-9 April 2003

Kyeyamba, H., S. Speelman, G. Van Huylenbroeck and W. Verbeke, 2007. Raising offtake from cattle grazed on natural rangelands in Sub-Saharan Africa: A transaction costs approach. Working Paper.

Lewis, A., 1954. Economic Development with Unlimited Supplies of Labour. The Manchester School of Economics and Social Studies:139-191.

Mazoyer, M., 2004. Mundialización Liberal y Pobreza Campesina: qué alternativa? In Francois Houtart (eds.), Globalización, Agricultura y Pobreza. Quito: Ediciones Abya-Yala.

McMurtry, J., 1998. Unequal Freedoms: the Global Market as an Ethical System. Connecticut: Kumarian Press.

Mittenzwei, K., 2004. Rules and Equilibria: a formal conceptualization of institutions with an application to Norwegian agricultural policy making. In: G. VanHuylenbroeck, W. Verbeke and L. Lauwers (eds.) Role of Institutions in Rural Policies and Agricultural Markets. The Netherlands: Elsevier.

North, D.C., 1990. Institutions, Institutional Change, and Economic Performance. Cambridge: Cambridge University Press.

Ostrom, E., 1990. Governing the Commons: The Evolution of Institutions for Collective Action. Cambridge: Cambridge University Press.

Ostrom, E., 1992. Crafting Institutions for Self-governing Irrigation Systems. Institute for Contemporary Studies, San Francisco: ICS Press.

Parker, R., 2000. La Ciencia de las Plantas. España: Paraninfo.

Powers, L. and R. McSorley, 2001. Principios Ecológicos en Agricultura. España: Paraninfo.

Prebisch, R., 1981. Capitalismo Periférico: crisis y transformación. Mexico: Fondo de Cultura Económica.

Reardon, T., C.P. Timmer and J.A. Berdegué, 2003. The Rise of Supermarkets in Latin America and Asia: implications for international markets for fruits and vegetables. In: A. Regni and M. Gehlar (eds.) Global Markets for High-Value Products. USDA-ERS: Agricultural Information Bulletin.

Renos, V., E. Sadoulet and A. de Janvry, 2003. Measuring Transactions Costs from Observed Behavior: Market Choices in Peru. CUDARE Working Paper 962, Department of Agricultural & Resource Economics, University of California Berkeley, http://repositories.cdlib.org/are_ucb/962 (accessed March 2006).

Risler, J.F. and M.G. Mellor, 1996. The Ecological Risks of Engineered Crops. Boston: MIT Press.

Schultz, T.W., 1964. Transforming Traditional Agriculture. New Haven: Yale University Press.

Vera, R., 2001. Sustainable Agriculture in the Lowlands and Subtropics? Trends and Bioeconomic Opportunities and Constraints. In: O. Solbrig, R. Paarlsberg and F. di Castri (eds.) Globalization and the Rural Environment, Cambridge: David Rockefeller Center for Latin American Studies, Harvard University Press.

Vermillion, D., 1999. Property rights and collective action in the devolution of irrigation system management. in: Proceedings of the International Conference: Collective Action, Property Rights, and Devolution of Natural Resource Management. Philippines: 21-25 June 1999.

Vermillion, D., 2004. Collective action and property rights for sustainable development irrigation. Collective Action and Property Rights. Brief 6 of 16. Washington: International Food Policy Research Institute.

Williamson, O.E., 1979. Transaction-Cost Economics: The Governance of Contractual Relations. Journal of Law and Economics 22: 233-61.

Williamson, O.E., 1985. The Economic Institutions of Capitalism. New York: Free Press.

World Bank, 1993. World Development Report. Oxford University Press for World Bank, New York.

World Bank, 2002. World Development Report (2002). Building Institutions for Markets. Oxford University Press for World Bank, New York.

World Bank, 2003. World Development Report (2003). Sustainable development in a dynamic world. Transforming Institutions, growth and quality. Oxford University Press for World Bank, New York.

Endogenous agricultural structures and institutions: insights from transition countries

Johan F.M. Swinnen

Abstract

There is an extensive literature on the optimality of farms and agricultural institutions, which received renewed attention with the reforms and restructuring processes in transition countries. In this chapter I argue, first, that the farm structures that have emerged from the transition process are much more diverse than expected ex ante. Second, I argue that this diversity importantly reflects initial conditions and external constraints, but also policy differences. Third, I argue that these differences are not necessarily a transitional (temporary) phenomenon, but are likely to have long-lasting impacts on the agricultural structures, among others because institutional innovations are emerging to address the constraints and opportunities posed by the current structures, and are as such 'locking-in' the existing structures in a long-run institutional framework.

1. Introduction

There is an extensive literature on the optimality of farms and agricultural structures more generally. This literature got much renewed attention with the reforms and restructuring processes in transition countries.

The transition experience provides a unique natural experiment in this area. In this chapter, I review some of the findings from several studies who have analyzed the determinants and effects of agricultural restructuring across countries from Central Europe to East Asia (the ECA region). I present several conclusions. First, the farm structures that have emerged from the transition process are much more diverse than expected ex ante. Second, this diversity reflects initial conditions and external constraints, but also policy differences. Third, these differences are likely to have long-lasting impacts on the agricultural structures, because institutional innovations are emerging to address the constraints and opportunities posed by the current structures, and are as such 'locking-in' the existing structures in a long-run institutional framework.

The chapter is organized as follows. The first section gives a brief overview of the literature on the optimality of farm structures. The second section describes how farm structures have evolved and emerged in the transition countries. I then provide a series of hypotheses to explain these structural developments, drawing on theory and empirical studies. In the last part of the chapter I discuss the emergence of contracting and vertical coordination in transition countries and what these imply for the future structural developments.

2. A brief review on the optimality of farm structures

Most of the literature on farm efficiency finds that there are relatively few economies of scale in farm operations, albeit it with some important modifications. The main argument relates to relative imperfections in the labour markets versus the capital and product markets (Eswaran and Kotwal, 1985; Pollak, 1985). The essence of the argument goes as follows.

Farming is characterized by important supervision costs because in most circumstances farm workers' true efforts are not easily observable, due to the specific characteristics of agricultural production. Such imperfections imply that wage workers have limited incentives to exert effort, and either need to be supervised at a cost or be offered contracts that provide higher incentives, such as sharecropping.

Family members have higher incentives to provide effort than hired labour. They share in output risk and can be employed with no or less supervision costs. This is the main advantage of family farming over wage-labour based farming.

These advantages may be offset by disadvantages of family farms in accessing credit and other markets. It is well known that rural credit markets are notoriously imperfect and that especially poor and small farmers are constrained in formal credit markets. Larger and richer farms may have easier access to credit, either because their initial wealth is larger (for self-financing) or because their transaction costs in credit markets are lower. Another reason is access to output or input markets. Small farmers in remote areas may not be able to sell their products to urban markets, or they may get lower prices from traders. Small farmers may be less likely to access (quality) inputs for their production. Hence, in such cases, imperfections of the input, output, and credit markets have the opposite effect of labour market imperfections in determining the optimal farm size.

Therefore, the optimality is largely an empirical issue. Several studies find that there is an inverse U-function between size and efficiency (Feder, 1985). Efficiency grows with size for the smallest farms, but beyond a certain size, typically coinciding with larger family farms, there is a declining relation between size and efficiency. However, not surprisingly, these relative effects, and hence the 'optimum' depends on the nature of the farm activity (livestock, crops, ...), available technology, relative factor abundance, market imperfections, and existing regulations and institutions.

What do the transition experiences contribute to these insights ?

3. The development of farm structures in transition countries

Ex ante, two sets of arguments were forwarded on the development of farm structures in transition countries after land reform and liberalization. The first argument posited that farm workers in transition countries had little human capital, managerial expertise and entrepreneurship for managing a farm that they would not start farming on their own. This argument predicted that the large farms would continue to operate, albeit in an adjusted mode reflecting different environments.

The second argument predicted the opposite. It argued that, since communist-designed collective and state farms were very inefficient, liberalization and removal of state control would lead to their collapse, and therefore there would be a total shift to family farms (or 'individual farming').

When looking at *what happened in reality* it appears that none of these arguments was correct. Or both were, depending on the way one looks at it (Table 1). In fact, a large variety of farm structures has emerged in the transition world, incorporating both extremes and everything in between (Rozelle and Swinnen, 2004; Lerman *et al.*, 2004). In most transition countries, farms vary widely in size and organization, from small household plots, to family farms, and to large co-operatives or farming companies. One extreme is Slovakia where almost all land is used by large farming corporations. The other extreme is in countries such as Albania, Azerbaijan, China, etc. where almost all land is used by small individual farms. Most countries have a mix of large and small farms (Figure 1).

Some countries have *large differentiations by region or commodity*. For example, in Kazakhstan the northern regions are dominated by vast grain producing farming corporations, sometimes using hundreds of thousands of hectares, while the cotton areas in the south are dominated by very small household farms. In Russia, around 60% of agricultural output is produced by household plots. However, the vast majority of the land is used by large farms, sometimes organized in huge agro-holdings (Figure 2). One example of such agro-holding is the Orel Niva holding, which controls 337,000 hectares of land and employs 16,000 workers. It processes 200-300,000 tons of wheat. Its activities include 102 large farms, 28 processing plants, 100 trade organizations, 32 service enterprises, etc. (Gataulina *et al.*, 2006).

Another remarkable variation between countries is in the *labour shedding on the farms*. In some countries, farms absorbed labour, while in other countries farms massively shed labour, even in the early years of transition (Figure 3). We found that these labour adjustments where strongly correlated with the farm restructuring process (Swinnen *et al.*, 2005). Individual farms often absorbed labour, while large farms, where they survived and faced hard budget constraints, such as in Central Europe, laid off surplus labour. In several cases, such as in Slovakia, Czech Republic, Hungary and Estonia, this contributed to a massive outflow of labour from agriculture. In other cases, it contributed to an inflow of labour, such as in Central Asia and Romania.

Table 1 Restructuring of farming organization and general reform indicators.

	Individual land use			Individual production		Agr. reform
	Pre-reform	After 5 yrs	After 10 yrs [a]	Pre-reform	After 7 yrs	After 10 yrs
East Asia						
China	5-10	98	99	n.a.	n.a.	n.a.
Vietnam	5	99	99	n.a.	n.a.	n.a.
Laos	54	99	99	n.a.	n.a.	n.a.
Myanmar	99	99	99	n.a.	n.a.	n.a.
Central Asia						
Mongolia	n.a.	n.a.	n.a.	n.a.	n.a.	n.a.
Kazakhstan	0	5	24	28	38	5.6
Kyrgyzstan	4	34	37	34	59	6.4
Tajikistan	4	5	9	23	39	4.2
Turkmenistan	2	3	8	16	30	2.0
Uzbekistan	5	13	14	28	52	2.0
Transcaucasus						
Armenia	7	95	90	35	98	7.2
Azerbaijan	2	5	n.a.	35	63	6.2
Georgia	12	50	44	48	76	6.0
European CIS						
Belarus	7	16	12	25	45	1.8
Moldova	7	12	20	18	51	6.0
Russia	2	8	13	24	55	5.6
Ukraine	6	10	17	27	53	4.0
Baltics						
Estonia	4	41	63	n.a.	n.a.	8.4
Latvia	4	81	87	n.a.	n.a.	8.4
Lithuania	9	64	85	n.a.	n.a.	7.6
Central Europe						
Czech Rep.	1	19	26	n.a.	n.a.	8.6
Hungary	13	22	54	n.a.	n.a.	8.8
Poland	76	80	84	n.a.	n.a.	7.8
Slovakia	2	5	9	n.a.	n.a.	7.6
Balkans						
Albania	3	95	n.a.	n.a.	n.a.	6.8
Bulgaria	14	44	56	n.a.	n.a.	7.6
Romania	14	71	82	n.a.	n.a.	6.6
Slovenia	83	90	94	n.a.	n.a.	8.0

Source: Rozelle and Swinnen, 2004.

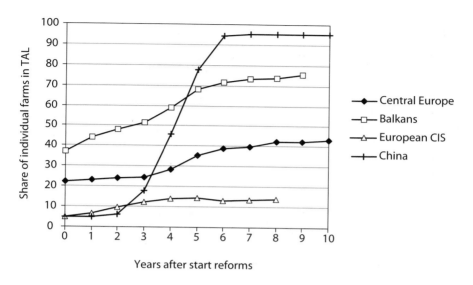

Figure 1. Share of agricultural land used by individual farms (%) (Rozelle and Swinnen, 2004).

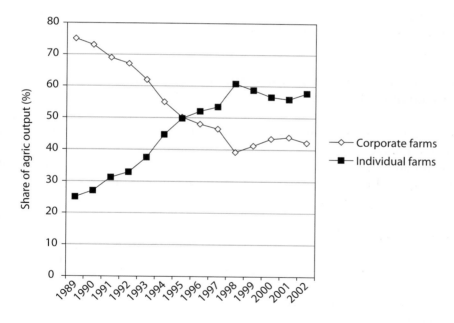

Figure 2. Agricultural output by farm organization in Russia, 1989-2002 (%) (Rozelle and Swinnen, 2004).

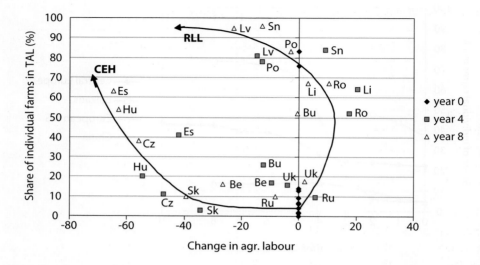

Figure 3. Labour adjustment and farm restructuring in transition (Swinnen et al., 2005).
Notes:
- CEH is average of Czech, Estonia, Hungary, RLL is average of Romania, Latvia, Lithuania
- 8 years after reforms is 1997 for CEEC, except Slovenia (1996), and 1998 for FSU

4. Technology, endogenous farm restructuring, and the nature of productivity gains

Looking across transition countries from Central Europe to East Asia, indicates a link between technology, policy and performance (Swinnen and Rozelle, 2006). Although gains in productivity have come both from property rights reforms and organizational restructuring, the relative importance of each component differs between countries reflecting technology and policy differences (Macours and Swinnen, 2002). In countries with -intensive technologies the shift from large-scale collective farming to small-scale individual farming caused dramatic gains in technical efficiency with relatively small losses in scale efficiency. In capital and land intensive regions, gains in productivity, if any, came primarily from large farms shedding with privatization of the farms.

These different sources of productivity gains are not coincidental. Technology has an important impact on the relative efficiency of different farm organizations, and thus on the incentives for farm restructuring. Technology affects both the costs and benefits of the shift to individual farming, as summarized in Figure 4. An important factor in the optimal scale of farming is transaction costs in management. Large operations in agriculture face transaction costs because of principal agent problems and monitoring costs in contracting, which are typically large in agriculture (Pollak, 1985). Hence, individual farming will improve effort

Table 2. Selected initial condition indicators for transition countries[1].

	Share of agr. in empl. (%)	GNP/capita (PPP$ 1989)	/land (pers./ha)	Agr. land in indiv. farms (%)	CMEA export (% of GDP)	Years central planning
East Asia						
China	69.8	800	0.672	5-10	0.01	42
Viet Nam	70.2	1100	2.298	5	0.05	21
Laos	n.a.	n.a.	n.a.	54	n.a.	16
Myanmar	66.2	n.a.	0.970	99 [a]	n.a.	38
Central Asia						
Mongolia	32.7	2100	0.002	0	0.17	n.a.
Kazakhstan	22.6	5130	0.008	0	0.18	71
Kyrgyzstan	32.6	3180	0.054	4	0.21	71
Tajikistan	43.0	3010	0.185	4	0.22	71
Turkmenistan	41.8	4230	0.015	2	0.34	71
Uzbekistan	39.2	2740	0.109	5	0.24	71
Transcaucasus						
Armenia	17.4	5530	0.218	7	0.21	71
Azerbaijan	30.7	4620	0.203	2	0.33	70
Georgia	25.2	5590	0.217	12	0.19	70
European CIS						
Belarus	19.1	7010	0.105	7	0.45	72
Moldova	32.5	4670	0.269	7	0.25	51
Russia	12.9	7720	0.044	2	0.13	74
Ukraine	19.5	5680	0.118	6	0.25	74
Baltics						
Estonia	12.0	8900	0.072	4	0.27	51
Latvia	15.5	8590	0.085	4	0.31	51
Lithuania	18.6	6430	0.098	9	0.34	51
Central Europe						
Czech Rep.	9.9	8600	0.122	1	0.10	42
Hungary	17.9	6810	0.131	13	0.10	42
Poland	26.4	5150	0.258	76	0.17	41
Slovakia	12.2	7600	0.139	2	0.10	42

Source: Macours and Swinnen, 2002.

Table 2. Continued.

	Share of agr. in empl. (%)	GNP/capita (PPP$ 1989)	/land (pers./ha)	Agr. land in indiv. farms (%)	CMEA export (% of GDP)	Years central planning
Balkans						
Albania	49.4	1400	0.627	3	0.02	47
Bulgaria	18.1	5000	0.132	14	0.15	43
Romania	28.2	3470	0.204	14	0.03	42
Slovenia	11.8	9200	0.116	83	0.07	46

[1] Pre-reform indicators are from 1978 for China, 1981 for Viet Nam, 1986 for Laos, 1989 for the CEECs and Myanmar and 1990 for the FSU and Mongolia.

[a] Own estimation

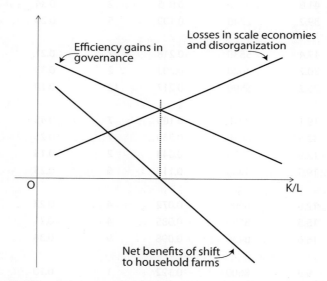

Figure 4. Costs and benefits of the shift to individual farming (Swinnen and Rozelle, 2006).

and a farmer's control over farm activities and this will lead to efficiency gains. However, the importance of these efficiency gains vary with specialization and technology (Allen and Lueck, 1998). Since the greatest improvement in efficiency from farm individualization is attributable to rising effort from better incentives, the benefits will be relatively greater for systems in which incentives plays a greater role.

However, there are also costs that are incurred when collective or corporate farms are broken up into individual farms. In many cases there are two major types of costs. First, there is one set of costs that could arise due to the loss in scale economies. As in the case of the incentive effects, the impact on scale economies will be sensitive to the nature of the technology. The economy of scale losses may be considerable in the case of capital intensive production systems, systems in which we would expect economies of scale to be relatively significant since there are many fixed expenses and many large assets used in farming activities. In countries in which farming is intensive and few capital inputs are used, however, such losses could be minimal.

Second, there also may be costs associated with disorganization that will occur with the restructuring of farms. The costs will arise from the mismatch that can occur between the farm's needs for inputs, services and equipment and the infrastructure that has been set up to provide those inputs and services. Initially designed for large scale farming, the inputs and services that the nation's agricultural input supply chain are set up to provide are not always suitable for individual farms. Hence, newly formed individual farms may require an entirely different set of inputs, services and equipment. The disorganization and economies of scale costs could be high (initially) if such inputs, services, and equipment play an important role in the local farming systems. Again, this is affected by technology. These disruption costs are more likely to be lower in intensive systems than in more advanced, integrated and capital-intensive agricultural systems.

The importance of technology (resource intensity) in the growth of individual farming is illustrated empirically by Figure 5, which shows a strong positive relationship between the pre-reform intensity of farming and the importance of individual farming five years after the start of transition. As such, the farm restructuring process, in particular the growth of individual farming, is at least partially endogenous in this transition process.

In countries with -intensive technologies there is a strong shift from large-scale collective farming to small-scale individual farming and with it strong gains in technical efficiency with relatively small losses in scale efficiency, as we documented above. For example, in countries such as China, Vietnam, Albania, Armenia, Georgia, and Romania, significant gains in productivity came mostly from the shift to household farming when land was distributed to rural households. In all these countries the man/land ratio was over 0.2 persons per hectare and total factor productivity increased strongly during early transition (between 4 per cent and 9 per cent annually) when individual farming grew from 8 per cent of total land use on average to 84 per cent on average (Rozelle and Swinnen, 2004).[55]

---

[55] A further argument that can be made on this, pushing the endogeneity argument even further, is that in labour intensive economies, ultimately a land reform process emerged that was conducive to farm individualization. This land reform procedure, i.e. distribution of land in kind to households, often came about only after changes were made to the existing policies, such as in Azerbaijan, Kyrgyz Republic, Moldova, etc., reflecting changes in governments and political economy pressures (Swinnen and Rozelle, 2006).

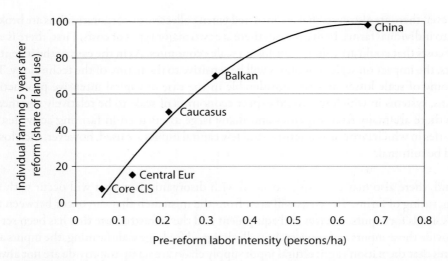

Figure 5. Pre-reform technology and the growth of individual farming (Swinnen and Rozelle, 2006).

In contrast, in capital and land intensive regions, large-scale corporate farming remained important and productivity gains came primarily from large farms shedding with privatization of the farms. For example in the Czech Republic, Slovakia and Hungary, countries in which farming was more capital and land intensive (man/land ratio of 0.14 or less), gains in productivity came primarily from large farms shedding with privatization of the farms. During the first five years of transition, use declined by 44 per cent on average in these three countries (Figure 3), yielding an annual increase in productivity of 7.5 per cent on average, while individual farms used only 15 per cent of the land.

5. Land reform and farm fragmentation and concentration

Different land reform resulted in different structures. The most important land reform choices were: restitution, distribution in kind (in actual plots), distribution of land shares, and a combination (sequence), e.g. first distribution in shares, then in kind. These differences have important implications for restructuring.

The dominant land reform procedure in Central Europe, the Balkans, and the Baltic countries was restitution of land to the former owners that had lost their rights during the collectivization movement in the past. If the original owners were not alive, reformers restored ownership rights to their closest heirs. Typically land reform laws restituted land to the historical boundaries. If restitution to the original boundaries was not possible, former owners received rights to a plot of land of comparable size and quality.[56] Distribution of land has been done by allocating physical plots (such as in Albania) or in shares (such as in Russia and Kazakhstan) or first in shares and later in physical plots (such as in Azerbaijan).

There is a complex impact of the land reforms on farm restructuring. Here we limit the discussion to a few important effects. I first argue that land restitution, such as implemented in most of Central and Eastern Europe contributed to the consolidation of large farms, instead of to its fragmentation – as was mostly argued. Second, I argue that the distribution of land in-kind contributed to individualization, and third, that shares distribution has contributed to farm consolidation and, more recently, to a concentration of land ownership, and thus inequality.

5.1. Land distribution, market constraints and farm fragmentation

Fragmentation of land due to the restitution process is often cited as a constraint on farming. However, evidence suggests that labour market constraints may be a more fundamental cause of fragmentation.

First, there is *no evidence that fragmentation of ownership causes fragmentation of land use.* In fact, quite the opposite. Fragmentation of ownership is very strong in Central Europe (eg Slovakia, Czech Republic and Hungary) while this has not led to a fragmentation of land use, quite the contrary – farm land is very consolidated through rental agreements.

Mathijs and Swinnen (1998) explain how the nature of transaction costs in land markets actually led to a consolidation of land. Restitution in many countries gave land back to individuals that were no longer active in agriculture, most commonly to either former farmers or their heirs. Except for the case in some of the poorer countries, the new landowners did not return to farming and primarily were interested in renting their land. Because the search and negotiations costs of identifying individuals that were willing to rent the land were so high, the easiest way for the new land owners to find a renter was to contact those that were already using the land. Consequently, in most cases the new leasees became those that had been involved with farming on the large pre-reform farms.

Transaction costs also favoured the large farmers from the point of view of their search for land to rent. Almost all of those that farmed after reform were those that were active in agriculture prior to reform. Most were farm workers or cooperative members. Since land was restituted to people outside agriculture, if they wanted to stay in farming, they were forced to search for the owners of the land and strike a rental contract. However, since the

---

[56] In several countries restitution was combined with other land reform programs, for example, voucher privatization (Hungary), distribution of state land (Romania) or the leasing of state-owned land (Czech Republic).

management of the large farms was closely involved in the restitution process, they had an information advantage in identifying the new owners. Transaction costs on both the supply and demand side gave an advantage to large farms. As a result, after restitution, farm size did not fragment as much as had been feared. Although a small farming class did emerge everywhere, many large farms did not disappear and the agricultural sector in several CEE countries remained characterized by a dual farm structure.

Second, a closer look at the fragmentation of land use across Eastern Europe suggests that *fragmentation has less to do with the land market than with the market.* Figure 6 illustrates that land use fragmentation is strongly correlated with the employment structure of the economy. In the mid 1990s there was an almost perfectly linear relationship between the share of land used by very small farms and employed in agriculture. Land use fragmentation was a problem mostly in countries where too many rural households had to rely on agriculture.

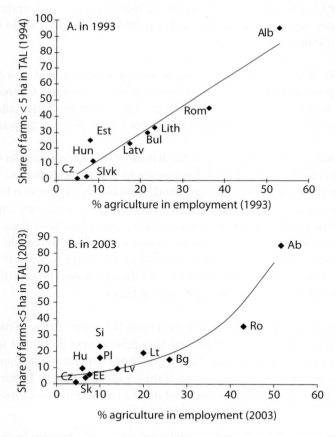

Figure 6. Farm fragmentation and the share of agriculture in employment. A. in 1993 (Mathijs and Swinnen, 1998); B. in 2003 (Swinnen and Vranken, 2005).

In-country survey data also confirm that within countries fragmentation of land and small plots are essentially associated with old, often retired, and part-time farmers. For example in Hungary, both larger family farms and large corporate farms in Hungary use large and consolidated land plots. Commercial farms rent a large share of the land they operate (Vranken and Swinnen, 2006). Hence, this evidence is consistent with the earlier conclusion that if there is a fragmentation problem, it is primarily caused by market constraints.

In several countries (e.g. Moldova, Bulgaria, Albania) rural households have tried to cope with labour market constraints by migrating to urban areas or to other countries (Macours and Swinnen, 2005). Migration and the associated remittances have contributed to the growth of farming, and to a lesser extent of rental markets, by allowing households to obtain a more productive labour/land ratio, by reducing credit constraints, and by stimulating the supply of land in labour intensive agricultural systems (Swinnen and Vranken, 2005).

5.2. Distribution in kind and farm individualization

The distribution of land in specific plots (boundaries) created much stronger property rights for the new owners than with share distribution. The distribution of shares has often implied uncertain property rights and high transaction costs.[57] The stronger rights (with distribution in plots) has caused a stronger growth of family farms as it was easier for these new farms to access their land. For example, the distribution of land in kind lead to the rapid growth of family farming in China in the late 1970s, in Vietnam in the 1980s, in Albania in the early 1990s, in Azerbaijan, Kyrgyz Republic, and Moldova in the second half of the 1990s. In all countries, within a few years after the start of the land reform, a large share of agricultural land shifted to family farms. In contrast, where land shares were distributed (e.g. Russia, Kazakhstan and pre-2000 Ukraine), the shift of land use to family farms was much less.

5.3. Land shares and the growth of mega-farms

Another effect of the share distribution is that it has allowed a concentration of land ownership, much more so than the other types of land reform. Shares were exchanged without being linked to specific plots. In several cases, workers transferred their land shares to the corporate farm, for example in exchange for employment. When farms were sold, often after bankruptcy, the land shares were now part of the farm assets, and investors who took over the farm became land owners. This process led to the concentration of land ownership, e.g. in parts of Russia and Kazakhstan, with vertically integrated companies

---

[57] Individuals usually had to declare their intention to start up their own farm in order to take physical possession of their land. However, the barriers to exit were severe as leaving the farm was often discouraged by farm managers and local officials. In addition, in several countries, the share distribution system was accompanied by continued soft budget constraints for the large farms (eg in Ukraine, Russia and Kazakhstan), further reducing incentives for restructuring of the farms.

owning now hundreds of thousands of hectares of land. In contrast, land distribution in plots and restitution[58] has led to relatively egalitarian land ownership distributions.

6. Summary: why is there such a difference in farm structures between transition countries?

Differences in farm structures are due partly to different reform choices and partly due to exogenous factors, in particular relative factor endowments, technology, scale economies, market imperfections and existing institutions.

First, relative factor endowments are important. These differ enormously across the ECA region. The pre-reform land/labour ratio in, for example, Russia and Kazakhstan is many times higher than in, for example, Albania, Azerbaijan, Moldova or Romania. In relatively labour intensive agricultural systems, the benefits of shifting to family farms (from corporate farms) are larger while the disruption costs are lower. That is why we observe a strong correlation between factor intensities and the growth of family farming: corporate farms remain much more important in land and capital intensive farming.

Second, scale economies matter but vary by commodity. For example, grain production tends to have more economies of scale than, e.g. dairy or vegetable production. Therefore, within a country one may observe strong differences in farm organizations. The most extreme example is Kazakhstan where the northern grain belt is dominated by huge farms, while in the southern part of the country one finds much more small farms, e.g. in cotton production.

Third, imperfections in output and input markets and existing institutions are particularly important in ECA countries, where there are substantial market imperfections and where traditional institutions, for e.g. product marketing and input supplies, have been designed to serve large scale farms. In the absence of such institutions for small scale farms, it is not surprisingly that large scale farms have remained more prominent in ECA than would have been predicted based on models from outside ECA where institutions were much more targeted to smaller farms. In fact, in a survey of a series of studies on the relative efficiency of large corporate farms and smaller family farms, Gorton and Davidova (2004) conclude that there is no clear evidence of corporate farms being inherently less efficient than family farms and that even when family farms are on average more efficient, some corporate farms also perform as well as the best family farms. In countries with a more supportive institutional

---

[58] The ownership distribution following the restitution process depends on the pre-collectivization ownership distribution. This distribution was relatively egalitarian as it was typically preceded by a Communist-imposed land reform which distributed land from large landowners and institutions (such as the Church) to landless peasants and small farmers. The main exception is Albania where the pre-collectivization was very inegalitarian (feudal). This was one of the reasons why restitution was heavily opposed in Albania and the government distributed the land equally to rural households (Swinnen *et al.*, 1997).

environment for small-scale farming, the family farms are more efficient relative to large corporate farms as in countries where small family farms are a relatively new phenomena.

Furthermore, non-traditional farm structures have turned out to be well fitted for this non-traditional farm environment. For example, in Romania the most efficient farm organization for resource-constrained small farmers are 'family societies' in which farmers collectively share in the provision of mechanized services (Sabates-Wheeler, 2002). In East Germany, 'partnerships' (small groups of farmers in that pooled their effort in certain production and marketing tasks) outperformed all other forms of farm organization between 1992 and 1997 (Mathijs and Swinnen, 2001). In Russia the most successful household farms refrain from registering as 'private farms,' instead choosing to remain connected in some fashion to large farm enterprises. Such producers use their connections to gain access to inputs, marketing channels and other services in an environment where traditional markets, if any, function poorly (O'Brien and Wegren, 2002). Even in Turkmenistan, producers have begun to shift to family-based leasing within the nation's highly regulated environment in order to be able to access basic inputs, services and output channels through the state marketing order system (Lerman and Brooks, 2001).

The most extreme version of large corporate farms are in the grain producing regions of Kazakhstan and Russia. There, huge farms have developed as part of vertically integrated agribusiness companies, sometimes owning and operating more than 100,000 hectares have emerged. Scale economies are more important in extensive grain production than in, e.g. vegetable or dairy production. However, the main reasons appear to be transition-specific (Gray, 2000):

- *Access to inputs*: in a very tight capital market these companies control a large part of the liquid financial resources in the regions concerned making it possible for them to farm when many other farm types are not assured access to inputs. They have access to bank lending, apart from their own liquid resources, on the basis of non-agricultural assets with higher collateral value. Their vertical ownership in the grain market allows them to purchase inputs at the source (e.g., the refinery for fuel) and to avoid barter terms.
- *Access to output markets*: in northern Kazakhstan, land is not the most critical input in the farming process. It is not surprising that the organization of farming in the north is evolving in a way in which land ownership is almost irrelevant. The new successful farms comprise a set of property and contractual relationships in which land ownership is a peripheral issue. The greatest difference between the large-scale investor-led farms and smaller individual farms and partnership farms lies in the difficulties experienced by the smaller farms in marketing their output.
- *Bargaining power with* the (local) *authorities*, who still intervene in many ways. The oblast authorities continue to play a highly interventionist role in agricultural commodity markets, in spite of the greatly reduced role of the state in procurement. In practice such interventions are open to abuse, with favoured (large) operators allowed to export grain to neighboring oblasts or to Russia while smaller producers are prohibited, often until all

outstanding debts for inputs are paid. Moreover, there was a widespread practice until the mid 1990s whereby local authorities continued to require farms, even after they became producer cooperatives, to engage in production activities which were well-known to be loss-making, especially livestock production. The continued dominance of the seed industry based on state farm production (in the grain sector) tends to perpetuate the single channel system and places farms under the control of the local authorities which continue to determine the physical flows of seed grains (especially when it crosses oblast boundaries). Most farms continue to depend on the local authorities to supply key inputs and for finance for these inputs through the issue of local authority guarantees for the provision of seed and fuel by supplies on a barter basis against the season's production. The increased size and financial wealth of the large, integrated, grain companies protects them against these state interventions.

In summary, corporate farms and 'non-traditional' large farming organizations are more likely to be (relatively) efficient in the specific institutional environment and structural conditions of transition. However, the extensive use of land by corporate farms in several transition countries is also influenced by significant transaction costs in the land market, monopoly power in the regional land markets, and property rights imperfections.

7. Contracting and vertical coordination with market imperfections

Farm contracting and vertical coordination in supply chains has grown strongly since the mid 1990s in transition countries. At the end of the 1990s, in the Czech Republic, Slovakia and Hungary, 80% of the corporate farms, who dominated farm production in these countries, sold crops on contract, and 60-85% sold animal products on contract; numbers which are considerably higher than on commercial farms in the US or the EU. A survey by White and Gorton (2006) of agri-food processors in five CIS countries (Armenia, Georgia, Moldova, Ukraine and Russia) found that food companies which used contracts with suppliers grew from one-third in 1997 to almost three-quarters by 2003.

Part of this vertical coordination includes the provision of farm assistance programmes to the farms. These farm assistance programmes include a variety of measures, such as credit, prompt payments, transportation, physical inputs, and quality control. However also investment loans and bank loan guarantees are provided in several cases.

A recent large study by the World Bank comes to the following conclusions (Swinnen, 2006; World Bank, 2005):
- In the *dairy* sector, extensive production contracts have developed between dairy processors and farms, including the provision of credit, investment loans, animal feed, extension services, bank loan guarantees, etc.
- In the *sugar* sector, marketing agreements are widespread, but also more more extensive contracts, including also input provisions, investment loan assistance, etc.

- In both the dairy and sugar sectors, the extent of supplier assistance by processors also goes considerably beyond some of the trade credit and input assistance provided by agribusiness to farms in some developing countries.
- In *cotton*, cotton gins typically contract farms to supply seed cotton and provides them with a variety of inputs. This model, which is common in Central Asia, resembles that of the gin supply chain structure in developing countries, such as in Africa. However, the extent of contracting and supplier assistance seems to be more extensive in Central Asia, with credit, seeds, irrigation, fertilizer, etc. being provided by the gins.
- In *fresh fruits and vegetables*, the rapid growth of modern retail chains with high demands on quality and timeliness of delivery is changing the supply chains. New supplier contracting, which is developing rapidly as part of these retail investments, include farm assistance programs, which are more extensive than typically observed in Western markets. They resemble those in emerging economies, but appear more complex in several cases.
- In *grains* there is extensive and full vertical integration in Russia and Kazakhstan, where large agro-holdings and grain trading companies own several large grain farms in some of the best grain producing regions.

7.1. The problem of exclusion

A key concern is that the process of vertical coordination will exclude a large share of farms, and in particular small farmers. There are three important reasons for this. First, transaction costs favour larger farms in supply chains, since it is easier for companies to contract with a few large farms than with many small ones. Second, when some amount of investment is needed in order to contract with or supply to the company, small farms are often more constrained in their financial means for making necessary investments. Third, small farms typically require more assistance from the company per unit of output.

The concern of the exclusion of small farmers is voiced often and raised in many studies, in particular also in the emerging literature on the impact of the growth of modern supply chains, which emphasize the shift to larger preferred suppliers and the exclusion of small farms.

However, what *empirical evidence* do we have on the exclusion of small farms from vertically integrating supply chains ? We looked at empirical evidence from transition countries and from emerging and developing countries. Interestingly, the two sets of empirical evidence show a largely consistent picture. Our studies and interviews with companies generally confirm the main hypotheses coming out of global observations:
1. transaction costs and investment constraints are a serious consideration;
2. companies express a preference for working with relatively fewer, larger, and modern suppliers;
3. but empirical observations show a very mixed picture of actual contracting, with much more small farms being contracted than predicted based on the arguments above.

Johan F.M. Swinnen

In fact, surveys in Poland, Romania and CIS find no evidence that small farmers have been excluded over the past six years in developing supply chains (Dries and Swinnen, 2004; Van Berkum, 2006). In the CIS, the vast majority of companies have the same or more small suppliers in 2003 than in 1997 (White and Gorton, 2006). In terms of the supplier assistance, better and more assistance seems to go to larger farms, although there is significant variation with the type of assistance.

While vertical coordination does not seem to exclude small farmers from the supply chains and all major companies contract with small farmers, more sophisticated supplier assistance programs tend to be more available for larger farms. Often, supplier programs differ to address the characteristics of these varying farms. For example, in case studies of dairy processors investment support for larger farms include leasing arrangements for on-farm equipment, while assistance programs for smaller dairy farms include investments in collection units with micro-refrigeration units.

Hence, despite the apparent disadvantages noted earlier, the empirical evidence suggests that vertical coordination with small farmers is widespread. Furthermore, our empirical evidence indicates that companies in reality work with surprisingly large numbers of suppliers and of surprisingly small size.

7.2. Why contracting with small farmers?

There are several reasons. First, the most straightforward reason is that companies sometimes have no choice. In some cases, small farmers represent the vast majority of the potential supply base. This is, for example, the case in the dairy sector in Poland and Romania, where the vast majority of farms only have a few cows, and in many other sectors in transition countries.

Second, company preferences for contracting with large farms are not as obvious as one may think. While processors may prefer to deal with large farms because of lower transaction costs in e.g. collection and administration, contract enforcement may be more problematic, and hence costly, with larger farms. Processors repeatedly emphasized that farms' 'willingness to learn, take on board advise, and a professional attitude were more important than size in establishing fruitful farm-processor relationships'.

Third, in some cases small farms may have substantive cost advantages. This is particularly the case in labour intensive, high maintenance, production activities with relatively small economies of scale.

Fourth, processors may prefer a mix of suppliers in order not to become too dependent on a few large suppliers.

Finally, processing companies also differ in their willingness to work with small farms. Some processing companies continue to work with small local suppliers even when others do not. These companies have been able to design and enforce contracts which both the small firms and the companies find beneficial. This suggests that small-scale farmers may have future perspectives when effectively organised.

8. Endogenous contracting and 'the farm assistance paradox'

The evidence presented here suggests an interesting paradox. With the demand of modern supply chains, small farmers in ECA may not be able to make the necessary upgrades by themselves without support packages by processors or agribusiness. If there are sufficient (quality) supplies available for processors from large farms, processors will not be willing to introduce such support packages. If there are not sufficient supplies, support will be forthcoming. Hence, we have the paradoxical situation that small poor farms may be best off (in the perspective of 'supply chain driven development') if they are in an environment which is dominated by small poor farms.

There is some empirical evidence for this hypothesis. Companies seem to be most likely to reach out to small farms when they face a supplier base which is dominated by small farmers not able to supply the commodities they want, and least likely when there is a heterogeneous farm structure with some farms able to deliver the desired supplies. For example, some international dairy companies and foreign investors target larger farms as their preferred suppliers and only reach out to smaller suppliers if they need them to secure supplies.

Yet, 'large' is a relative concept: in countries like Poland, Romania, and many CIS countries dairy production is dominated by household production. As a consequence, there is evidence that in Hungary processors only work with farms of over 20 cows, in Poland with farms who have more than 5 cows, and in Romania with farms with 1 cow. Or, as Van Berkum (2006) puts it: 'In Romania, large farms are farms with more than five cows'.

These developments have major implications for the development of agricultural structures in these countries. As private-sector-driven institutions develop to address these different supplier bases, these institutions will in the longer run have an important impact on the resulting and evolving agricultural structures, with the initial structure having an important impact on the one evolving in the medium term. Hence, the existing differences are not necessarily a transitional (temporary) phenomenon, but are likely to have long-lasting impacts on the agricultural structures, because institutional innovations which are emerging to address the constraints and opportunities posed by the current structures, are 'locking-in' the existing structures in a long-run institutional framework.

That said, further restructuring and at least some consolidation is expected to be essential everywhere to sustain such agricultural systems where institutions and farm structures are mutually endogenous.

References

Allen, D.W. and D. Lueck, 1998. The Nature of the Farm. Journal of Law & Economics, 41: 343-86.

Dries, L. and J. Swinnen, 2004. Foreign Direct Investment, Vertical Integration and Local Suppliers: Evidence from the Polish Dairy Sector. World Development 32:1525-1544.

Eswaran, M. and A. Kotwal, 1985. A Theory of Contractual Structure in Agriculture. American Economic Review 75: 352-367.

Feder, 1985. The Relation between farm size and productivity: the role of family labour, supervision, and credit constraints. Journal of Development Economics (18) 2-3: 297-313.

Gataulina, E.A., V.Y. Uzun, A.V. Petrikov and R.G. Yanbykh, 2006. Integration Processes in Agriculture-Industrial Complex: Agro-Firms and Agro-Holdings in Russia. In: J. Swinnen (ed.) Case Studies on the Dynamics of Vertical Coordination in the Agri-food Supply Chains of Eastern Europe and Central Asia. World Bank Publications.

Gorton, M. and S. Davidova, 2004. Farm productivity and efficiency in the CEE applicant countries: a synthesis of results. Agricultural Economics 30: 1–16.

Gray, J., 2000. Kazakhstan: a review of farm restructuring. World Back Technical Paper No 457, Washington D.C. The World Bank.

Lerman Z. and K. Brooks, 2001. Turkmenistan: An assessment of leasehold-based farm restructuring, World Bank Technical Paper No. 500, The World Bank, Washington DC.

Lerman, Z., C. Csaki and G. Feder, 2004. Agriculture in Transition: Land Policies and Evolving Farm Structures in Post-Soviet Countries. Lanham MD, Lexington Books.

Macours, K. and J. Swinnen, 2002. Patterns of Agrarian Transition. Economic Development and Cultural Change 50: 365-395.

Macours, K. and J. Swinnen, 2005. Agricultural Labour Adjustments in Transition Countries: The Role of Migration and Impact on Poverty. Review of Agricultural Economics.

Mathijs, E. and J. Swinnen, 1998. The Economics of Agricultural Decollectivization in East Central Europe and the Former Soviet Union. Economic Development and Cultural Change, 47:1-26.

Mathijs, E; and J. Swinnen, 2001. Production Organization and Efficiency during Transition: An Empirical Analysis of East German Agriculture. The Review of Economics and Statistics 83: 100-107.

O'Brien, D. and S. Wegren (eds.), 2002. Rural Reform in Post-Soviet Russia, Woodrow Wilson Center Press and Johns Hopkins University Press: Washington DC.

Pollak, R.A., 1985. A Transaction Cost Approach to Families and Households. Journal of Economic Literature 23: 581-608.

Rozelle, S. and J. Swinnen, 2004. Success and Failure of Reforms: Insights from Transition Agriculture. Journal of Economic Literature, XLII (June): 404-456.

Sabates-Wheeler, R., 2002. Farm Strategy, Self-Selection and Productivity: Can Small Farming Groups Offer Production Benefits to Farmers in Post-Socialist Romania?, World Development, 30: 1737-1753.

Swinnen, J. (ed.), 2006. Case Studies on the Dynamics of Vertical Coordination in the Agri-food Supply Chains of Eastern Europe and Central Asia. World Bank Publications.

Swinnen, J. and L. Vranken, 2005. The Development of Land Rental Markets in Transition. ECSSD, The World Bank.

Swinnen, J. and S. Rozelle, 2006. From Marx and Mao to the Market: The Economics and Politics of Agricultural Transition. Oxford University Press.

Swinnen, J., A. Buckwell and E. Mathijs (eds.), 1997. Agricultural Privatization, Land Reform and Farm Restructuring in Central and Eastern Europe. London: Ashgate Publ.

Swinnen, J., L. Dries and K. Macours, 2005. Transition and Agricultural. Agricultural Economics 32: 15-34.

Swinnen, J., L. Dries and K. Macours, 2005. Transition and Agricultural Labour, Agricultural Economics, 32(1): 15-34.

Van Berkum, S., 2006. Dynamics in vertical coordination in the Romanian dairy sector. In: J. Swinnen (ed.) Case Studies on the Dynamics of Vertical Coordination in the Agri-food Supply Chains of Eastern Europe and Central Asia, World Bank Publications.

Vranken, L. and J. Swinnen, 2006. Land rental markets in transition: Theory and evidence from Hungary, World Development 34(3): 481-500.

White, J. and M. Gorton, 2006. Vertical Coordination in CIS Agrifood Chains. In: J. Swinnen (ed.) Case Studies on the Dynamics of Vertical Coordination in the Agri-food Supply Chains of Eastern Europe and Central Asia. World Bank Publications.

World Bank, 2005. The Dynamics of Vertical Coordination in the Agri-food Supply Chains of Eastern Europe and Central Asia. ECSSD, World Bank Publications.

Experience and challenges in rural microfinance

Henk A.J. Moll

Abstract

The current discussion on microfinance deals predominantly with institutions operating in urban or densely populated rural areas. This chapter focuses on microfinance in rural areas that are agrarian based and less densely populated because access to microfinance services in these areas is still problematic. The position of microfinance as a useful service to rural households is outlined in the context of the monetisation of the rural economy and its gradual integration into national and international markets. The relative neglect of physical infrastructure in rural areas, generally low levels of education and limited health services as well as under-developed markets and a dependence on an unpredictable agriculture are reflected in low and irregular incomes for a large proportion of the population. The provision of financial services under these circumstances requires microfinance institutions to pay specific attention to the organisation structure and mode of operation at the retail unit level, and on the design of financial products that suit the regional circumstances. The chapter concludes with a discussion of the role microfinance can play in rural development in the long term and the incorporation of microfinance institutions into national and international financial systems.

1. Introduction

Microfinance has firm roots in agricultural credit and the 'Spring review of credit programs to small farmers' (Donald, 1976) can be seen as marking the beginning of the international discussion on small-scale finance and development. The discussion has broadened and developed substantially since Donald's review with important theoretical insights being contributed by Bell (1988); Hoff and Stiglitz (1993) and Krahnen and Schmidt (1994) amongst others. The body of knowledge and understanding on microfinance has also been strengthened by contributions from the worlds of finance, banking, management and political science. Case studies have complemented these theoretical insights by documenting practical experiences of working with microfinance in all parts of the world (Ledgerwood, 1999). 'Microfinance' has gradually became an umbrella term for a wide ranging discussion and - as the distinction between urban and the rural situations faded under the impact of generalisation - critical differences have been overlooked. As a result current discussions are based on experiences with microfinance in predominantly urban and densely populated rural areas.

This chapter deals with microfinance in less-favoured rural areas characterised by agrarian based economies and low population density. High transportation and transactions costs and co-variant risk are serious obstacles to establishing financial services in such areas. The purpose of this chapter is to provide an overview of experiences and challenges in an effort to expand the financial frontier of microfinance. It begins with a discussion of the position of microfinance in the process of rural transformation in developing countries. Conditions generally associated with the rural environment are analysed and their implications for the provision of financial services assessed. The abilities, limitations and desire of rural households to engage in relationships with microfinance institutions are examined in detail and - together with the arguments developed earlier in the chapter – this enables an identification of the specific challenges associated with establishing effective rural microfinance institutions. The chapter concludes by examining the future of microfinance in the long term. It stresses the need for microfinance institutions to shift their objectives and the factors that determine their possible incorporation into national financial markets.

2. The rural transformation

The position of microfinance in rural areas must be viewed in the context of the rural transformation taking place in many developing countries. At the start of this transformation process agriculture forms the mainstay of the rural economy. Agricultural production is a small-scale household activity and is primarily concerned with ensuring family food security. Surpluses are offered for sale on local, regional or national markets and the proceeds are used to meet the household's cash needs. Technologies used in agricultural production at this stage are based on the households' resources land, labour, livestock and home-made tools. Very few external services, inputs and purchased capital goods will be used.

In many developing countries rural economies have developed beyond the early situation sketched above and the process of transformation is well underway. New agricultural production technologies have been introduced, adapted and used by rural households. These new technologies lead to higher productivity of the households' resources land and labour, require more and more diversified inputs and services, and generally result in a larger quantity of marketed output. The Green Revolution accelerated the process of technological innovation and made the transformation clearly visible in many rural areas, especially in Asia.

Agriculture thus developed backward linkages with specialised services providers including extension and veterinary agents and suppliers of fertilizer and hybrid seed. As production increased forward linkages were also established with marketing, processing and transport agents. Two central aspects of this process have been the incorporation of small-scale agricultural producers into input and produce markets and the monetisation of agricultural production. A further significant development has been the expansion of non-agricultural enterprises into rural industry and service sectors. Initially these enterprises are often agriculture related, but as they develop they gain their own dynamic and their links with agriculture fade. As rural cash incomes grow, the scope for retail trade in food, clothing, building materials and other consumer goods increases.

The process of rural transformation outlined above is not driven by technical developments alone. There are significant linkages with the national context in which these developments take place. First, employment opportunities in the urban sector or expectations of employment generally lead to a rural exodus. It is usually the young, relatively better educated people who leave the village for the urban areas or for employment abroad. As a result the rural labour force is reduced with often adverse effects on agricultural as well as non-agricultural production. At the same time, however, successful migrants transfer money to their families in the rural areas when their expectations materialise. These remittances increase the purchasing power of the rural household and enable investment in enterprises (Wouterse, 2006). When migrants return home they introduce new ideas about consumption and production. However, transfers the other way round may also occur when employment in towns is interrupted, e.g. the closure of mines in South Africa, or when other calamities are encountered.

The development of demand for agricultural products in the urban areas is a second major linkage. Increasing demand for staple foods like grains and tubers enables production to expand and stimulates the development and acceptance of more productive technologies, generally with positive effects on rural employment and incomes. The development potential of increases in basic food production for rural areas is, however, limited as land holdings are generally small and the production value per unit of land low. Increasing demand for higher quality foodstuffs such as fruits, vegetables, dairy products, eggs and meat from a thriving urban population offers more scope for a substantial and continuous increase in rural incomes. Demand for these products enables rural agricultural producers to enter agricultural enterprises that offer a higher rate of return per unit of land and labour and thus improved income opportunities through diversification and specialisation. This is especially important where land holdings are small and rural labour is relatively abundant. In addition, as marketing and processing linkages develop more employment is created.

A third linkage is the incorporation of the rural areas into national or even global spheres of influence. Radio, television and (mobile) phones as well political parties and even regional conflicts reduce rural isolation and shape the ideas, expectations and behaviour of both rural and urban populations.

The availability and introduction of new agricultural production technologies and rural-urban linkages do not result in a uni-directional and steady process of rural transformation. Transformation requires and means quantitative and qualitative changes in institutions to support an increasingly complicated fabric of relationships between a growing numbers of actors who perform specialised tasks (Hayami and Ruttan, 1985). These changes are sometimes slow and hardly noticeable and in other periods rapid when bottlenecks are overcome.

Microfinance plays a significant role in institutional adaptation and change. In this context, microfinance is a non-specific institution because a wide range of actors - agricultural producers, traders and shopkeepers, local manufacturers, wage-earners and consumers – with differing needs - require finance to maintain and safeguard their pattern of consumption or to enable them to invest. The focus of microfinance is on the majority of rural households that never had regular access to the financial services provided by formal financial institutions such as banks. Many explanations have been offered for their exclusion including the information gap between commercial banks and rural entrepreneurs; the absence of collateral to support credit transactions; low income levels leading to small transactions and the pervasive and persistent idea that low income households cannot save.

New insights gained from experiences in the 1970s and 1980s showed that many of these explanations were false. Innovative approaches showed that financial services could reach low income households successfully (Moll, 1989; Yaron, 1992). Currently, a wide diversity of microfinance institutions including rural banks, cooperatives and NGOs operate in the rural areas of Africa, Asia and Latin America. However, many communities especially those living in less populated and generally less-favoured areas, remain outside the financial frontier of microfinance (Von Pischke, 1991).

3. The rural environment and microfinance

3.1. Agriculture and the rural economy

Agriculture is still the core business of many rural societies and this affects the rural economy in specific ways. First, regionally specific agro-ecological conditions determine the extent to which crop, livestock, forestry, and fishery enterprises can be developed. This leads to regional specialisation or at least to the concentration of a limited number of enterprises that give each region its unique character. Second, the agricultural calendar with its regular and recurring planting, maintenance and harvesting periods leads to similar cash flows for the farmers and agents involved in backward and forward linkages. Third, variations in crop yields as a result of weather conditions, or price changes of agricultural products may cause the incomes of a large proportion of the rural population to fluctuate unpredictably. These fluctuations can be substantial and can lead either to crisis situations or periods of relative affluence. The consequence of dependency on agriculture is a rural economy characterised by a predictable seasonality in activities and financing requirements and unpredictable and possibly substantial fluctuations in rural incomes from year to year. In addition, there may be long-term trends that negatively affect the returns expected from major agriculture enterprises. This will have a deep and serious impact on rural economies that lack the capacity or resources to switch to other, more attractive enterprises.

3.2. Physical environment

Population density in rural areas can range from a few persons per km^2 to several hundred persons per km^2. This is much lower than in urban areas where there are thousands of people per km^2. In rural areas the issue of how many potential microfinance clients live within an acceptable distance from the microfinance retail point is an important one. For this reason physical infrastructure such as roads, bridges, and (mobile) phone networks are of direct concern for rural microfinance institutions because they determine the transport cost and time involved for both staff and clients. In a wider sense physical infrastructures such as market places, warehouses, electricity and water supply and sewage systems are also important for the rural economy and influence the extent to which rural people can save, invest and establish links with microfinance institutions. National opportunities and policies determine the volume of public funds allocated to providing physical infrastructure in rural areas. Priorities set at the national level, therefore, have a direct effect on the rural economy as a whole.

3.3. Institutional environment

The markets of agricultural inputs, products and services such as financial services, extension services and animal health services are the core issues of the agricultural policies in many countries. There is widespread consensus regarding the importance of well-functioning markets for the development of agriculture and the rural areas in general. Transaction cost literature provides new and valuable insights into the functioning and shortcomings of markets and a wealth of case studies. However, there is a wide variation in the way insights are interpreted and remedies proposed. The debate on liberalisation - *state or market* or *state and market* continues in many countries (Tilburg *et al.*, 2000; Jayne *et al.*, 2002) leading to policy changes and the reversal of policy changes. Despite discussions and policy changes farmers in many countries still do not have reliable access to inputs and produce markets with negative consequences for agricultural production and farmers' incomes.

Land markets in rural areas typically operate in a dualistic manner with the formal law on the one hand and the traditional practice on the other hand. Security of individual tenure or regulated access to communal resources based either on law or on tradition, provides incentives for maintaining or improving the productive capacity of land and water resources. Experience shows that the absence of security or ineffective regulation of common property leads to land degradation and threatens production in the long run. Land policies have potentially a wide political dimension, especially in former colonies, but there may be substantial differences between formal policies and their implementation in rural areas.

Rural financial markets consist of relationships between buyers and sellers of financial assets who are active in the rural economy. These relationships are based on transactions that include borrowing, lending and the transfer of financial assets: debt claims, verbal or written promises to pay and ownership claims as well as rights of access, use or control (Von Pischke *et al.*, 1983). The concept of rural financial market covers formal institutions, informal intermediaries and enterprises and households demanding financial services as well as financial relationships based on reciprocity. Studies of rural financial markets have shown a limited presence of formal institutions such as banks on the one hand and a rich variety of informal financial arrangements between individuals and groups on the other. A major characteristic of the rural financial market is segmentation: financial institutions provide services to only one specific group of households (Moll *et al.*, 2000). Thus, commercial banks offer a range of services to traders and salaried workers; cooperatives offer savings services and possibly some credit to their members, and informal financial institutions provide credit to a small group of people living in the same neighbourhood. This segmentation results in a large proportion of the rural people being excluded from formal financial services. Hoff and Stiglitz (1993) focus on imperfect information as an explanation for this exclusion and for other shortcomings prevailing in the rural credit market.

Public services in health and education are usually at lower levels and with less depth in rural areas than in urban areas with as consequences a lower overall education level and more serious negative effects when health is impaired. Health risks combined with the absence of formal (health) insurance and a low level of wage employment with social security provisions means that large sections of rural society depend on informal institutions when problems and difficulties arise. These informal, mostly local, institutions can be effective in as far as idiosyncratic risks are concerned but they are inadequate when it comes to dealing with co-variant risk.

Other aspects of the institutional environment such as personal and property protection and access to the judiciary system are often the same for both rural as urban dwellers. For example, the staff of microfinance institutions – whether urban or rural based – are equally dependent on the degree of police presence available when they transport cash or try to protect their offices against burglary. Insufficient protection by police will lead to higher costs and - in extreme cases - to financial institutions refusing to provide services in risk-prone areas.

3.4. Conclusions

This general overview of the main features of the rural environment leads to a number of conclusions that have specific consequences for the supply of financial services in rural areas. Loans should be designed to fit the agricultural enterprises of a region both in terms of timing and the amount of finance required. Specific attention should be given to the co-variant risks associated with the agricultural enterprises and to the fact that a large proportion of clients in a given area will have a similar cash flow pattern. The latter requires attention to ensure adequate liquidity management so seasonal liquidity shortages or excesses are avoided. These three consequences are especially relevant in areas where agriculture is the dominant economic sector. The cost of transport and information are high in rural areas. This in combination with the small size of individual microfinance transactions leads to high transaction costs for microfinance institutions and their clients. Establishing microfinance institutions in rural areas means adding to the number of formal and informal institutions operating in the rural financial market. In order to become an accepted and respected partner, the microfinance institution will have to compete with established institutions.

This overview of the rural environment pinpoints the importance of public services in the health, education, infrastructure and law enforcement sectors. The policy environment in which agricultural markets operate are extremely important because public services and policies impact on the rural economy as a whole and therefore have a direct effect of the way in which financial arrangements are structured and operate.

4. Rural households and enterprises: the demand for microfinance services

Figure 1 provides a general overview of the relationships between rural households, the markets for goods and production factors.

Resources such as land, labour, livestock and implements are stated as productive assets as they have a production value for the household and are related to enterprises. Non-

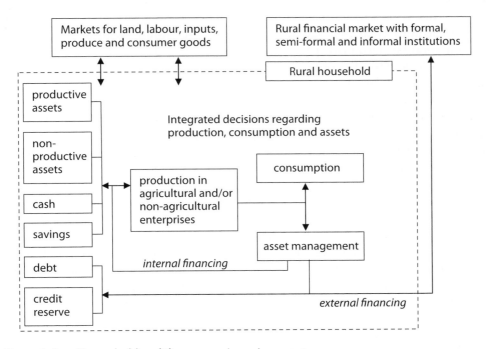

Figure 1. Rural household and the economic environment.

productive assets such as gold, jewellery and consumer durables have no production value (although they do have a consumption value) but have a liquidity value which is dependent on accessible exchange or market possibilities and the transaction cost involved in their conversion into cash. The third type of assets - cash - can be used directly for production and consumption purposes without transaction costs. Debts and claims result from financial relationships with the various institutions operating within rural financial markets. Potential financial relationships lead to the existence of a credit reserve, i.e. the volume of credit that lenders are prepared to supply on the basis of their perception of the borrowing capacity of the household. This credit reserve, stated as 'credit' by Baker and Bhargava (1974), can be used to secure loans without actually disposing of productive or non-productive assets and must, therefore, be considered a liquid asset.

Assets can be utilised for economic enterprises inside or outside agriculture. A division of rural households according to their (main) enterprise has little relevance as household members are often engaged in enterprises in agricultural, trading, service and small-scale industrial sectors, either as entrepreneur or as employee and often on a seasonal or part-time bases. Studies show that as much as 50% of 'farm' household income is derived from non-agricultural enterprises (Haggblade *et al.*, 2002). Some members of rural households, for example, may have migrated to the urban areas but continue to maintain their household 'membership' and their claim to household resources through remittances and periodical visits. Household income, usually partly in kind and partly in cash, is used for consumption requirements. Temporary income shortfalls or excesses are dealt with either by an internal rearrangement of assets (internal financing), or by establishing financial relationships with the rural financial market (external financing). Continuing shortfalls in income will eventually lead to a loss of assets whereas continuing income excess will result in the expansion of the resource basis.

Rural households face at least two major concerns in their daily struggle to survive. First, they must safeguard consumption in the face of seasonal or irregular incomes and shocks. Households develop a range of strategies that include ex ante measures such as saving in various forms and ex post coping mechanisms such as borrowing, increasing the amount of labour invested in their own enterprises or seeking employment elsewhere. The widespread desire of households to save has been extensively discussed from many points of view. To mention a few: Carroll (1997) developed a buffer-stock saving version of the Life Cycle/Permanent Income Hypothesis; Morduch (1995) discusses a range of risk coping mechanisms including saving and borrowing; Kochar (2004) includes ill-health in the analysis of savings and portfolio choices; Moll (2005) discusses the role of livestock in saving and insurance and Ndirangu (2007) elaborates the effects that shocks arising from ill-health and weather variability have on savings behaviour and labour input.

A second major concern refers to the financing of enterprises. Short-term lending can be used to finance production inputs if resources fall short of requirements and medium-term lending is required for investing in the expansion of enterprises.

In conclusion, it can be stated that rural households have diverse financing requirements and a portfolio of financing options through internal rearrangements of assets as well as external relationships. However, financing requirements in terms of amount and tenor are often larger than the internal possibilities and the external (informal) options available. In such situations access to microfinance services provides a welcome addition to the financial options already available. Reliability of access and flexibility in tenor, size and the repayment conditions attached to loan contracts determine whether microfinance services are more attractive to rural households than the other financing options available.

5. The challenge of rural microfinance institutions

Rural microfinance institutions may operate as independent unit banks without a branch network such as the People's Credit Banks in Indonesia which operate within one district (Sunarto, 2007) or as a multi-tier organisation with a head office in the capital, provincial offices and retail offices in rural areas. Both types of microfinance organisations have their advantages and disadvantages which are largely similar for both urban and rural institutions. Therefore, the discussion in this section focuses on the retail unit of the microfinance institution operating in rural areas. It is at this point that decisions are made by both microfinance institution and rural households regarding the establishment and maintenance of financial relationships. The structure of the discussion is depicted in Figure 2.

The starting point is the retail unit of the microfinance institution where two internal policy issues - organisation structure and method of operation - determine its capacity to supply financial products. Rural households have been discussed in the previous section

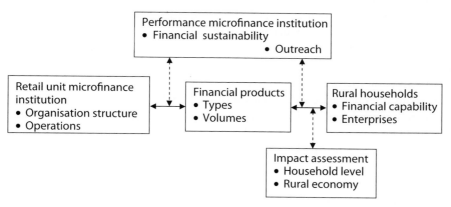

Figure 2. The relationship between rural microfinance institution and rural households.

and their abilities and wishes are used in the discussion of financial products in this section. The performance of the microfinance institution is measured along two central objectives – financial sustainability and outreach – while the effect of the financial products on the rural households is captured under impact assessment.

5.1. Organisation structure and operations

Rural microfinance institutions operating in less densely populated areas face high delivery costs per transaction because of travel costs incurred between retail unit and client. These costs are in addition to those incurred in dealing with the standard microfinance problems of information asymmetry and the difficulties of handling small transactions efficiently. Managing high delivery costs requires responses at the level of organisation structure and in the way operations are conducted at the retail unit where staff and clients have the most direct interaction. A variety of organisational structures and operations have been developed to meet different circumstances.

5.1.1 Retail office with counter services and field staff
The retail unit is located in a rural centre and consists of a permanent office with some staff members in charge of counter services and with field staff equipped with motorbikes or cars. Clients visit the office and transactions are dealt with in the office. Field staff may visit clients on their premises to appraise and monitor activities and for contract enforcement.

This option is suitable in rural areas where clients are concentrated around a rural centre and there is a public transport system that enables them to visit the retail office. The BRI Unit Desa in Indonesia operates with this combination of organisational structure and mode of operation in the densely populated areas of Java and Bali, but also in the much less densely populated islands east of Bali (Ibrahim, 2004; Robinson, 2004; and, Mauer, 2004).

5.1.2. Mobile office with counter services and field staff
The mobile retail unit is housed in a van that moves to different locations according to a fixed schedule. Some staff members are in charge of the counter services and others operate with motorbikes or cars. Clients visit the mobile banking unit at designated opening times and transactions are dealt with in the mobile unit. Field staff may visit clients at their premises to appraise and monitor activities and to enforce contracts. In situations where the volume of transactions does not justify a permanent office, a mobile unit is an interesting option. If transactions expand, the mobile unit can be replaced by a permanent retail office. The Equity Bank in Kenya operates with mobile offices for one, two or three days a week in one location in order to explore market conditions in new areas.

5.1.3. Retail office without counter service and with field staff
The field staff operates from an office which may simply be one room in an office building at a central location. Various ways of operating are possible with this organisational structure:

- Field staff visit individual clients at a specified time at their premises and transactions are dealt with on the spot.
- Field staff visit groups of clients at a specified time and at a specified place and transactions are dealt with on the spot.
- Field staff provides counter services at pre-determined places and times and clients come as and when necessary.

These possible combinations are suitable in situations where clients are dispersed over a wide area either because population density in the region is low or because operations have just started. Examples of these types of operations can be found in many places and include the Grameen Bank and the many microfinance institutions that have followed this model as well as the Kenya Women Finance Trust. There are many case studies that can be used as a basis for an extensive discussion on advantages and disadvantages of group approaches vis-à-vis individual relationships between client and organisation (see for example Giné and Karlan, 2006; and, Dellien *et al.*, 2005).

Technological developments have a positive affect on the provision of financial services to rural areas. The presence of mobile phone networks in rural areas greatly facilitates contacts between clients and microfinance institution. The use of mobile phones for information exchange is the first obvious application, but mobile phones in combination with personal identification codes are also being used to facilitate money transfers.

A second development is the use of portable computers by field staff in their contacts with clients. Financing options with repayment schedules can be shown and discussed on the spot and transactions can be administered efficiently (Barton, 2004). Other developments such as the providing members or clients with swipe or smart cards will further improve information exchange and transactions efficiency and open up new possible combinations of input supply, financing and marketing (Campaigne, 2004). These recent technological developments can contribute to mitigating the problems of inadequate information exchange and help rural lenders and borrows by reducing the burden of transaction costs.

5.2. Financial products

The development of financial products capable of addressing the expected and unexpected financing requirements of rural households is a serious challenge for rural microfinance institutions (Evans and Ford, 2003). The overview of the rural environment provided in Section 3 and the discussion of the rural household in Section 4 indicate that a 'one-size-fits-all' type of product is totally inadequate. To meet potential demand a combination of several services is needed. Some examples are outlined below.

- Saving accounts without restrictions with regard to size of deposits and frequency of withdrawal (Maurer, 2004).

- Seasonal credit in line with the agricultural season, which could mean loans for a period of six to nine months. These loans - with one repayment after a relative long period - are more exposed to moral hazard and enforcement problems than the usual trade loans with weekly instalments. Maintaining contact with borrowers during the lending period is, therefore, necessary to reduce the possibility of moral hazard and to assess repayment possibilities in case enforcement measures are required.
- Short-term emergency loans. Absence of insurance turns credit into a coping mechanism that may enable rural households to deal with shocks without reducing their productive assets and future earning capacity.
- Short-term loans to meet cash-flows problems. These loans are usually very productive because the marginal value of capital can be high: no credit means an opportunity lost. This applies to trade as well as to agricultural enterprises.
- Investment loans for a period of one to three years. This type of loan can finance the expansion of production capacity inside and outside agriculture. Whether or not these loans are actually required will depend on the specific circumstances involved: reliable access to short-term loans might reduce the need for investment loans as own capital can be directed towards investment while access to short-term loans is kept (costless) as a financing option.
- Money transfer services. Links with national and international money transfer services bring these services to rural households.
- Insurance. An emerging number of insurance programmes in health, agricultural and life insurance have been developed for the rural population.

The provision of several products has advantages for microfinance institutions. Being able to meet the diverse requirements of rural households makes the relationship more valuable and act as an additional incentive when it comes to repaying loans. Serving more types of clients means that risk is spread over different types of people and economic sectors. Combining services can also lead to economies of scope as both staff and funds can be more efficiently utilised.

5.3. Financial products and contract farming

Agricultural production can give rise to problems for both financial institutions and farmers, but it can also offer specific opportunities. The production chains of some agricultural commodities provide unique possibilities for the provision of credit. The commodity chains of milk, cotton, sugar and tea in particular require that the primary products - milk, seed cotton, sugar cane and tea leaves respectively - are processed in comparatively large-scale factories. Moreover, the time between milking or harvesting and processing must be short to maintain product quality.

This technology-determined linkage between the production of primary product and processing enables processors to provide finance and inputs to smallholder farmers at little or no monitoring and enforcement costs because product delivery is secured. Co-operatives and state and private enterprises in many different countries under a wide variety of circumstances organise production, transport and processing in these chains in an integrated way with smallholder farmers either as members or contractors (Sartorius *et al.*, 2004). In this way membership or contracts with processing enterprises offer smallholders the possibility of participating in modern production technologies and, at the same time enables processors to align processing with product supply. The efficiency gained by integrated production and processing offers advantages to all parties concerned provided monopoly positions are either avoided or not used for unilateral benefit.

5.4. Performance

Performance is generally measured according to the financial and social objectives of financial sustainability and outreach. The basic data needed for financial analysis are provided in annual financial statements with the profit and loss statement showing income and expenditures and the balance sheet indicating assets, liabilities and equity. Prior to financial analysis, this basic data might require adjustments as far as loan portfolio quality and the value of fixed assets are concerned. Adjustments are especially relevant for finance institutions operating either without or under minimal financial legislation as standard rules for writing off loans and depreciation of assets might be neglected.

The aim of the –adjusted - profit and loss statement analysis is to assess financial sustainability (Yaron, 1992). The first step in sustainability analysis is to determine the net operating margin: income minus actual operating and financing costs. If this margin is positive the institution complies with the first condition of sustainability: operational sustainability. The second step deals with whether or not the real value of the equity can be maintained. In this step the imputed financial cost is determined with either the inflation rate or the opportunity cost of capital as a bench mark. If the net operating margin is sufficient to cover the imputed cost of capital, the institution is financially sustainable.

Outreach is generally measured in terms of numbers and types of clients and the scale and depth of outreach (Ledgerwood, 1999). Financial institutions generally have data on the number of savers and lenders together with the volumes of savings and credit products provided. Annual data on new clients and clients leaving are generally not provided although it is particularly interesting. Data collected over a period of years will show whether or not an organisation is able to extend its services to more clients. These data can be related to the number of households present in a certain area to obtain an insight into the proportion of rural households reached. If qualitative data are available on clients, participants or rural households in general it is possible to determine which segment(s) of rural households are actually served by individual institutions. The latter information is relevant to assess the claims of finance institutions that they do indeed reach 'the poor'.

5.5. Impact assessment

Impact assessment has become an intrinsic part of the international discussion on microfinance currently being waged by donor agencies, microfinance institutions and national governments. The claim that microfinance interventions can play a significant role in the struggle against poverty and the willingness of donors and governments to fund microfinance initiatives has inevitably lead to the requests for proof of its effectiveness. Impact evaluation in this case goes beyond an analysis of depth of outreach because the purpose of the assessment is to determine the effects of microfinance services on the situation of 'the poor'.

The discussion of the rural households in Section 4 provides a first insight into the methodological problems of such impact assessments: microfinance lending must be distinguished from other actual or potential sources of finance, and the effects of microfinance lending on production and income must be established. An overview of the methodological and practical issues involved here has been provided by Hulme (1997).

Leaving methodological issues aside, impact studies have brought forward three positive effects of microfinance interventions. First, access to microfinance leads to an increased capacity to deal with risk because it provides access to savings or credit when emergency strikes. This may mean that farm households no longer have to sell productive assets such as machinery, inventory, land and livestock to meet emergency needs and their income generating capacity in the longer term will, therefore, not be negatively affected. Second, microfinance can ensure that even in difficult circumstances farm families are able to maintaining an adequate diet and remain able to work (Pitt and Khandker, 1998). This is of major importance as wage labour is often the main source of income in low-income households. Third, increased opportunities to invest in productive enterprises have been observed.

The discussion on impact focuses almost entirely on the effects of the relationships between microfinance institution and clients over a period of several years. A long-term perspective on the effects of microfinance for the development of rural areas is provided in the next section.

6. The development of microfinance in rural areas

6.1. Establishment and expansion

Microfinance institutions entering rural areas generally meet with a large demand for financial services. Part of this initial demand may be based on false assumptions, but information spreads quickly and soon policies and products are understood. During the establishment period the emphasis of both institution and clients is on access and less importance is attached to interest rates. In this period interest rates on loans of 25 to 40% on an annual basis are acceptable for clients and necessary for institutions that still operate with little information on clients and enterprises. After having operated for some time a microfinance institution will have built up a portfolio of trusted clients, increased its insight into local rural enterprises, and expanded the volume of services. All these developments will enable institutions to reduce interest rates.

Other microfinance institutions with a different client focus and other types of operation may start in the same area and show the same pattern of development towards more clients and new products. As a result competition will gradually develop. This competition – together with the increased volume of information on rural clients and their enterprises – will induce all microfinance institutions to become more efficient and develop innovative and attractive products to attract new clients and retain old ones. A well-known strategy, for example, is to start with group loans and then expand to individual loans for those group members who have developed specific requirements and shown particular abilities. These individual clients have a track record as trustworthy client and if they are not offered the possibility of obtaining individual loans they may approach other microfinance institutions or commercial banks.

The establishment and expansion of microfinance in rural areas thus has quantitative and qualitative dimensions: more clients and more types of services. For microfinance institutions this requires continued internal assessment of organisation structures and operations and adjustments when necessary for the institution to flourish and remain in business.

In a wider sense, microfinance institutions can also play a pioneering role in the rural institutional landscape. First, both saving and credit products can be used as insurance substitutes because formal insurance is absent. Second, but possibly more important, microfinance institutions are often the first formal, external institution that deal with people on a contractual basis with mutual confidence and mutual duties as basic conditions. This differs fundamentally from government and non-government agencies that provide health, education or religious services free of charge or on the basis of voluntary contributions. The contractual dimension in the relationship between microfinance institution and the rural population requires specific attention when microfinance institutions are being set up to avoid problems being created because of gaps in information and expectation.

6.2. Impact in rural areas

Extrapolating the effects of microfinance on individual households to rural areas as a whole will provide some indication of the overall effect of providing wider access to financial services. Increasing the capacity of the individual to deal with shocks will reduce the effect of co-variant shocks on rural communities at least when a substantial proportion of the population is within the financial frontiers of the microfinance organisation. Further, increased saving in the form of financial assets will mean a shift away from storing wealth in non-financial assets that have zero or low productivity. Financial savings become available for investment in agriculture, in agriculture-related trade and processing as well as a host of other enterprises and this can be expected to lead to technological progress and rural employment. The availability of capital for new, trustworthy clients with productive enterprises will have a fundamental affect on economic relationships in rural areas. Rajan and Zingales (2003) state: 'a healthy financial system can be a powerful anti-monopoly tool, providing the lubrication for the emergence of competitors that can undermine the power of incumbent firms'. In this way microfinance can positively affect economic life in rural areas and an expanded outreach will make these effects more visible (Moll, 2006).

At a later stage, when remunerative investment opportunities in rural areas are more limited and the volume of savings overtakes the volume of credit, excess capital can be channelled via microfinance institutions and the national banking system to urban areas where large-scale industries and services offer extensive investment opportunities. In this way, rural savers will benefit from their investments and the children of the rural savers might find the urban jobs they are looking for when they migrate.

The process of financial development at national level has been studied by Shaw (1973) and McKinnon (1973) and described as financial deepening. The shift towards saving in the form of financial assets, the subsequent intermediation by the banking system and investments by borrowers are central to their theory of financial development. In this way microfinance in rural areas can lead to financial deepening by providing rural households with access to savings and credit through local intermediation. Empirical studies of the relationship between 'financial deepening' and economic development will become possible in rural areas where microfinance institutions have been operating for a prolonged period and have involved a substantial and increasing proportion of rural households in their activities.

7. Microfinance in rural areas and beyond

This overview of microfinance in rural areas has been concerned with the specific conditions that prevail in rural areas and the effect these have on the demand for and supply of microfinance services. Millions of rural households are still outside the reach of microfinance, but experience, partly reflected in the case studies quoted above, shows that in a wide variety of situations microfinance institutions and rural households have developed productive financial relationships. This means that microfinance institutions have found combinations of organisation structure, operations and financial products that comply sufficiently with the possibilities and priorities of rural households. In this sense recent experience provides some guidance for the establishment of retail units in new rural areas. Once microfinance institutions have overcome the initial problems of operating and established a foothold, the usual performance parameters – outreach and sustainability – should be re-assessed.

Outreach must be considered in both quantitative and qualitative terms: more clients, new clients that are more difficult to reach, and more diversified financial services that not only meet the needs of current clients but also address the needs of new and emerging clients. Expanding outreach in this way requires continuous attention to the specific conditions of the rural area concerned and the potential clients. Organisation structures, operations and new products will have to be designed with this context in mind. This overview aims to provide some guidance as to how this process can be managed.

Sustainability in the sense that all cost can be met by expected returns is not sufficient in the long run. The focus should in fact be on financial stability. Shocks, whether they are man-made or the result of natural disasters, are part of life and microfinance institutions have an obligation towards their clients to continue operations even in adverse circumstances. Stability is partly an internal matter and discussions along the lines of 'What do we do if .....' should result in provisions and scenarios capable of dealing with foreseeable shocks. External relationships are also important and here rural microfinance institutions find themselves in the company not only of other microfinance institutions but of all the other financial institutions in the country as well. Linkages established with national and international financial markets can be effective when rural communities are faced with adverse conditions.

Outreach and sustainability in this wider sense are not competitive but complementary: the more successful a microfinance institution is in overcoming the specific difficulties related to providing services in rural areas, the more worthwhile it becomes to defend and safeguard established financial relationships.

References

Baker, C. B. and V. K. Bhargava, 1974. Financing small-farm development in India. Australian Journal of Agricultural Economics 18: 101-118.

Barton, S., 2004. PDA technology in microfinance ACCION's experiences implementing Porta Credit. In: A. Pakpahan, E.M. Lokollo and K. Wijaya (eds.) Developing micro banking: creating opportunities for the poor through innovation. PT Bank BRI, Jakarta. pp. 309-322.

Bell, C., 1988. Credit markets and interlinked transactions. In: Chenery, H. and T.N. Srinivasan. Handbook of development economics Volume I., North-Holland, Amsterdam, pp. 763-772

Campaigne, J.F., 2004. DrumNet: a project of Pride Africa. In: Pakpahan, A., E.M. Lokollo and K. Wijaya (eds.), Developing micro banking: creating opportunities for the poor through innovation. PT Bank BRI, Jakarta. pp. 295-308.

Carroll, C.D., 1997. Buffer-stock saving and the life cycle/permanent income hypothesis. Quarterly Journal of Economics 112: 1-55.

Dellien, H., J. Burnett, A. Gincherman and E. Lynch, 2005. Product diversification in microfinance: introducing individual lending. Women's World Banking, New York.

Donald, G., 1976. Credit for small farmers in developing countries. Colorado: Westview Press, 286p.

Evans, A. and C. Ford, 2003. A technical guide to rural finance: exploring products. WOCCU Technical Guide 3, Madison:World Council of Credit Unions

Giné, X. and D. Karlan, 2006. Group versus individual liability: a field experiment from the Philippines. Center Discussion Paper No. 940, Economic Growth Center Yale University, New Haven.

Haggblade, S., P. Hazell and T. Reardon, 2002. Strategies for stimulating poverty-alleviating growth in the rural nonfarm economy in developing countries. International Food Policy Research Institute, Washington D.C., 82p.

Hayami, Y. and V.W. Ruttan, 1985. Agricultural development: an international perspective. The Johns Hopkins University Press, Baltimore, 506p.

Hoff, K and J.E. Stiglitz, 1993. Imperfect information and rural credit markets: puzzles and policy perspectives. In: K. Hoff, A. Braverman and J.E. Stiglitz (eds.) The economics of rural organization, theory, practice, and policy. Oxford University Press, Oxford, 590p.

Hulme, D., 1997. Impact assessment methodologies for microfinance: a review. Paper prepared in conjunction with the AIMS Project for the Virtual Meeting of the CGAP Working Group on Impact Assessment Methodologies.

Ibrahim, M., 2004. Strategic issues in developing microfinance to promote micro, small and medium enterprises. Paper presented at BRI International Seminar on Microbanking System, Nusa Dua, Bali, Indonesia.

Jayne, T.S., J. Govereh, A. Mwanaumo, J.K. Nyoro and A. Chapoto, 2002. False promise or false premise? The experience of food and input market reform in Eastern and Southern Africa. World Development 30: 1967-1985.

Kochar, A., 2004. Ill-health, savings and portfolio choices in developing countries. Journal of Development Economics 73: 257.

Krahnen, J.P. and R.H. Schmidt, 1994. Development finance as institution building. A new approach to poverty-oriented banking. Westview Press, Boulder, 154p.

Ledgerwood, J., 1999. Microfinance handbook: an institutional and financial perspective. The World Bank, Washington D.C., 286p.

Maurer, K., 2004. The role of BRI Units in capital accumulation and rural savings mobilization. In: A. Pakpahan, E. Lokollo and K. Wijaya (eds.) Developing micro banking: creating opportunities for the poor through innovation. PT Bank BRI, Jakarta. pp. 207-220.

McKinnon, R., 1973. Money and capital in economic development. Washington D.C. The Brookings Institution.

Moll, H.A.J., 1989. Farmers and finance: experience with institutional savings and credit in West Java. Wageningen Agricultural University,135 p.

Moll, H.A.J., R. Ruben, E.W.G. Mol and A.A. Sanders, 2000. Exploring segmentation in rural financial markets: an application in El Salvador. Savings and Development 24: 33-54.

Moll, H.A.J., 2005. Costs and benefits of livestock systems and the role of market and nonmarket relationships. Agricultural Economics 32: 181-193.

Moll, H.A.J., 2006. Microfinance and rural development: a long term perspective. Journal of Microfinance 7: 13-31.

Morduch, J., 1995. Income smoothing and consumption smoothing. The Journal of Economic Perspectives 9: 103-114.

Ndirangu, L., 2007. Households' vulnerability and responses to shocks: evidence from rural Kenya. Thesis Wageningen University, in press.

Pitt M.M. and S.R. Khandker, 1998. The impact of group-based credit programs on poor households in Bangladesh: does the gender of the participants matter? Journal of Political Economy 106: 958-996.

Rajan, R. and L. Zingales, 1996. Financial dependence and growth. University of Chicago, unpublished. Quoted in Khan and Senhadji, 2003.

Robinson, M., 2004. Why BRI has the world's largest sustainable microbanking system and what commercial microfinance means for development. In: A. Pakpahan, E. Lokollo and K. Wijaya (eds.) Developing micro banking: creating opportunities for the poor through innovation. PT Bank BRI, Jakarta. pp. 295-308.

Sartorius, K., J. Kirsten and M. Masuku, 2004. Contract farming models and small farm vertical co-ordination partnerships in Southern Africa. In: C.J. van Rooyen, O.T. Doyer, L. D'Haese and F. Bostyn (eds.) Readings in agribusiness, a source book for agribusiness training. University of Pretoria and the Flemish Interuniversity Council.

Shaw, E., 1973. Financial deepening in economic development. New York. Oxford University Press.

Sunarto, H., 2007. Understanding the role of bank relationships, relationship marketing, and organisational learning in the performance of People's Credit Bank Evidence from surveys and case studies of Bank Perkreditan Rakyat and clients in Central Java, Indonesia. Thesis Free University, Amsterdam, 306p.

Tilburg, A. van, H.A.J. Moll and A. Kuyvenhoven, 2000. Agricultural markets beyond liberalisation. Kluwer Academic Publishers, Boston Dordrecht London.

Von Pischke, J.D., D.W. Adams and G. Donald, 1983. Rural financial markets in developing countries: their use and abuse. The Johns Hopkins University Press, Baltimore, 441 p.

Von Pischke, J.D., 1991. Finance at the frontier: debt capacity and the role of credit in the private economy. EDI Development Studies, Bank, Washington D.C., 427 p.

Wouterse, F.S., 2006. Survival or accumulation: migration and rural households in Burkina Faso. PhD thesis Wageningen University.

Yaron, J., 1992. Assessing development finance institutions: a public interest analysis. World Bank Discussion Paper 174. Washington D.C.

www.woccu.org: World Council of Credit Unions, Madison.

Growing importance of land tenancy and its implications for efficiency and equity in Africa

Stein Holden

Abstract

This chapter summarises recent evidence on the functioning of land markets in developing countries, with a special focus on Africa. Land markets are becoming more prevalent in Africa, and their impact on economic growth and income distribution in the future are of great importance. The review of the empirical evidence on the efficiency and equity implications of land markets in Africa suggests that there are complex interactions between land markets, the distribution of resources and the characteristics of non-land factor markets including credit and insurance markets. Future land policies should pay careful attention to these complex interactions and the dynamics of change. Finally, while there is some evidence that land rental markets tend to be beneficial for the poor, the picture is more mixed when it comes to land sales markets as distress sales by the poor could magnify existing inequalities.

1. Introduction

Scarcity is the basis for economic value and for the emergence of markets. However, the economic literature also points towards many other factors that may contribute positively or negatively towards whether land markets emerge. These factors include, in addition to land scarcity; unequal distribution of land, imperfect markets for non-land factors of production, seasonality, shocks and imperfections in inter-temporal markets (credit and insurance), government policies and projects, and traditional institutions that substitute for, stimulate or hinder land market formation (Deininger, 2003, Holden, *et al.*, 2006).

Africa has for long been seen as a land-abundant continent. However, land scarcity is increasing in parts of Africa, like the East African Highlands and parts of Southern and Central Africa, due to population increase and population concentration. Population concentration or unequal access to land has partly been introduced by colonial powers (e.g. in South Africa, Rhodesia and Nyasaland by the British Empire) and partly by feudal power structures, e.g. in historical Ethiopia. Partly such unequal distribution of land persists today and partly it has been reduced through various forms of land reforms. Urbanisation and globalization processes are another important reason for population concentration and increasing and competing demands for land. These processes also cause increasing pressures on traditional systems of tenure that partly tend to lag behind in their adjustments to the new pressures. This also puts pressures on governments to introduce land reforms to facilitate growth, provide security of tenure and reduce poverty.

Land reforms are again high on the international policy agenda as evidenced by the increasing number of land reform projects in the project portfolio of the World Bank and the recent establishment of the (High-Level) Commission for the Legal Empowerment of the Poor. Hernando de Soto has also contributed to this by attracting international donors' interests towards providing the poor with formal property rights to mobilize their 'dead capital' as a basis for investment loans. However, the route towards well-functioning credit markets based on the formal property rights of the poor, seems long and thorny in Africa. Also, De Soto's explanation *'why capitalism triumphs in the West and fails everywhere else'* (De Soto, 2000), fails to explain how China has succeeded in having an impressive economic growth for many years without private property rights to land.

I do not address the issue of private property rights directly here but concentrate on the role of land markets, and in particular whether they are informal or formal. What are the historical and current roles of land markets in form of land sales and land rental markets in Africa? What are the efficiency and equity implications? To what extent do land markets in contribute to growth and poverty reduction? Can land reforms and land markets be shaped to become more pro-poor?

This chapter provides a theoretical and material basis for understanding the emergence of land markets and for why particularly tenancy markets are on the increase in parts of the world while land sales markets are more restricted in their occurrence, especially in developing countries. Furthermore it reviews some of the empirical literature on land tenancy with a focus on Africa where there has been relatively less research on these markets as compared to e.g. in Asia (Otsuka *et al.*, 1992; Otsuka and Hayami, 1988). The focus of the empirical review is on the efficiency and equity implications of land markets in Africa. Finally, I draw some policy implications out of the lessons from theory, material conditions and empirical evidences.

2. Roles of land markets

2.1. Land scarcity and factor market characteristics

The fundamental characteristic of land as a resource is that it is immobile in the physical sense due to the extremely high costs of moving land as compared to other resources. Non-land factors of production have therefore to be brought to the land to initiate production, and then, after a necessary time delay from planting and till harvest time, outputs have to be harvested and transported from the land to the places for consumption and sale. This spatial dispersion of production leads to high transaction costs, especially in land-abundant economies where other factor markets also tend to be imperfect (Binswanger and Rosenzweig, 1986). This spatial dispersion and immobility of land also ties up land users and typically make them less able to migrate and access information, often creating information asymmetries, thin markets and imperfect competition environments in remote rural areas.

The quantity and quality of infrastructure, transportation services and market information access are crucial for the functioning of rural markets and the degree of integration of rural producers into markets for inputs, outputs and consumer goods.

Land scarcity is a necessary but not sufficient condition for the emergence of land markets. There must be sufficient heterogeneity among agents in the economy such that a single price of land attracts willing sellers and buyers simultaneously. Imbalances in ratios between factors of production across producers and the basic characteristic that these factors of production also serve as complements rather than substitutes in the production process (due to low elasticities of substitution) create opportunities for efficiency gains through trade in factors of production, as long as the transaction costs involved are not too high.

On the other hand, with perfect markets for all other factors of production and outputs, there is no need for a market for land as one missing market does not lead to any loss of efficiency, like in the neoclassical farm household model (Singh *et al.*, 1986). In such a world, the residual profit would be the land rent and the shadow value of land would be independent of who owns the land, and would only depend on land characteristics and exogenous prices.

For various reasons, markets for non-land factors of production are imperfect. These reasons include the immobility and spatial dispersion of land and poor infrastructure that already have been pointed out, but also the specific characteristics of some of these other factors of production. First, rural labour markets tend not to function well and hired labour does not represent a perfect substitute to family labour in farm production (Feder, 1985). This is because of the moral hazard problems due to drudgery aversion of workers and risks of opportunistic behaviour leading to a higher supervision requirement for hired labour. Second, hiring of labour may require cash outlays before the benefits of the work can be harvested and sold. Cash shortages and limited access to credit markets may thus also limit hiring of labour and consequently production efficiency. Third, the dominance of and seasonality in rain-fed agriculture due to limited investment in irrigation in Africa, causes synchronised production with related patterns of input demand and output supply where timing of operations becomes crucial for productivity and for profitability. Covariate risk in production also affects market risk when marketing constraints causes output prices to be negatively correlated with realized production outcomes. Risk in production and risk aversion may affect *ex-ante* and *ex-post* adjustments in behaviour in ways that undermine production efficiency as well as the functioning of input and output markets. Risk also introduces a need for insurance markets or substitutes to such markets. Such substitutes may be in form of credit markets, buffer stocks of essential commodities or livestock, or family networks. Even land itself plays an important role as a safety net in many poor economies where the non-farm economy is poorly developed and access to non-farm sources of income is constrained. The right to have access to land has therefore even been seen as a basic human right and has been implemented by law with a strong emphasis on an egalitarian distribution of land, e.g. in Ethiopia. Fourth, imperfect information and transaction costs cause pervasive imperfections in inter-temporal markets (credit and insurance) in poor rural economies even though some of the new innovative approaches in microfinance. can reduce some of these asymmetries and costs.

Mechanisation leading to economies of scale has played a modest role in tropical agriculture expect in certain types of export-oriented production, but also here the economies of scale arise mainly from the processing and marketing side and not so much from the production side itself (Deininger and Binswanger, 1999). With increasing land fragmentation due to high population pressures economies of scale may also become important on very small farms when mechanisation in form of use of oxen for ploughing are an essential part of the production technology since oxen are indivisible and a pair of oxen is required for ploughing, and timing of the operation is crucial for land productivity. Markets for oxen ploughing services tend therefore to function poorly (Bliss and Stern, 1982). This is also related to the vulnerability of animals to mismanagement and the critical timing of operations where owners of oxen give priority to their own fields before they eventually are willing to provide ploughing services to others.

A consequence of the above facts is that land rental markets tend to work better than oxen rental market, making land more mobile in an economic sense than oxen, even though oxen are more mobile in a biophysical sense (Tikabo and Holden 2007; Tikabo *et al.*, 2007).

The fact that non-land factor markets are imperfect creates a potential for land markets to play a role to enhance production efficiency in order to equalize factor ratios across farms. The efficiency-enhancing effects of such markets depend on the degree of imbalances in the factor ratios before the land transaction, the elasticity of substitution among factors of production and the transaction costs involved in alternative factor markets.

2.2. Land sales vs. land rental markets

It is very important to distinguish between land sales and land rental markets. In many countries land rental markets tend to be much more active than land sales markets, not only because land sales markets are prohibited by law. One of the advantages of land rentals are that they can be used for temporary adjustment of factors of production as a response to shocks. Land sales have much more long-term consequences and require much larger amounts of money which may depend on access to credit markets and typically disfavour poor people on the demand side for that reason. On the supply side of land sales markets the poor may be forced to sell their land due to shocks or economic hardships ('distress sales') and such sales tend to take place at unfavourable prices due to the circumstances, while the wealthy are capable of taking advantages of such situations and accumulate land. This is also one of the reasons why land sales markets are prohibited in many countries.

But how significant is this problem? Under what conditions is it a problem and are necessarily prohibitions of land sales the best solution? Distress sales may take place because they are the best option of the poor in that specific situation because other alternatives are inferior or non-existent. This is a likely consequence of imperfect credit and insurance markets. The policy solution of prohibiting land sales may even make the poor worse off in their acute situation unless alternative solutions to selling of land are provided.

Another reason for prohibiting land sales may be to avoid rapidly increasing landlessness and excessive migration to urban areas. The question, however, is also whether and how many would sell their land if they were given the opportunity, and whether they necessarily would sell all their land or just a part of it. A third reason may be for the state to keep control over the land and be able to access it easily and without having to pay too much compensation to those who lose their land. There may also be development and growth benefits from such a property rights regime, like in China.

The widespread prohibition of land sales also causes land rental markets to play a more important role, even though there have been many attempts to regulate and even prohibit such markets as well. One of the reasons for such regulations has been the historical experiences with exploitative contracts utilised by rich and powerful landlords to maximise their returns from poor tenants that lack or have very unfavourable exit options. Secondly, sharecropping has been prohibited because of the belief in Marshallian inefficiency as a necessary outcome of such contractual arrangements (Marshall, 1890). More recent theoretical contributions (Cheung, 1969; Stiglitz, 1974) provided a rationale for the existence and persistence of sharecropping and triggered a large theoretical inquiry into the logic of such contracts under varying assumptions and circumstances. On the other hand, empirical studies to assess the efficiency impacts started later. In Asia the search for Marshallian inefficiency has provided mixed results (Otsuka *et al.*, 1992). Still, there are few very good empirical studies that control properly for endogeneity of contract choice, technology choice and unobservable heterogeneity of the agents in the rental market.

2.3. Land markets and efficiency of land use

Economies of scale related to mechanisation of agriculture in the Western world has caused the optimal and viable farm sizes to grow. At the same time, owners of small farms have tended to hesitate with selling their land, causing a rapid increase in land renting. Adjustments through land sales typically take more time due to higher transaction costs in the land sales market, risk and uncertainty and imperfections in inter-temporal markets.

At the other extreme end, in some of the most densely populated rural areas in Asia and Africa, land fragmentation has reached levels where further fragmentation would lead to inefficiencies and farms have become too small to provide a subsistence income for the owners. Land renting and migration may here also be an important element of a survival strategy both for the landless and those who look for off-farm survival opportunities. Land renting provides an exit option for some households and an entry opportunity for other households and may thus be a step on the ladder out of poverty.

What are the short-term and longer-term efficiency and growth consequences of land rental markets? If we ignore the issue of contract choice for the moment, will a land rental market ensure that rented land is treated in the same way as owner-operated land? It is to expect that those who are willing and able to rent in land are also able to use the land more efficiently than the landlord who is willing to rent it out at the agreed price. But do the short-term productivity effects ensure that the rented land is conserved and invested on such that its future productivity is ensured and enhanced? There are many studies on how tenure security may affect investment (Otsuka, 2007) but there are very few studies that study the management of rented vs. owner-operated land, an exception being Dubois (Dubois, 2002). With short-term contracts there are reasons to question whether rented land is managed in a sustainable way.

3. Historical tenancy markets in Africa

Before we go to a condensed review of studies of land markets in Africa with emphasis on their efficiency and equity impacts, we give a couple of examples of the historical situation in parts of Africa that were or were not exposed to colonial rule, and the role of tenancy in those cases.

3.1. Ethiopia

Tenancy is far from new in Africa. Tenancy was an important element of the feudal system, like in Ethiopia before the land reform in 1974, and it was also an important element of the colonial system to extract surpluses from the rural economies, like in Malawi. Particularly in Southern and Central Ethiopia, land distribution was skewed and tenants operated a large share of the land. Various sources reported by Rahmato have estimated that more than 40% of the land was operated by tenants and tenants represented more than one third of the total population in the country very good historical data do not exist (Rahmato, 1984). Sharecropping was the dominant contract form, contracts were oral and the rent was one third or one half of the output, depending on the land quality.

Tenure insecurity was prevalent for tenants; '*The threat of eviction, rather than the act in itself, was the potent weapon in the hands of the landlords, and the tenant over whom the danger of unemployment and destitution hung like the sword of Democles had no alternative but to accommodate all the demands of his landlord..*' (Rahmato, 1984, p.25-26). The large landlords did not usually live in the rural areas and did not deal with their many tenants themselves but had agents who took care of this business for them.

It is therefore not strange that the term 'landlord' today is linked to this feudal history and that Ethiopians find it strange to use the same term for female-headed households that rent out their land because they lack oxen and the other necessary resources to farm their land themselves. The land reform that lead to a very egalitarian land distribution was complemented with more or less frequent land redistributions to maintain the egalitarian distribution and to provide land to new households. The reform drastically changed the power balance between 'landlords' and 'tenants' although the basic contractual form in terms of share tenancy and use of oral contracts is pretty much the same. All land was made state property and land rentals as well as hiring of labour were illegal under the Derg regime (1974-91). After the change in government in 1991 land short-term land renting and hiring of labour were legalised while land redistributions were mostly stopped with a few exceptions. Recent land reforms in form of providing households with inheritable user certificates may also have contributed to tenure security and enhancement of land rental markets.

3.2. Malawi

Land tenancy played an important role in the colonial history of Malawi. Africans became tenants on European controlled farms in the Shire highlands in a tenancy system called *thangata* (meaning 'to assist' in Chichewa). The system implied an exchange of labour for land. The landlords provided land to the tenants while the tenants were expected to work for the landlords for a given period. The Africans perceived the system as a symbol of European oppression while the settlers viewed it as reciprocal assistance. Abusive practices were common and contributed to the resistance against the system.

Europeans started to settle in the Central Region in the 1920s where land was more abundant. Unlike in the Southern Region where it was shortage of land that created tenancy, it was a demand for cash income that attracted tenants. The tenancy system that was established was called 'Visiting Tenancy System' and may be seen as a form of contract farming and was fairly lucrative for the tenants compared to subsistence farming.

In the post-colonial period tenants became an important component in the expansion of burley tobacco estates in Central Region of Malawi. The number of tenants increased from 20,000 in 1980 to 105,000 in 1989. Tenants rather than workers cultivated most of the land on the new estates. Several studies found that the average annual net returns of tenants were better than those of estate workers but tenants also faced higher risk and their incomes were therefore more skewed and many of them received no income.

There is evidence of exploitative tenancy contracts contributing to high turnover rates (>30%) for tenants. These included high interest rates on credit of 40-140% that tenants obtained from their landlords while the estate landlords were favoured with subsidised access to credit at negative real interest rates. Furthermore, the tenants were paid only a fraction of the tobacco auction price, and this fraction was set as a maximum fraction and not a minimum fraction, and was determined by the Tobacco Association of Malawi (TAMA). This maximum fraction declined from 41% in 1983 to 19% in 1988. This regulation mechanism favoured the estate owners at the expense of the tenants. It made them able to underpay the tenants but this was also done through under-grading of tobacco purchased from tenants, delayed payment and payment in instalments. One study found that tenants only received cash payment of 10% of the value of the crop after the credit had been paid on one estate.

The tobacco tenancy model as it was practiced in the 1980s seems not to be a good instrument for helping tenants out of poverty. The removal of the restrictions on smallholders' right to grow tobacco in the 1990s may have made tenancy contracts less exploitative but no recent good studies on tenancy on tobacco estates have been carried out. The New Land Policy in Malawi does not address tenancy issues and this is a serious deficit in the policy. Land rental markets are also emerging in other parts of the country and it is of high policy relevance to investigate how these tenancy markets may help land-poor households to access land and how the land policy reforms should be shaped to make land rental markets pro-poor (Holden, et al., 2006).

4. Importance of land markets in Africa: implications for efficiency

4.1. Allocative efficiency of land markets

It is worth looking at some of the basic references on analysis of allocative efficiency of land markets (Bell and Sussangkarn, 1988; Bliss and Stern, 1982; Rahmato, 1984; Skoufias, 1995). Bliss and Stern studied the land rental market in Palanpur village in India using a theoretical model to assess whether the land adjustment in the rental market was complete ('desired') or less than desired due to transaction costs in the market. They found evidence of incomplete adjustment.(Bell and Sussangkarn, 1988, Bliss and Stern, 1982, Skoufias, 1995)

Bell and Sussangkarn (1988) suggested that the market for tenancies may not clear in Walrasian fashion because of adverse selection and moral hazard problems, as some households that want to lease land may be rationed. They suggested that this is due to high transaction costs, particularly for landlords. Their study was in an Indian setting with absentee landlords, a well functioning labour market and many landless households. Similarly, Bliss and Stern (1982, p.54-65) also argued that sharecropping involved moral hazard problems and a strong likelihood of rationing. This is because sharecropping does not involve a market-clearing price in the land rental market, leading to an asymmetry between landlords and tenants and to rationing on the tenant side of the market.

Skoufias developed a model with transaction costs that he tested on a data set from India with censored tobit models on each side of the market (Skoufias, 1995). He found signs of significant asymmetries across the two sides of the market and transaction costs in the market.

Kevane was the first to test the allocative efficiency model of Bliss and Stern in Africa, in Western Sudan, where he found the adjustment in the land rental market to be efficient (Kevane, 1997). In a study in Eritrea, Tikabo and Holden (2007) found significant transaction costs in the land rental market causing only partial adjustment. They also found significant asymmetries in the market and selection bias on the tenant side of the market where different factors appear to affect whether to participate or not vs. the degree of participation. Teklu and Lemi analysed rental market behaviour on the landlord side in Southern Ethiopia using a Heckman selection model (Teklu and Lemi, 2004). They also found an adjustment coefficient that indicated significant transaction costs (significantly less than one) in relation to land adjustment and tested and rejected the single stage model while they found significant selection bias. Households with more labour force and with more healthy heads of households (predicted body mass index) were found to lease out less land.

Pender and Fafchamps (2005) also tested for transaction costs and the allocative efficiency of the land rental market using a dataset from 1994 in Arsi in Eastern Ethiopia. They use a Censored Least Absolute Deviations (CLAD) model for the test since they had to reject the tobit model because of lack of normality. They did not find significant inefficiency in the land rental market but this could be due to the relatively small sample size (only 72 uncensored observations) and the fact that the CLAD model is demanding with respect to sample size. They found that oxen endowment significantly affected land renting activity showing that the oxen factor market was imperfect and that the land rental market helped to equalize factor ratios across households.

Benin *et al.* (2005) found evidence of expansion in land rental market activity in the Amhara Region of Ethiopia during the 1990s in a community survey with group interviews in about 100 communities where the respondents were asked to assess the changes that had taken place (Benin *et al.*, 2005).

Holden and Ghebru (2005) analysed the adjustment in the land rental market in Tigray in northern Ethiopia with a focus on the role of kinship and trust in relation to access and participation in the market. Building on Fafchamps (Fafchamps, 2004) they developed a theoretical model for entry and participation in the land rental market as a landlord or as a tenant. The basic idea is that trust among contracting partners build up gradually over time and may depend on kinship relationships (Holden and Ghebru, 2005). They found a high share of kinship contracts in Tigray and that tenants appeared to have easier access to the market in communities with a high share of kinship contracts but adjustment for those inside the market appeared to be no more efficient for contracts among kin than among non-kin. Overall they found significant transaction costs in the market and indications of non-convex transaction costs creating entry barriers while those inside the market appeared to face lower transaction costs. Their findings also appeared to contradict the basic assumption of constant returns to scale that is an underlying assumption in the Bliss and Stern (1982) and Skoufias' (1995) models. It is likely that economies of scale occur when farm sizes become very small and a pair of oxen is required for ploughing due to the lumpiness of this essential factor of production, and due to the difficulties of establishing an efficient market for its services.

Deininger and Mpuga (2003) assessed the performance of land rental and land sales markets in Uganda. Based on a national sample they found that about 29% of the land had been purchased and that about 7% of the land was rented, while about 13% of the households were renting in land. The adjustment coefficients on area owned were significantly below one in absolute value demonstrating transaction costs in the market and the coefficients were close to but significantly different from zero for the area leased out (Deininger and Mpuga, 2003).

4.2. Contract choice and production efficiency

Sharecropping caught the interest of the classical economists as an inefficient institutional arrangement (Marshall, 1890). However, seminal contributions by Cheung (1969) and Stiglitz (1974) provided new light and renewed interest in analysing the efficiency-implications of sharecropping (Cheung, 1969; Stiglitz, 1974). A large theoretical literature has developed since then and many empirical studies have been carried out, especially in Asia. Many of the studies in Asia have been reviewed by Otsuka (Otsuka *et al.*, 1992; Otsuka and Hayami, 1988). They do not find any general evidence of Marshallian inefficiency and explain this by the endogeneity of contract choice; sharecropping may occur mainly in cases where it works efficiently while alternative contracts will be preferred where sharecropping would be inefficient. Another problem is that many of the studies are flawed by methodological weaknesses.

Still, there are some studies that identified significant Marshallian inefficiency and also use credible methodologies. The most famous among these is the study by Shaban in India (Shaban, 1987), who found a 16% difference in yields on sharecropped and owneroperated land of the same tenants. He compared productivity on tenants' owner-operated and sharecropped-in plots while controlling for plot and household characteristics. Otsuka, however, suggests that the Marshallian inefficiency found in India is due to policy distortions ('Land-to-the-tiller'-policies) causing sharecropping contracts to be short-term and therefore inefficient (Otsuka, 2007).

There are relatively few studies of the efficiency of sharecropping in Africa. However, we will give a brief overview of the existing studies.

Laffont and Matoussi found significant evidence of Marshallian inefficiency in a study in Tunisia (Laffont and Matoussi, 1995). Due to financial constraints some tenants preferred sharecropping to fixed-rent contracts and such contracts were associated with lower land productivity[59]. One may therefore attribute the lower efficiency on sharecropped plots to the financial constraint rather than to the sharecropping contract per se as it is not known whether a prohibition of sharecropping in such an environment necessarily would enhance production efficiency. This may instead have reduced the allocative efficiency in the land rental market which sharecropping contracts may have enhanced by relaxing the financial constraint.

There are three studies with the same dataset from Arsi in Ethiopia but with different theoretical and methodological approaches.

Pender and Fafchamps (2006) develop a theoretical model of tenant and landlord behaviour taking transaction costs, Marshallian inefficiency, and monitoring costs into account. They demonstrate that positive monitoring costs will cause the landlord to restrict area rented to tenants such that the tenant's yield on the leased-in plot should be higher than the landlord's yield on his own plot. In their empirical analysis they fail to find evidence of this, however, possibly due to the small sample size which would have required the yield difference to be as high as 40%. Similarly, they also failed to find evidence of Marshallian inefficiency when comparing input levels and output values of owner-operated and rented-in plots of tenant households, using household fixed effects to control for unobservable household characteristics. Here also the sample size of only 64 fixed-rent plots, 31 sharecropped plots and 149 owner-operated plots may be the main reason for lack of significant differences.

Earlier tests of the same dataset by Gavian and Ehui (1999) found total factor productivity to be lower on contracted land (fixed rent, sharecropped, borrowed or gifted land) than on owner-operated land but they found no significant difference in input use (Gavian and Ehui, 1999).

Using a stochastic frontier approach on the same dataset, Ahmed *et al.* (2002) found significant inefficiency on sharecropped land but not so on land under fixed-rent contracts. This demonstrates that different methods applied to the same dataset has lead to different conclusions. However, the household fixed effects approach used by Pender and Fafchamps (2006) must be considered to be the preferred approach as it controls for unobservable heterogeneity of households. The requirement of the stochastic frontier models that only strictly exogenous variables should be included in the first stage, while input use and crop choice were included in this stage in the analysis by Ahmed *et al.*, variables that are clearly endogenous, may give reason to be cautious when interpreting the results. It is also not clear why technical inefficiency should be higher on rented land than on owner-operated land

---

[59] The study did not control for possible systematic differences in land quality between rented plots under different contracts and owner-operated plots.

since Marshallian inefficiency from theory should lead to a reduction in input use (i.e., allocative inefficiency rather than technical inefficiency).

Among other studies in Ethiopia, Holden *et al.* (2001) found land productivity to be significantly higher on rented land than on owner-operated land (Holden *et al.*, 2001). However, this study mixed owner-operated, rented-in and rented-out plots and did therefore not correct for unobservable household characteristics. The finding could, however, possibly be explained by significant transaction costs as suggested by the model of Pender and Fafchamps (2006).

Kassie and Holden (2006) developed a sharecropping model with threat of eviction and kinship, where kinship could reduce the problem of Marshallian inefficiency but also reduce the incentive effect of threat of eviction. They tested the model with a dataset from Western Gojjam in Ethiopia. Using household fixed effects models and comparing productivity on tenants' own plots with that on rented-in plots they found higher productivity on sharecropped land than on owner-operated land of tenants. Sharecropped plots of non-kin tenants also received significantly more fertiliser than other plots. They also found non-kin sharecropped plots to be associated with higher productivity than kin sharecropped plots. The results were consistent with the threat of eviction effect dominating over the Marshallian inefficiency effect and kinship reducing the threat of eviction effect.

In Eritrea, Tikabo and Holden (2007) found land productivity to be significantly higher for owner-tenants than for pure owner-operators and owner-landlords showing that land is transferred from less efficient to more efficient land users (Tikabo and Holden, 2004). The fact that the productivity differential persisted after participation in the land rental market indicates that there are significant transaction costs in the land rental market. Using household fixed effects, they also found significantly higher land productivity on plots with costsharing contracts than on owner-operated and pure sharcropped plots of tenants, while there was no significant difference between the owner-operated and sharecropped plots.

In the Amhara Region of Ethiopia Bezabih and Holden found higher land productivity on land rented out by male-headed landlord households than by female-headed landlord households. The difference was attributed to lower tenure security and lower bargaining power of female-headed households. Female (widows or divorced) heads of households may sometimes be forced to rent out their land to their in-laws such that coersion may be involved (Bezabih and Holden, 2006). Coersion may therefore cause inefficiency but also has equity implications.

5. Importance of land markets in Africa: implications for equity

Are land markets good or bad for equity? Economic theory does not give a clear answer to this. Land sales markets and land rental markets may also have different effects on equity. It is also appropriate to ask: equity of what? We may look at different effects of land markets; a) the distribution of land in absolute terms, b) the relative distribution of factors of production, and c) the distribution of welfare. We will look at the available empirical evidence on these matters in Africa.

5.1. Evidence on changes in the land distribution

The historical experience has implied that land rental markets have transferred land from land-rich households to land-poor and often landless households in situations where the initial land distribution was inequitable due to the historical power relations. This pattern may for several reasons not necessarily hold today. For example, mechanisation in agricultural production has caused an increase in optimal farm sizes (e.g. in Norway, Japan, parts of India) and this has caused small farms to rent out their land to larger farms that can take the advantages of economies of scale.

It has also frequently been stated that land sales markets cause a concentration of land on the hands of the wealthy but how common are distress sales? We will look at some of the empirical evidence from Africa. André and Platteau (1998) found that inequity of land had increased over a period of five years (1988-93) in Western Rwanda where two-thirds of the land sales were distress sales to meet subsistence and medical expenses (André and Platteau, 1998). Dubuisson (1998, cited in Baland *et al.*, 2000) found that land markets reduced inequity in the land distribution in a study in Benin as land was purchased by land-poor households from land-rich households (Dubuisson, 1998). Similarly, Place and Migot-Adholla (1998) found that landless households purchased land in Kenya (Place and Migot-Adholla, 1998). Baland *et al.* (2000) studied the distributive impact of land markets in 36 villages in East and Central Uganda. They concluded that both the land sales and land rental markets corrected initial inequality in land endowments. It appeared that what households failed to adjust in their land size through the land sales market they compensated by additional adjustment through the land rental market to come closer to their desired land area. Deininger and Mpuga (2003) found that land rental markets, but not land sales markets, transferred land to more efficient and relatively poor producers in Uganda, making land available for the landless. They found no evidence of land sales markets leading to concentration of land. Holden *et al.* (2006) provided evidence of land-poor households accessing land through emerging land rental markets in Malawi, where the land distribution is inequitable.

After the land reform in Ethiopia in 1975 and through later land redistributions there was an equitable distribution of land within villages. However, with the growing number of young landless households, the skewness in the distribution of land has again increased. While land sales markets are prohibited, the land rental markets are very active, driven by inequities in non-land factors of prodution, like oxen and labour endowments. Oxen- and labour-poor households, often female-headed household, are forced to rent out their land to households with more of these endowments. This typically leads to a more skewed distribution of operated holdings than own holdings (Holden and Ghebru, 2005; Kassie and Holden, 2006). Similarly, in Eritrea Tikabo *et al.* (2007) found that land moved through the rental market to households with oxen and more labour endowments and causing a more skewed distribution of operated land holdings.

5.2. Evidence on changes in relative factor ratios

If land markets imply that factors of production are reallocated to more efficient producers this should lead to less skewed distribution of relative factor ratios across farms after the land transactions than before the transactions. As discussed above the issue whether land is redistributed to other factors of production or vice versa, depends on the costs and benefits involved with the redistribution. With perfect markets for non-land factors of production land markets would not contribute to enhance efficiency. Only when transaction costs are lower in land markets than in other factor markets would there be efficiency gains from land market transactions. The initial distribution of land and non-land factors of production and the size of the transaction costs would determine the degree of factor adjustment in the factor markets. We review some of the empirical evidence from Africa.

Tikabo *et al.* (2007) found an egalitarian distribution of land owned in Eritrea where mean area of owned land was smaller for landlord households (3.28 tsimdi) than for tenant households (3.47 tsimdi). Landlord households were characterised by having lower endowments of oxen and male labour than tenant households. Landlord households therefore had larger amounts of land per adult labour unit in the household and the land rental market redistributed land such that operated land holding per adult labour unit was evened out. Similarly, the land rental market contributed to even out the factor ratio between oxen and land. The overall consequence was that households with more endowments of non-land factors of production also carried out more of the agricultural production because they were more efficient producers and because the transaction costs appeared to be lower in the land rental market than in the markets for labour and oxen ploughing services.

Holden and Ghebru (2005) also found in their study in Tigray, Ethiopia, that households without oxen and with little male labour force were more likely to rent out land and households with more oxen and male labour were more likely to rent in land, in line with the idea that land markets serve to even out factor ratios across households.

5.3. Evidence on welfare effects

In principle, if a household has the voluntary choice between participating in a market and not, and we believe the household maximises its utility to be better off in the cases when it chooses to participate in the market. The appropriate counterfactual would be the welfare situation of the same household if it were rationed out of that market. Therefore, comparing with households that voluntarily do not participate in the market is a wrong. This simple example illustrates in a nutshell the difficulty of measuring welfare effects of land markets.

In some cases, land market participation may be the best choice in a second-best world, where short-term needs for food or to cover medical expenses take presedence over long-term needs due to lack of functioning credit and insurance markets. In the short run this may be a question of life or death while in the longer run the land market 'causes' the household to fall into poverty.

Land markets may provide an opportunity for landless or land-poor households to access land and thus a basis for survival and poverty reduction according to the agricultural ladder hypothesis. The agricultural ladder hypothesis stems from England in the 19th century (Wehrwein, 1958). Spillman (1917) identified four different ladders based on more than 2000 farms in Illinois, Iowa, Kansas, Nebraska and Minnesota in the USA (Spillman, 1917, 1919). The steps in the ladders included; working on the home farm with or without wages (F), working as hired labour on other farms (H), tenancy (T) and ownership (O) with the following alternative ladders; F-H-T-O, F-H-O, F-T-O, and F-O. Hallagan (1978), based on the agricultural ladder hypothesis, introduced a model for self-selection into land rental contracts based on the characteristics of tenants, allowing landlords to screen their tenants (Hallagan, 1978).

How relevant is the agricultural ladder hypothesis to understand the development pathways and the role of land markets in Africa today? Can land markets serve as a ladder out of poverty? One pathway could be if the landless could move from being wage labourers to become tenants and then, if land sales were legal, they could buy land and become owner-operators. The other pathway could be that they could sell their land and migrate to another place, with better opportunities, with sufficient funds from their sale to establish a new business. The new business may not necessarily be in agriculture. What evidence of such pathways do we see in Africa and how successful are those using these ladders?

Or do land markets represent the snakes in the 'snakes and ladders game' that the poor are forced to play? If land sales and land rentals are in form of distress sales and rentals at very unfavourable conditions because of lack of alternative safety nets, the land markets may primarily serve as a form of expensive insurance.

The empirical evidence appears quite limited. It is particularly difficult in surveys to trace households that have outmigrated and few studies do this. The study by Baland *et al.* (2000) reveals that in Uganda, where both land rental and land sales markets are coexisting, that households may sell their land in one area and migrate to another area, possibly to buy land there. The study by Deininger and Mpuga (2003) indicate that the 'ladder'-effect may be more important than the 'snake'-effect in Uganda. The study by André and Platteau (1998) indicate that the 'snake'-effect was important in Rwanda before the genocide.

Land rental markets appear to be emerging in Malawi as an opportunity for landless or near landless households (Holden *et al.*, 2006). More research is needed, however, to determine how important this pathway is and to assess how successful households chosing it are.

In a study in the Amhara Region in Ethiopia, using community level data based on group internviews where the group was asked to explain the changes that have taken place during the 1990s, Benin *et al.* (2005) found that land rental market activity has increased. They found that a decline in the use of fixed-rent contracts was associated with increasing landlessness and preference for sharecropping among the young landless households, in line with the agricultural ladder hypothesis.

The perspective of land as a safety net is central in the policies of some countries, like Ethiopia, where all residents in a rural community, according to the land proclamation, have a right to access land to meet their subsistence needs. However, with the growing population pressure and prohibition of further land fragmentation, the role of land as a safety net is slowly eroding and enhancing vulnerability of the land-poor. The number of young landless households is growing rapidly and their ability to access land through the land rental market may depend on their farming capital in form of oxen, farm implements and the characteristics of the rental market (Holden and Ghebru, 2005; Kassie and Holden, 2006). Further studies on this are recommended using panel data to trace the alternative pathways and the related welfare effects.

In Eritrea, Tikabo and Holden (2007) found that poor female-headed households tended to rent out their land to wealthier, usually male-headed households. A similar pattern of poor landlords and wealthier tenants has also been found in Ethiopia (Bezabih and Holden, 2006; Holden and Ghebru, 2005; Kassie and Holden, 2006) and in Madagascar (Bellemare, 2006). In Eritrea and Ethiopia it appears that the land rental market serves as an important opportunity for poor households that are too poor to farm their land efficiently themselves. By sharecropping out their land to households that are better endowed with non-land factors of production they can get an income from their land in form of food which provide them with the food security that they need.

Also the study by Laffont and Matoussi in Tunisia revealed that the tenants had higher levels of working capital than landlords (Laffont and Matoussi, 1995). They also found that the amount of working capital of landlords and tenants affected contract choice. Tenants with more working capital and landlords with less working capital were more likely to have fixed-rent contracts rather than sharecropping contracts. This may indicate that sharecropping is a response to credit constraints on the tenant side while fixed-rent contracts may be a sign on distress rentals due to credit constraints on the landlord side. Wealth of tenants appeared to play a less significant role in contract choice than the financial constraints. Tikabo and Holden (2003) found that wealthier tenants and poorer landlords were more likely to have fixed-rent contracts and less wealthy tenants and more wealthy landlords were more likely to have cost-sharing contracts, in line with the credit constraint hypothesis. However, if poverty also is correlated with risk aversion such that the wealthier are less risk averse, this could also explain the same pattern of contract choice and matching in the tenancy market. The study was unable to distinguish between these possible alternative explanations.

6. Conclusions

Land markets are growing in importance in Africa and will have to play an important role to stimulate economic growth and reduce poverty in the future. This chapter briefly reviews some of the empirical evidence on the efficiency and equity implications of land markets in Africa. Many questions are left unanswered demonstrating the need for further research. The review reveals that there are complex interactions between land markets, the distribution of resources and the characteristics of non-land factor markets including credit and insurance markets. Future land policies should pay careful attention to these complex interactions and the dynamics of change before attempts are made to regulate these markets since history has shown that well-intended interventions often have had unintended and counter-productive effects.

Much of the research highlights that land rental markets often are beneficial for the poor and policies that facilitate land rental markets should therefore in general be good for poverty-reduction. When it comes to land sales markets, the evidence is more mixed and we do not draw any strong conclusions. It is possible that land sales markets are not as bad as their reputation and that they may benefit the poor if proper land policies are in place. However, further research should be conducted to study these markets. Land policies in general should be formulated to enhance efficient utilization of land while protecting the land rights of the poor. Where policy distortions hinder efficient utilisation of land, such distortions should be attempted removed while minimising the potential conflicts involved.

References

Ahmed, M.M., B. Gebremedhin, S. Benin and S. Ehui, 2002. Measurement and sources of technical efficiency of land tenure contracts in Ethiopia. Environment and Development Economics 7: 507-527.

André, C., and J.P. Platteau, 1998. Land Tenure under Unbearable Stress: Rwanda Caught in the Malthusian Trap. Journal of Economic Behavior and Organization 34: 1-47.

Baland, J.M., F. Gaspart, F. Place and J.-P. Platteau, 2000. The Distributive Impact of Land Markets in Uganda. mimeo, CRED, Université de Namur, Belgique.

Bell, C. and C. Sussangkarn, 1988. Rationing and adjustment in the market for tenancies: The behavior of landowning households in Thanjavur district. American Journal of Agricultural Economics.

Bellemare, M., 2006. Testing between Competing Theories of Reverse Share Tenancy. Duke University.

Benin, S., M. Ahmedb, J. Penderc and S. Ehuid, 2005. Development of Land Rental Markets and Agricultural Productivity Growth: The Case of Northern Ethiopia. Journal of African Economies 14: 21-54.

Bezabih, M. and S.T. Holden, 2006. Tenure Insecurity, Transaction Csots in the Land Lease Market and their Implications for Gendered Productivity Differentials. Brisbane, Australia, International Association of Agricultural Economists.

Binswanger, H. P. and M. Rosenzweig, 1986. Behavioural and Material Determinants of Production Relations in Agriculture. Journal of Development Studies 22: 503-539.

Bliss, C.J. and N.H. Stern, 1982. Palanapur: The Economy of an Indian Village. Delhi and New York: Oxford University Press.

Cheung, S.N.S., 1969 The Theory of Share Tenancy. Chicago: University of Chicago Press.

De Soto, H., 2000. The Mystery of Capital. Why Capitalism Triumphs in the West and Fails Everywhere Else. New York: Basic Books.

Deininger, K., 2003. Land policies for growth and poverty reduction. A World Bank Research Report. The World Bank.

Deininger, K. and H. Binswanger, 1999. The evolution of World Bank's land policy: principles, experience, and future challenges. The World Bank Research Observer 14: 247-76.

Deininger, K. and P. Mpuga, 2003. Land markets in Uganda: Incidence, impact and evolution over time, Proceedings of the 25th IAAE meeting (Durban, RSA).

Dubois, P., 2002. Moral Hazard, Land Fertility and Sharecropping in a Rural Area of the Philippines. Journal of Development Economics 68: 35-64.

Dubuisson, J.F., 1988. Evolution du régime foncier et intensification de l'agriculture en Afrique sub-saharienne: le cas d'un village du Sud Bénin. M.A. Thesis, Faculté des Sciences économique, sociales et de gestion, University of Namur.

Fafchamps, M., 2004. Market Institutions in Sub-Saharan Africa. Massachusettes & London: The MIT Press, 2004.

Feder, G., 1985. The relation between farm size and farm productivity: the role of family labor, supervision and credit constraints. Journal of Development Economics 18: 297-313.

Gavian, S. and S. Ehui, 1999. Measuring the Production Efficiency of Alternative Land Tenure Contracts in a Mixed Crop-Livestock System in Ethiopia. Agricultural Economics 20: 37-49.

Hallagan, W., 1978. Self-selection by contractual choice and the theory of sharecropping. Bell Journal of Economics 9: 344-54.

Holden, S., B. Shiferaw and J. Pender, 2001. Market imperfections and land productivity in the Ethiopian highlands. Journal of Agricultural Economics 52: 53-70.

Holden, S.T. and H. Ghebru, 2005. Kinship, Transaction Costs and Land Rental Market Participation. Ås. Working Paper.

Holden, S.T., R. Kaarhus and R. Lunduka, 2006. Land Policy Reform: The Role of Land Markets and Women's Land Rights in Malawi. Norwegian University of Life Sciences. Working Paper.

Kassie, M. and S.T. Holden, 2006. Sharecropping Efficiency in Ethiopia: Threats of Eviction and Kinship. Brisbane, Australia, International Association of Agricultural Economists.

Kevane, M., 1997. Land tenure and rental in Western Sudan. Land Use Policy 14: 295-310.

Laffont, J.J. and M.S. Matoussi, 1995. Moral hazard, Financial Constraints and Sharecropping in El Oulja. Review of Economic Studies 62: 381-399.

Marshall, A., 1890. The Principles of Economics. Vol.I. London: MAcMillan.

Otsuka, K., 2007. Efficiency and Equity Effects of Land Markets. In: P. Pingali (ed.), vol. 3, Elsevier, pp. 2678-2709.

Otsuka, K., A.H. Chuma and Y. Hayami, 1992. Land and Labour Contracts in Agrarian Economies: Theories and facts. Journal of Economic Literature 30: 1965-2018.

Otsuka, K. and Y. Hayami, 1988. Theories of share tenancy: a critical survey. Economic Development and Cultural Change 37: 32-68.

Pender, J. and M. Fafchamps, 2006. Land Lease Markets and Agricultural Efficiency in Ethiopia. Journal of African Economies 15(2): 251-284.

Place, F. and S.E. Migot-Adholla, 1998. Land Registration and Smallholder Farms in Kenya. Land Economics 74: 360-373.

Rahmato, D., 1984. Agrarian Reform in Ethiopia. Scandinavian Institute of African Studies.

Shaban, R.A., 1987. Testing between competing models of sharecropping. Journal of Political Economy 95: 893-920.

Singh, I., L. Squire and J. Strauss, 1986. Agricultural Household Models: Extensions, Applications and Policy. Washington, D.C., The World Bank and The John Hopkins University Press.

Skoufias, E., 1995. Household Resources, Transactions Costs, and Adjustment Through Land Tenancy. Land Economics 71: 42-56.

Spillman, W.J., 1919. The Agricultural Ladder. American Economic Review: 170-179.

Spillman, W.J., 1917. How Farmers Acquire Their Farms. Washington, D.C., pp. 87-90.

Stiglitz, J.E., 1974. Incentives and Risk Sharing in Sharecropping. Review of Economic Studies 41: 219-255.

Teklu, T. and A. Lemi, 2004. Factors affecting entry and intensity in informal rental land markets in Southern Ethiopian highlands. Agricultural Economics 30: 117-128

Tikabo, M.O. and S. Holden, 2004. Factor market imperfections and the land rental market in the highlands of Eritrea: Theory and evidence. Department of Economics and Resource Management, Norwegian University of Life Sciences.Discussion paper 12-2004

Tikabo, M.O. and S.T. Holden, 2007. Land Tenancy and Efficiency of Land Use in the Highlands of Eritrea. Department of Economics and Resource Management, Norwegian University of Life Sciences. Working Paper

Tikabo, M.O., S.T. Holden and O. Bergland, 2007. Factor Market Imperfections and the Land Rental Market in the Highlands of Eritrea: Theory and Evidence. Working Paper.

Wehrwein, C.F., 1958. An Analysis of Agricultural Ladder Research. Land Economics 34: 329-337.

From the green revolution to the gene revolution: how will the poor fare?

Prabhu Pingali and Terri Raney

Abstract

The past four decades have seen two waves of agricultural technology development and diffusion to developing countries. The first wave was initiated by the Green Revolution in which an explicit strategy for technology development and diffusion targeting poor farmers in poor countries made improved germplasm freely available as a public good. The second wave was generated by the Gene Revolution in which a global and largely private agricultural research system is creating improved agricultural technologies that flow to developing countries primarily through market transactions. The Green Revolution strategy for food crop productivity growth was based on the premise that, given appropriate institutional mechanisms, technology spillovers across political and agro-climatic boundaries can be captured. A number of significant asymmetries exist between developed and developing, e.g.: agricultural systems, market institutions and research and regulatory capacity. While farmers in some developing countries are benefiting from transgenic crops, these asymmetries raise doubts as to whether the Gene Revolution has the same capacity to generate widespread spillover benefits for the poor. A strong public sector – working cooperatively with the private sector – is essential to ensure that the poor benefit from the Gene Revolution.

1. Introduction

The past four decades have seen two waves of agricultural technology development and diffusion to developing countries. The first wave was initiated by the Green Revolution in which improved germplasm was made available to developing countries as a public good through an explicit strategy for technology development and diffusion. The second wave was generated by the Gene Revolution in which a global and largely private agricultural research system is creating improved agricultural technologies that are flowing to developing countries primarily through market transactions in the form of transgenic constructs imported and adapted for local conditions. Recent evidence shows that farmers in developing countries can benefit from these transgenic innovations (Raney, 2006), but asymmetries between developed and developing countries in research capacity, regulatory capacity and market institutions pose formidable challenges in bringing the potential of the Gene Revolution to bear on the problems of poor farmers in poor countries.

The Green Revolution was responsible for an extraordinary period of growth in food crop productivity in the developing world over the last forty years. Productivity growth has been significant for rice in Asia, wheat in irrigated and favorable production environments worldwide and maize in Mesoamerica and selected parts of Africa and Asia. A combination of high rates of investment in crop research, infrastructure and market development, and appropriate policy support fueled this land productivity. These elements of a Green Revolution strategy improved productivity growth despite increasing land scarcity and high land values (Pingali and Heisey, 2001).

The transformation of global food production systems defied conventional wisdom that agricultural technology does not travel well because it is either agro-climatically specific, as in the case of biological technology, or sensitive to relative factor prices, as with mechanical technology (Byerlee and Traxler, 2002). The Green Revolution strategy for food crop productivity growth was explicitly based on the premise that, given appropriate institutional mechanisms, technology spillovers across political and agro-climatic boundaries can be captured. Hence the Consultative Group on International Agricultural Research (CGIAR) was established specifically to generate spillovers particularly for nations that are unable to capture all the benefits of their research investments. What happens to the spillover benefits from agricultural research and development in an increasingly global integration of food supply systems?

Over the past decade the locus of agricultural research and development has shifted dramatically from the public to the private multinational sector. Three interrelated forces are transforming the system for supplying improved agricultural technologies to the world's farmers. The first is the strengthened and evolving environment for protecting intellectual property in plant innovations. The second is the rapid pace of discovery and growth in importance of molecular biology and genetic engineering. Finally, agricultural input and output trade is becoming more open in nearly all countries. These developments have created a powerful new set of incentives for private research investment, altering the structure of the public/private agricultural research endeavor, particularly with respect to crop improvement (Pingali and Traxler, 2002).

At the same time, developing countries are facing increasing transactions costs in access to and use of technologies generated by the multinational sector. Many developing countries lack the regulatory frameworks and trained personnel needed for assessing the safety of transgenic crops for consumers and the environment. Existing international networks for sharing technologies across countries and thereby maximizing spillover benefits are becoming increasingly threatened. The urgent need today is for a system of technology flows which preserves the incentives for private sector innovation while at the same time meeting the needs of poor farmers in the developing world.

2. Green revolution R&D: access and impact

The major breakthroughs in yield potential that kick started the Green Revolution in the late 1960s came from conventional plant breeding approaches. Crossing plants with different genetic backgrounds and selecting from among the progeny individual plants with desirable characteristics, repeated over several cycles/generations, led to plants/varieties with improved characteristics such as higher yields, improved disease resistance, improved nutritional quality, etc. The yield potential for the major cereals has continued to rise at a steady rate after the initial dramatic shifts in the 1960s for rice and wheat. For example, yield potential in irrigated wheat has been rising at the rate of 1 percent per year over the past three decades, an increase of around 100 kilograms per hectare per year (Pingali and Rajaram, 1999).

Prior to 1960, there was no formal system in place that provided plant breeders access to germplasm available beyond their borders. Since then, the international public sector (the CGIAR) has been the pre-dominant source of supply of improved germplasm developed from conventional breeding approaches, especially for self-pollinating crops such as rice and wheat and for open pollinated maize. CGIAR managed networks of international nurseries for sharing crop improvement results evolved in the 1970s and 1980s, when financial resources were expanding and plant IPR laws were weak or nonexistent.

The international flow of germplasm has had a large impact on the speed and the cost of NARSs crop development programs, thereby generating enormous efficiency gains (See Evenson and Gollin, 2003, for a global assessment of gains from the international exchange of major food crop varieties and breeding lines). Traxler and Pingali (1999) have argued that the existence of a free and uninhibited system of germplasm exchange that attracts the best of international materials allows countries to make strategic decisions on the extent to which they need to invest in plant breeding capacity. Small countries behaving rationally choose to free ride on the international system rather than invest in large crop breeding infrastructure of their own (Maredia *et al.*, 1994).

Evenson and Gollin (2003) report that even in the 1990s, the CGIAR content of modern varieties was high for most food crops; 36% of all varietal releases were based on CGIAR crosses. In addition, 26% of all modern varieties had a CGIAR-crossed parent or other ancestor. Evenson and Gollin (2003) suggest that germplasm contributions from international centers helped national programs to stave off the 'diminishing returns' to breeding that might have been expected to set in had the national programs been forced to work only with the pool of genetic resources that they had available at the beginning of the period.

2.1. Impacts of food crop improvement technology

Substantial empirical evidence exists on the production, productivity, income, and human welfare impacts of modern agricultural science and the international flow of modern varieties of food crops. Evenson and Gollin (2003) provided detailed information for all the major food crops on the extent of adoption and impact of modern variety use, they also show the crucial role played by the international germplasm networks in enabling developing countries to capture the spillover benefits of investments in crop improvement made outside their borders. The adoption of modern varieties during the first 20 years of the green revolution-aggregated across all crops-reached 9% in 1970 and rose to 29% in 1980. By the 1990, adoption of MVs had reached 46%, and by 1998 63%. Moreover, in many areas and in many crops, first generation modern varieties have been replaced by second and third generation modern varieties Evenson and Gollin (2003).

Much of the increase in agricultural output, over the past 40 years, has come from an increase in yields per hectare rather than an expansion of area under cultivation. For instance, FAO data indicate that for all developing countries, wheat yields rose by 208% from 1960 to 2000; rice yields rose 109%; maize yields rose 157%; potato yields rose 78%; and cassava yields rose 36% (FAOSTAT; FAO, 2005). Trends in total factor productivity (TFP) are consistent with partial productivity measures, such as rate of yield growth Pingali and Heisey (2001) provide a comprehensive compilation of TFP evidence for several countries and crops.

The returns to investments in high-yielding modern germplasm have been measured in great detail by several economists over the last few decades. These studies found high returns to the Green Revolution strategy of germplasm improvement. The very first studies that calculated the returns to research investment were conducted at IRRI for rice research investments in the Philippines (Flores-Moya *et al.*, 1978) and at CIAT in Colombia (Scobie and Posada, 1978). More detailed evidence on the high rates of return to public-sector investments in agricultural research was provided by the International Service for National Agricultural Research (ISNAR) (Echeverría, 1990) and the International Food Policy Research Institute (IFPRI) (Pardey *et al.*, 1992). For detailed synthesis of the numerous studies conducted across crops and countries, see Evenson (2001) and Alston *et al.* (2000). Alston *et al.* concluded from a review of 289 studies that there was no evidence that the rate of return to agricultural research and development has declined over time.

Widespread adoption of modern seed-fertilizer technology led to a significant shift in the food supply function, contributing to a fall in real food prices. The primary effect of agricultural research on the non-farm poor, as well as on the rural poor who are net purchasers of food, is through lower food prices.

Early efforts to document the impact of technological change and the consequent increase in food supplies on food prices and income distribution were made by Hayami and Herdt (1977), Pinstrup-Andersen *et al.* (1976), Scobie and Posada (1978), and Binswanger (1980). Pinstrup-Andersen argued strongly that the primary nutritional impact for the poor came through the increased food supplies generated through technological change.

Several studies have provided empirical support to the proposition that growth in the agricultural sector has economy-wide effects. One of the earliest studies showing the linkages between the agricultural and non-agricultural sectors was done at the village level by Hayami *et al.* (1978). Hayami provided an excellent micro-level illustration of the impacts of rapid growth in rice production on land and labor markets and the non-agricultural sector. Pinstrup-Andersen and Hazell (1985) argued that the landless labor did not adequately share in the benefits of the Green Revolution because of depressed wage rates attributable to migrants from other regions. David and Otsuka (1994), on the other hand, found that migrants shared in the benefits of the Green Revolution through increased employment opportunities and wage income. The latter study also documented that rising productivity caused land prices to rise in the high-potential environments. For sector level validation of the proposition that agriculture does indeed act as an engine of overall economic growth see Hazell and Haggblade (1993); Delgado *et al.* (1998); and Fan *et al.* (1998).

Although the favorable, high-potential environments gained the most in terms of productivity growth, the less favorable environments benefited as well through technology spillovers and through labor migration to more productive environments. According to David and Otsuka, wage equalization across favorable and unfavorable environments was one of the primary means of redistributing the gains of technological change. Renkow (1993) found similar results for wheat grown in high- and low-potential environments in Pakistan. Byerlee and Moya (1993), in their global assessment of the adoption of wheat MVs, found that over time the adoption of MVs in unfavorable environments caught up to levels of adoption in more favorable environments, particularly when germplasm developed for high-potential environments was further adapted to the more marginal environments. In the case of wheat, the rate of growth in yield potential in drought-prone environments was around 2.5 percent per year during the 1980s and 1990s (Lantican and Pingali, 2003). Initially the growth in yield potential for the marginal environments came from technological spillovers as varieties bred for the high-potential environments were adapted to the marginal environments. During the 1990s, however, further gains in yield potential came from breeding efforts targeted specifically at the marginal environments.

3. The changing locus of agricultural R&D: from national public to international private sector

In the decades of the 1960s through the 1980s, private sector investment in plant improvement research was limited, particularly in the developing world, due to the lack of effective mechanisms for proprietary protection on the improved products. This situation changed in the 1990s with the emergence of hybrids for cross-pollinated crops such as maize, etc. The economic viability of hybrids led to a budding seed industry in the developing world, but the coverage of seed industry activity has been limited to date, leaving many markets under served. The seed industry in the developing world was started by multi-national companies based in the developed world, and then led to the development of national companies (Morris, 1998). Despite its rapid growth, the private seed industry continued to rely, through the 1990s, on the public sector gene banks and pre-breeding materials for the development of its hybrids (Morris and Ekasingh, 2001; Pray and Echeverría, 1991). The break between public and private sector plant improvement efforts came with the advent of biotechnology, especially genetic engineering. The proprietary protection provided for artificially constructed genes and for genetically modified plants provided the incentives for private sector entry.

The large multi-national agro-chemical companies were the early investors in the development of transgenic crops. One of the reasons that agro-chemical companies moved into crop improvement was that they foresaw a declining market for pesticides (Conway, 2000). The chemical companies got a quick start in the plant improvement business by purchasing existing seed companies, first in industrialized countries and then in the developing world.

The amalgamation of the national private companies with the multi-national corporations makes economic sense (Pingali and Traxler, 2002). As Figure 1 illustrates, the process of variety development and delivery is a continuum that starts at the upstream end of generating knowledge on useful genes (genomics) and engineering transgenic plants to the more adaptive end of backcrossing the transgenes into commercial lines and delivering the seed to farmers. The products from upstream activities have worldwide applicability, across several crops and agro-ecological environments. On the other hand, genetically modified crops and varieties are applicable to very specific agro-ecological and socioeconomic niches. In other words, spillover benefits and scale economies decline in the move to the more adaptive end of the continuum. Similarly, research costs and research sophistication decline in the

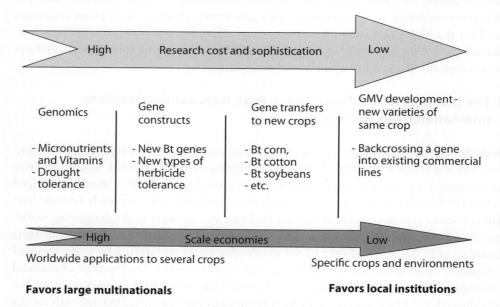

Figure 1. Organization of transgenic crop research and development.

progression towards downstream activities. Thus, a clear division of responsibilities in the development and delivery of biotechnology products has emerged, with the multinational providing the upstream biotechnology research and the local firm providing crop varieties with commercially desirable agronomic backgrounds (see Pingali and Traxler, 2002 for a more detailed discussion on this point).

The options available for public research systems to capture the spillovers from global corporations are less clear. Public sector research programs are generally established to conform to state or national political boundaries, and direct country-to-country transfer of technologies has been limited (Traxler and Pingali, 2002). Strict adherence to political domains severely curtails spillover benefits of technological innovations across similar agroclimatic zones. The operation of the CGIAR germplasm exchange system has mitigated the problem for several important crops, but it is not clear whether the system will work for biotechnology products and transgenic crops, given the proprietary nature of the technology.

Moreover, in the case of biotechnology innovations, the national and international public sector, do not have the resources to effectively create an alternative source of knowledge and technology supply. To understand the magnitude of private sector investment in agricultural research today, one need only look at its annual research budget relative to public research targeted to the developing country agriculture. The World's top ten multinational bioscience corporations' collective annual expenditure on agricultural research and development is nearly three billion U.S. dollars. In comparison the CGIAR, which is the largest international public sector supplier of agricultural technologies, spends less than 300 million U.S. Dollars annually on plant improvement research and development. The largest public sector agricultural research programs in the developing world, those of Brazil, China, and India, have annual budgets of less than half a billion dollars each (Byerlee and Fischer, 2001).

If we look at public expenditures for biotechnology alone, the figure comes out to be substantially smaller for the developing world as a whole. Public research expenditures on agricultural biotechnology (Table 1) reveal a sharp dichotomy between developed and developing countries. Byerlee and Fischer (2001) show that developed countries spend four times as much on public sector biotech research than developing countries, even when all sources of public funds – national, donor and CGIAR centers – are counted for developing countries. Developed countries also spend a higher proportion of their public sector agricultural research expenditures on biotech than developing countries or the CGIAR centers (Byerlee and Fischer, 2001).

The rapid growth of private sector investment in biotech research in developed countries means that private research expenditures now exceed public sector expenditures, both in absolute terms and as a share of total agricultural R&D expenditures. Comprehensive data on private sector biotech research in developing countries are not available, although partial

Table 1. Estimated crop biotechnology research expenditures (million US dollars).

	Biotech R&D (million $/year)	Biotech as share of sector R&D
Industrialized countries	1900-2500	
Private sector*	1000-1500	40
Public sector	900-1000	16
Developing countries	165-250	
Public (own resources)	100-150	5-10
Public (foreign aid)	40-50	n.a.
CGIAR centres	25-50	8
Private sector	n.a.	n.a.
World total	2065-2730	

* Includes an unknown amount of R&D for developing countries
Source: Byerlee and Fischer, 2001.

evidence suggests that the private sector is less developed than in industrialized countries. Data for seven Asian countries show that private agricultural research is equivalent to about 10 percent of total public sector agricultural research (Pray and Fuglie, 2000). If this ratio is applied to all developing countries (probably an upper bound), we can estimate private sector research expenditures at about $1 150 million per year. If we assume that the share of biotech in total private R&D is the same as in developed countries (40 percent, also an upper bound), this gives us an estimate of $458 million in annual private sector biotech research in developing countries. A lower-bound estimate could build on the fact that the share of total public sector research devoted to biotech in the developing countries is about 8 percent. If we apply that ratio to the private sector expenditures, we get a lower-bound estimate of $92 million for private biotech in developing countries. Although it is likely that private sector investment is somewhere between these estimates, even the upper bound estimate is only about one-third the level of private sector investment in developed countries.

4. Trends in biotechnology research, development and commercialization in the developing world

Only a few developing countries have highly sophisticated biotech research programs, but growing numbers have the capacity to adopt and adapt innovations developed elsewhere. Only three countries – China, India and Brazil – have extensive research programs in all areas of agricultural biotechnology, including advanced genomics and gene manipulation techniques (FAO BioDeC). According to IFPRI (2004) 15 developing countries have significant and growing capacity for biotechnology research, but these are typically more advanced countries such as Brazil, Argentina, and Egypt. Most of the least developed countries, however, have no documented research experience with genetically modified organisms and many have limited capacity for agricultural research of any kind (FAO BioDeC).

Many developing countries, including some higher-income countries, also lack regulatory capacity in the areas of intellectual property rights protection and biosafety procedures (food safety and environmental protection. Subsistence agriculture remains the dominant agricultural system in much of the developing world. The agricultural sectors of several more advanced developing countries exhibit a dualistic agricultural structure in which a few large very modern commercial farms coexist with many small subsistence farms. As a result, commercial markets for agricultural inputs and outputs remain weak, especially in the least developed countries, suggesting that the potential for commercial joint ventures between multinational biotech companies and local seed companies is limited to the more advanced developing countries.

4.1. Commercial cultivation of GM crops in the developing world

James (2007) reports that transgenic crops were commercially planted on 102 million hectares in 2006, in 22 countries, 12 of them developing (Figure 2). Developing countries accounted for 38 percent of the total global transgenic crop area in 2006, having increased consistently from 14 percent of the total 12.8 million hectares planted in 1997. Argentina, Brazil, India and China are the largest developing country producers (Figure 3).

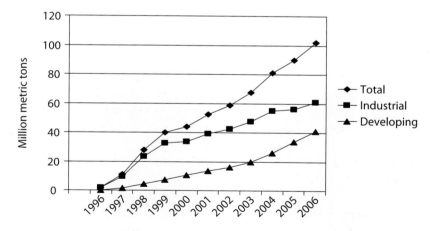

Figure 2. Transgenic crop area, 2006.

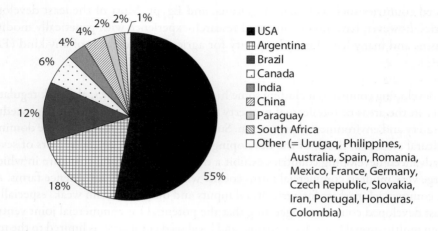

Figure 3. GM crops by country, 2006.

Herbicide tolerance and pest resistance remain the main GM traits that are currently under commercial cultivation and the main crops are soybean, maize, canola and cotton (Figure 4). In addition to these major crops, China also produces small quantities of virus resistant tomatoes and peppers, delayed ripening tomatoes and flower-color altered petunias.

Comprehensive data are not available on the origin of the GM varieties being planted in developing countries; however, the available evidence suggests that most of the GM varieties grown in developing countries in 2006 were developed by multinational companies

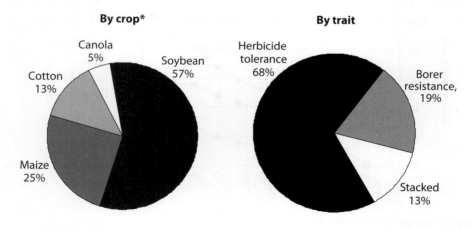

*Other = papaya, squash, alfalfa, rice

Figure 4. GM crops by crop and trait, 2006.

for the developed country markets of North America or were adapted locally from imported varieties. China is the exception. China is the only developing country to have commercialized transgenic varieties developed in the public sector from a locally developed genetic transformation construct. (Pray and Naseem, 2003). In other countries, commercial GM crops are imported or locally adapted from imported varieties.

This evidence suggests that globalization and the international transfer of technology have been essential factors in promoting the commercial spread of GM crops in developing countries. It also suggests that developing countries that lack strong public sector research systems and/or strong commercial seed sectors will be handicapped in the adoption of transgenic varieties.

4.2. Economic impacts of transgenic crops in developing countries

A recent survey of *ex post* farm-level studies of the impact of transgenic insect-resistant (IR) cotton shows that where these crops are available in developing countries, farmers can benefit and the benefits can be pro-poor (Raney, 2006). Table 2 summarizes the results from the most comprehensive economic studies of the farm-level impacts of IR cotton in developing countries to date. Each of the studies was based on data from two or three seasons of commercial farm production.

The figures in Table 2 reflect the average percent difference between IR and conventional cotton for all farmers over all seasons covered in the study. Although the averages conceal a high degree of temporal and spatial variation, they clearly indicate positive overall results. Farmers who adopted the transgenic varieties experienced higher effective yields (due to less pest damage), higher revenue and lower pesticide costs. These factors more than compensated for higher prices paid for IR seeds so that net profits increased for adopters.

Table 2. Performance advantage of IR over conventional cotton, percentage.

	Argentina	China	India	Mexico	South Africa
Yield	33	19	34	11	65
Revenue	34	23	33	9	65
Pesticide costs	-47	-67	-41	-77	-58
Seed costs	530	95	17	165	89
Profit	31	340	69	12	299

Source: author's calculations based on: Argentina – Qaim and De Janvry (2003); China – Pray, *et al.*, 2002; India – Qaim, *et al.*, 2006; Mexico – Traxler and Godoy-Avila, 2004; and South Africa – Bennet *et al.*, 2004.

5. Accessing biotechnology knowledge and products for the poor

Unlike the green revolution technologies, transgenic technologies are transferred internationally primarily through market mechanisms, often through commercial relationships between the multinational bio-science firms and national seed companies. This system of technology transfer works well for commercially viable innovations in well-developed markets, but perhaps not for the types of innovations needed in developing countries: crops and traits aimed at poor farmers in marginal production environments. These 'orphan' technologies have traditionally been the province of public sector research. Given the dominance of private sector research in transgenic crop research and meagre resources being devoted to public sector research in most developing countries, it is unlikely that public sector research can play this role for transgenic crops.

The options available for public research systems in developing countries to capture the spillovers from global corporations are limited. Public sector research programmes are generally established to conform to state or national political boundaries, and direct country-to-country transfer of technologies has been rare (Pingali and Traxler, 2002). Strict adherence to political domains severely curtails spillover benefits of technological innovations across similar agroclimatic zones. The operation of the CGIAR germplasm exchange system has mitigated the problem for several important crops, but it is not clear whether the system will work for biotechnology products and transgenic crops, given the proprietary nature of the technology.

Although private sector agricultural research expenditures seem overwhelmingly large, the reality is that they are focused very narrowly on the development of biotechnology related plant varieties, and even that for a very small number of crops. A large part of the private sector investment is concentrated on just four crops: cotton, corn, canola, and soybeans. Private sector investment on the world's two most important food crops, rice and wheat, is insignificant in comparison. Moreover, all of the private sector investment is targeted towards the commercial production sector in the developed world, with some spillover benefits flowing to the commercial sector in the developing world. The public sector, with its increasingly meager budget, is left to take care of the research and technology needs of the subsistence farming sector, as well as being the only source of supply for conventionally bred seed as well as crop and resource management technologies.

Will the poor benefit from any of the technological advances that are taking place today in the private sector? Private sector investments in genomics and genetic engineering could be potentially very useful for addressing the problems faced by poor farmers, particularly those in the marginal environments. Knowledge generated through genomics, for example, could have enormous potential in advancing the quest for drought tolerant crops in the tropics. The question that needs to be asked is whether incentives exist, or can be created, for public/private sector partnerships that allow the public sector to use and adapt technologies developed by the private sector for the problems faced by the poor. Can licensing agreements be designed that will allow private sector technologies to be licensed to the public sector for use on problems of the poor? Pingali and Traxler (2002) suggest that the public sector may have to purchase the right to use private sector technology on behalf of the poor.

Pingali and Traxler (2002) suggest three possible avenues for public sector institutions in developing countries to gain access to transgenic technologies: (i) directly import private- or public-sector transgenic varieties developed elsewhere, (ii) develop an independent capacity to develop and/or adapt transgenic varieties, and (iii) collaborate on a regional basis to develop and/or adapt transgenic varieties. The second option is the most costly and requires the highest degree of national research capacity, while the first option depends on the availability of suitable varieties developed elsewhere. The third option would require a higher degree of cooperation across national boundaries than has typically characterized public sector research. Pingali and Traxler (2002) ask whether incentives exist or can be created for public/private partnerships that allow the public sector to use and adapt technologies developed by the private sector.

Before considering the prospects for partnerships for accessing technologies in the pipeline it is important to conduct a detailed inventory of all prospective biotechnology products characterized by crop and by agro-ecological environments. Followed by an ex ante assessment of the impact each of these technologies could have on the productivity and livelihoods of the subsistence producers. The above assessment would lead to the identification of a set of products with high pro-poor potential that public/private partnerships could be built around.

Even if public-private partnerships could be developed, will the resulting technologies ever get to the poor? Given that technologies that are on the shelf today (generated by conventional research methods) have not yet reached farmers' fields, there is no guarantee that the new biotechnologies will fare any better. Are there any policy interventions that will make the situation any better? Identifying small farmer constraints to technology access and use continues to be an issue that the development community ought to deal with. This suggests a need for a third wave of globalization to ensure that international spillovers from the Gene Revolution make their way to the poor. Investments in biotechnology research capacity for the public sector will only be worthwhile if the current difficulties in delivering conventional technologies to subsistence farmers can be reversed.

References

Alston, J.M., M.C. Marra, P.G. Pardey, and T.J. Wyatt, 2000. Research returns redux: A meta-analysis of the returns to agricultural R&D. Australian Journal of Agricultural and Resource Economics 44(2): 185-215.

Bennett, R., Y. Ismael, S. Morse and B. Shankar, 2004. Reductions in Insecticide Use from Adoption of BT Cotton in South Africa: Impacts on Economic Performance and Toxic Load to the Environment. J Agr Sci 142:665-674.

Byerlee, D., and P. Moya, 1993. Impacts of International Wheat Breeding Research in the Developing World, 1966-1990. Mexico, D.F.: International Maize and Wheat Improvement Center (CIMMYT).

Byerlee, D. and K. Fischer, 2001. Accessing Modern Science: Policy and Institutional Options for Agricultural Biotechnology in Developing Countries. BioDevelopments–International Institute Inc, Cornell, Ithaca, NY (IP Strategy Today, no. 1-2001).

Byerlee, D., and G. Traxler, 2002. The role of technology spillovers and economies of size in the efficient design of agricultural research systems. In: J.M. Alston, P.G. Pardey and M.J. Taylor (eds.) Agricultural science policy: changing global agendas. Baltimore, USA: Johns Hopkins University Press.

Binswanger, H.P., 1980. Income distribution effects of technical change: Some analytical issues. South East Asian Economic Review 1(3): 179-218.

Conway, G., 2000. Crop Biotechnology: Benefits, Risks and Ownership. Rockefeller Foundation, New York (Foundation News 03062000).

David, C., and K. Otsuka (eds.), 1994. Modern Rice Technology and Income Distribution in Asia. Boulder: Lynne Rienner.

Delgado, L.C., J. Hopkins and V.A. Kelly, 1998. Agricultural Growth Linkages in Sub-Saharan Africa. IFPRI Research Report No. 107. Washington, D.C.: International Food Policy Research Institute (IFPRI).

Echeverría, R.G., 1990. Assessing the impact of agricultural research. In: R.G. Echeverría (ed.) Methods for Diagnosing Research System Constraints and Assessing the Impact of Agricultural Research. Volume 2. The Hague: International Service for National Agricultural Research (ISNAR).

Evenson, R.E., 2001. Economic Impacts of Agricultural Research and Extension. In: B.L. Gardner and G.C. Rausser (eds.) Handbook of Agricultural Economics, Vol 1.,. North Holland.

Evenson, R.E. and D. Gollin, 2003. Assessing the impact of the green revolution: 1960-1980, Science, 300: pp. 758-762.

Fan, S., P. Hazell and S. Thorat, 1998. Government Spending, Growth, and Poverty: An Analysis of Interlinkages in Rural India. EPTD Discussion Paper No. 33. Washington, D.C.: International Food Policy Research Institute (IFPRI).

FAO, 2005. FAOSTAT.

Flores-Moya, P., R.E. Evenson and Y. Hayami, 1978. Social returns to rice research in the Philippines: Domestic benefits and foreign spillover. Economic Development and Cultural Change 26(3): 591-607.

Hayami, Y. and RW. Herdt, 1977. Market price effects of technological change on income distribution in semi-subsistence agriculture. American Journal of Agricultural Economics 59(2): 245-256.

Hayami, Y., M. Kikuchi, P.F. Moya, L.M. Bambo and E.B. Marciano, 1978. Anatomy of a Peasant Economy: A Rice Village in the Philippines. Los Baños: International Rice Research Institute (IRRI).

Hazell, P. and S. Haggblade, 1993. Farm-nonfarm growth linkages and the welfare of the poor. In: M. Lipton and J. van de Gaag (eds.) Including the Poor. Washington, D.C.: The World Bank.

IFPRI, 2004. To reach the poor: Results from the ISNAR-IFPRI Next Harvest study on genetically modified crops, public research and policy implications. EPTD Discussion Paper No. 116.

James, C., 2007. ISAAA Brief No 35. Global Status of GM Crops.

Lantican, M. and P.L. Pingali, 2003. Growth in wheat yield potential in marginal environments. In: Proceedings of the Warren E. Kronstad Memorial Symposium, 1-17 March 2001. Mexico City: International Maize and Wheat Improvement Center (CIMMYT).

Maredia, M.K., D. Byerlee and C.K. Eicher, 1994. The efficiency of global wheat research investments: implications for research evaluation, research managers and donors. Staff Paper No. 94-17. Department of Agricultural Economics, Michigan State University, USA.

Morris, M., 1998. Maize Seed Industries in Developing Countries. Lynne Rienner Publishers, Boulder, Colorado.

Morris, M. and B. Ekasingh, 2001. Plant breeding research in developing countries: what roles for the public and private sectors. In: D. Byerlee and R. Echeverria (eds.) Agricultural Research Policy in an Era of Privatization: Experiences from the Developing World. CABI, Wallingford, UK.

Pardey, P.G., R.K. Lindner, E. Abdurachman, S. Wood, S. Fan, W.M. Eveleens, B. Zhang and J.M. Alston, 1992. The Economic Returns to Indonesian Rice and Soybean Research. Jakarta and the Hague: Agency for Agricultural Research and Development (AARD) and International Service for National Agricultural Research (ISNAR). Unpublished report.

Pingali, P.L. and P.W. Heisey, 2001. Cereal-Crop Productivity in Developing Countries: Past Trends and Future Prospects. In: J.M. Alston, P.G. Pardey and M. Taylor (eds.) Agricultural Science Policy. IFPRI & Johns Hopkins University Press: Washington.

Pingali, P.L. and S.R. Rajaram, 1999. World Wheat Facts and Trends, 1998/99. CIMMYT Institute, Mexico, DF.

Pingali, P. and G. Traxler, 2002. Changing locus of agricultural research: will the poor benefit from biotechnology and privatization trends. Food Policy 27(2002): 223-238.

Pinstrup-Andersen, P., N. Ruiz de Londoño and E. Hoover, 1976. The impact of increasing food supply on human nutrition: Implications for commodity priorities in agricultural research. American Journal of Agricultural Economics 58(2): 131-42.

Pinstrup-Andersen, P. and P.B.R. Hazell, 1985. The impact of the Green revolution and prospects for the future. Food Review International 1(1): 1-25.

Pray, C.E. and R.G. Echeverria, 1991. Private sector agricultural research in less developed countries. In: P.G. Pardey, J. Roseboom and J. Anderson (eds.) Agricultural Research Policy: International Quantitative Perspectives. Cambridge University Press, Cambridge.

Pray, C.E. and K.O. Fuglie, 2000. Policies for Private Agricultural Research in Asian LDCs. Paper presented at the XXIV International Conference of Agricultural Economists. Berlin, Germany.

Pray, C.E., A. Courtmanche and R.Govindasamy, 2002. The importance of intellectual property rights in the international spread of private sector agricultural biotechnology. Presented at the 6th International Conference convened by ICABR, Ravello (Italy), from July 11-14, 2002.

Pray, C., J. Huang, R. Hu and S. Rozelle, 2002. Five years of Bt cotton in China – the benefits continue. The Plant Journal 31(4):423-430.

Pray, C.E., and A. Naseem, 2003. The Economics of Agricultural Research, ESA Working Paper 03-07, FAO.

Qaim, M. and A. de Janvry, 2003. Genetically modified crops, corporate pricing strategies, and farmers' adoption: the case of Bt cotton in Argentina. Amer J Agr Econ 85(4):814-828.

Qaim, M., A. Subramanian, G. Naik and D. Zilberman, 2006. Adoption of Bt cotton and impact variability: insights from India. Rev Agr Econ DOI:10.1111/j.1467-9353.2006.00272.x., 28(1):48-58.

Raney, T., 2006. Economic impact of transgenic crops in developing countries. Current Opinion in Biotechnology. Vol. 17, Issue 2, pp 1-5.

Renkow, M., 1993. Differential technology adoption and income distribution in Pakistan: Implications for research resource allocation. American Journal of Agricultural Economics 75(1): 33-43.

Scobie, G.M. and R.T. Posada, 1978. The impact of technical change on income distribution: The case of rice in Colombia. American Journal of Agricultural Economics 60(1): 85-92.

Traxler, G. and P.L. Pingali, 1999. International collaboration in crop improvement research: current status and future prospects. Economics Working Paper No. 99-11. Mexico City, International Wheat and Maize Improvement Center.

Traxler, G. and S. Godoy-Avila, 2004. Transgenic Cotton in Mexico. AgBioForum 7(1&2): 57-62.

Income shocks and governance

Richard Damania and Erwin H. Bulte

Abstract

Do income increases lead to better institutions, or do better institutions raise income? We revisit the ongoing debate about the causality between income and institutions by using an instrument for income when explaining governance quality (the reverse approach of the one commonly taken in the literature). Our instrument for short-term income shocks is not associated with governance proxies, suggesting the causality runs from institutions to income in the short run. This implies there is no mutually enforcing virtuous cycle of income growth and institutional improvements.

1. Introduction

The relation between institutions and income (growth) is a hotly contested issue in economics. Institutional quality, income and various other economic measures such as human capital tend to move together over time as societies develop, and unraveling the underlying causal relationships – what drives what? – has proven extremely challenging.

Roughly speaking there are two opposing perspectives. There is the view that institutions are a key driver of economic outcomes, and there is the counter view that countries will improve their institutions after they become richer. Both views have influential supporters, and both point to persuasive evidence to back their positions. Both views may be correct simultaneously – in the presence of positive feedback effects between income and institutional quality, there could be a virtuous cycle of economic and institutional development. Since promoting economic development and institutional reform in the developing world are both key objectives of the international community, it would clearly be useful to have a better understanding of what drives what. Unfortunately, much is unknown about the main determinants of economic growth and institutional quality.

The main objective of this chapter is to analyze whether there exists a causal relation from short-term income to institutional quality in sub-Saharan Africa. Specifically, we analyze whether positive (negative) income shocks cause improvements (deteriorations) in the quality of governance. As a corollary, we evaluate the scope for virtuous cycles of economic and institutional development in the short run – to what extent does kick starting economic growth immediately translate into institutional improvements (further feeding economic growth, *etcetera*)?

The plan of the chapter is as follows. In section 2 we briefly summarize the main issues and positions in the causality debate, and clarify our contribution. Section 3 presents our data and some intermediate results. In section 4 we present our main regression results: while 'naïve' OLS regressions suggest a positive correlation between income and institutional quality, the correlation disappears when we add appropriate controls or instrument for income. We place our findings in perspective in section 5.

2. The context and our contribution

Following the wave of 'market fundamentalism' we are now in the midst of a wave of 'institutions fundamentalism' (Rodrik, 2006). Increasingly it is believed that getting institutions right is a prerequisite for economic development. Yet, the exact role of institutional reform in development is ill understood. Partly this is due to the simple fact that the exact meaning of the word institutions is in the eye of the beholder. The lack of agreement on the meaning of the concept logically extends to a dispute about how institutions ought to be measured. They are relatively durable features of economies, but does this mean that they are reasonably permanent and relatively unchanging – as argued by Glaeser *et al.* (2004) – or that they are stock variables in a state of constant flux with dynamics governed by policy choices (as argued by Rodrik *et al.*, 2004)? In an overview piece, Williamson (2000) refers to the former interpretation as the 'institutional environment' (where changes over time are measured with a time scale that spans decades), and the latter as 'governance' (where changes are measured on a 1-10 years scale).

Upon adopting the governance perspective, various survey indicators of institutional quality have been developed that aim to capture the behavior of policy makers and bureaucrats (e.g. Knack and Keefer, 1995). There is abundant evidence that these variables are significantly, in both an economic and statistical sense, associated with income. To circumvent the potential problems of reverse causality and omitted variables, it has now become standard to 'instrument' for the institutional variable in growth regressions. As instruments, Hall and Jones (1999) have used various correlates of the extent of Western European influence around the world (e.g., the fraction of the population speaking a major European language today), Bockstette *et al.* (2002) have developed an index of 'state antiquity', and Acemoglu *et al.* (2001) exploited data on settler mortality in early colonial days. A follow-up literature has emerged that suggests institutions thus measured are a key driver for development, possibly trumping everything else (e.g. Easterly and Levine, 2003; Rodrik *et al.*, 2004, see also work by Temple and Johnson, 1998).

Glaeser *et al.* (2004) reject this approach to measuring institutions and focus on the institutional environment perspective. They argue the survey variables measure policy outcomes rather than durable 'constraints', and that they are too volatile to capture anything 'deep' about the economic structure (see also Durham 1999). Further, when using constitutional measures of institutional quality the correlation between income and institutions disappears. Instead, their results suggest that human capital is the deep underlying variable that steers development.[60] Since accumulation of human capital is primarily a matter of policy choices (and not of institutions interpreted as durable constraints on policy makers), Glaeser *et al.* believe that policies determine economic growth and that growth, subsequently, determines institutions as traditionally measured. This effectively reverses the causality.

There is common ground between the opposing approaches: 'they both emphasize the need for secure property rights to support investments in human and physical capital ...' (Glaeser *et al.*: 272). The difference is that one camp views such property rights security as a policy choice, and the other interprets it directly as a measure of the institutional setting. The governance perspective on institutions is arguably broader and more 'loose.' Institutions, then, are characteristics of communities that may come in different forms and shapes, encompassing various measures of social constructs that facilitate the interaction between humans. This implies that institutions may be transient and subject to constant change.

This is exactly what is captured in the governance data. For example, empirical work by Brautigam and Knack (2004) on the relation between aid and governance suggests two important features about governance proxies: (i) they change quite a bit over relatively short time periods, and (ii) they respond to economic incentives. This suggests it may be hazardous to explain current income with instruments based on 'ancient' conditions of early colonial days. This is particularly true if, as argued by Glaeser *et al.* such instruments are correlated with other important explanatory variables of growth (such as human capital) or impact on economic growth directly (e.g. swamps may provide ample breeding ground for mosquito's, but also limit agricultural potential). The debate on the nature of the causality between income and governance appears open.

Which brings us to our contribution. We exploit the volatility of governance data to do a panel analysis. Rather than instrumenting for institutions in an income regression, as discussed above, we instrument for income growth in a regression that aims to explain (changes in) institutional quality. Using (lagged) rainfall measures as instrumental variables for income growth, we evaluate whether income shocks set in motion a process of institutional change. We also determine the scope for virtuous cycles of economic-institutional development, with both types of development mutually enforcing each other.

A key benefit of our panel approach is that we control for time-invariant country-specific effects, eliminating a source of omitted variable bias that looms large as a potential problem in regressions of institutional quality. On the other hand, one may ask whether relatively high frequency data are appropriate for testing relationships governed by slow dynamics

---

[60] This insight is disputed by Acemoglu *et al.* (2005a), who argue that the effect disappears if one controls for time effects. Acemoglu *et al.* (2005a) argue there is no significant relationship from education to institutions – high levels of education are not a prerequisite for democracy.

(but a similar tradeoff emerges in many other papers – e.g. Forbes 2000). Can the short-run variability in income carry the identification load for issues with long-run dimensions such as institutions? This is an open question, but a crucial one for economists interested in making contingent statements on the relation between economic outcomes and institutions – at what time scales do we find evidence of institutional change in response to economic incentives?[61] Note that the 'virtuous cycle' perspective is based on the implicit assumption that institutional change will occur relatively rapidly.

Income shocks can affect institutions through various channels, directly or indirectly. For example, direct effects may occur if income shocks affect government revenues, affecting the ability of the principal to constrain corruption and the incentives of civil servants (agents) to engage in corruption (e.g., real wages impacting on morale, incentives to work or accept bribes – see Aidt, 2003; Mauro, 2004). Similarly, adverse income shocks may impact on priority setting in policy formulation and implementation, perhaps because such shocks are a driver of civil conflict (Miguel *et al.*, 2004), impacting on institutions indirectly. We can envisage alternative channels as well. The time step in our panel (one year) is also the lower bound, according to Williamson (2000), for the frequency interval during which governance structures are updated.

3. Data and model

The approach taken in this chapter seeks to instrument for GDP in an institutions regression, using an instrument that is truly exogenous and unaffected by institutional structures and policies. The empirical focus is on sub-Saharan Africa – an arid region dependent on agriculture and where agriculture is heavily reliant on the level of rainfall.[62] Unlike many other arid regions in the world (e.g. the Middle East, Pakistan and Central Asia), sub-Saharan Africa has a relatively low hydrological storage capacity and limited irrigation distribution system. Hence, the fortunes of agriculture vary with the idiosyncrasies of rainfall. This, combined with a relatively under-developed secondary and tertiary sector, implies that precipitation levels have a significant impact on economic performance. Following Miguel *et al.* (2004), we use growth in (current and lagged) rainfall as instruments for African income growth (see also Paxson, 1992). While most of the earlier literature on the relation between institutions and income has focused on the link between institutional quality and income levels, we focus on income growth instead – this gives us more variation for the relatively

---

[61] Brautigam and Knack (2004) do not use annual variation in the governance data. Instead, they explain the changes in governance quality over the study period. The reasons are that institutional response may only occur with a time lag (something we test and try to control for), and the absence of a suitable annual instrument for aid (something that does not affect us as we have an annual instrument for income).

[62] Of course not all countries are equally reliant on agriculture. However, rainfall may also indirectly impact other sectors via primary production (through various linkages). For this reason we focus on total income rather than income in the primary sector only.

short study period that we consider. Note that conceptually it does not matter whether we regress institutions on income levels, or changes in institutions on income growth.

3.1. The data

Table 1 presents summary statistics for the main variables used in the analysis. The data are for countries in Sub-Saharan Africa from 1981-1999. The rainfall data is from Miguel *et al.*, and is based on the Global Precipitation Climatology Project (GPCP) database. The GPCP data are computed from local rainfall station gauge measures and corroborated against satellite data of cloud cover and precipitation. These measures are generally regarded as the most reliable estimates of rainfall. There is considerable variation in rainfall in sub-Saharan Africa – both across countries and over time.

Our measure of institutional quality (governance) is calculated from the ICRG governance indices. While there are many competing measures of institutional strength, the ICRG data are particularly useful for our purposes because of their extensive coverage of Africa over a relatively long time span. Following Brautigam and Knack (2004) we construct an Institutional Quality index based on three key characteristics: corruption, bureaucratic quality and the rule of law – three characteristics that could potentially vary with income. The index is formed by taking annual averages and summing the monthly sub-indices.

The remaining data are well known and drawn from established sources. We consider the 42 African countries also analyzed by Miguel *et al.* (2004, p.749 for the complete list).

Table 1. Summary statistics.

	Mean	Standard deviation	Min	Max
Institutional quality	6.3	2.09	0.6	12.4
Ethnic fractionalization	0.71	0.19	0.18	0.95
Change in institutional quality	-0.04	0.65	-2.8	3.8
Political instability	0.21	0.41	0	1
Wars	0.16	0.37	0	2
Population density	43.2	54.6	1.5	336
Rainfall	1001.6	502	96	2587
GDP growth	-0.004	0.07	-0.4	0.67
Aid per capita	50.1	41.1	0.38	412
Democracy (Polity IV)	-4.35	21.9	-88	9
Terms of trade (ToT)	109	34	45	348

Population density is the number of people per square kilometer taken from WDI. A terms of trade index for goods and services is used and is also from WDI, with 1995 as the base year; the impacts of conflict are captured using Collier and Hoeffler's (2002) measure of wars and conflict; ethnic fractionalization is based on Fearon and Laitin (2003). Also included are measures of political instability, the dependence on oil is measured by a dummy variable for a country that earns more than 1/3 of its revenue from oil, the log of mountainous terrain is used to assess potential landscape and remoteness impacts. We also use a dummy variable for British and French colonial heritage and a dummy variable for new states in the first two years of its existence (as new states are likely to be more fractional, etc. – see Miguel (*op cit*) for details). These data are from Miguel *et al.* as well.

Income shocks may impact on governance quality through various channels, affecting the demand and supply for control of corruption, bureaucratic efficiency, and so on. However, income changes *per se* may also lead to changes in governance (e.g., in environments with larger income shocks the incentive to plan for the future and invest in 'rules of the game' may be accentuated or attenuated). Therefore we also include a measure of rainfall volatility in some regressions as a robustness check.

A simple regression between GDP growth and governance quality confirms the well-known result that institutional quality remains highly and significantly correlated with growth at even the 1% level: $\Delta Y = -0.02 + 0.002 \times (\text{governance})$. Of course this does not establish causality, which may run in the opposite direction. To this we turn below.

3.2. Estimation framework

We start by exploring the relation between income and rain, and regress income growth in country i in period t (ΔY_{it}) on growth in (lagged) rainfall (ΔR_{it}). We include country fixed effects (α_i) to capture time-invariant characteristics of countries associated with income, or a series of controls (X_{it}). Finally, we sometimes include country-specific time trends. Hence:

$$\Delta Y_{it} = \alpha_i + \beta_0 \Delta R_{it} + \beta_1 \Delta R_{it-1} + \gamma_i \text{ year}_t + e_{it}, \tag{1}$$

where e_{it} is a disturbance term. The results are in Table 2. They are significant and robust – current and lagged increases in rainfall are positively associated with income growth. These results are robust with respect to including fixed effects or a vector of controls. We find similar results when we regress (lagged) rainfall *levels* on GDP *levels* (as opposed to rainfall growth on GDP growth) – details are available from the authors on request. Following Miguel *et al.* (2004) we conclude that (lagged) changes in rainfall is a potential instrument of income growth. The low R^2s indicate that a good deal of unexplained heterogeneity remains. This should be kept in mind when interpreting the regression results below.

Table 2. Rainfall and income (first stage regression), ordinary least squares.

Explanatory variable	Dependent variable: growth in income			
	(1)	(2)	(3)	(4)
Growth in rainfall, t	0.059**	0.057**	0.058***	0.051***
	(0.018)	(0.018)	(0.13)	(0.13)
Growth in rainfall, t-1	0.04**	0.04**	0.03***	0.03***
	(0.014)	(0.014)	(0.13)	(0.13)
Initial income	-0.001	-0.007*		
	(0.022)	(0.004)		
Democracy	0.0098	0.002		
	(0.006)	(0.009)		
Ethnic fraction.	-0.006	-0.003		
	(0.012)	(0.03)		
Religion	-0.0025	-0.005		
	(0.011)	(0.029)		
Oil exporting	-0.005	0.01		
	(0.005)	(0.02)		
Terms of ttrade growth	0.002	0.003		
	(0.022)	(0.023)		
Log(population)	-0.002	-0.002		
	(0.022)	(0.008)		
Fixed effects	No	No	Yes	Yes
Country time trends	No	Yes	No	Yes
R^2	0.03	0.09	0.04	0.13
Chi squared	26.43***	60.3***		
F statistic			7.88***	2.86***
Observations	660	660	742	742

Hubert robust standard errors in parenthesis for OLS estimates.
* denotes statistical significance at the 10% level.
** denotes statistical significance at the 5% level.
*** denotes significance at the 1% level.

We now turn to the second stage of the regression analysis. Throughout we examine whether income growth explains either (i) levels of institutional quality, IQ_t, and (ii) changes in institutional quality, ΔIQ_t:

$$IQ_{it} = \alpha_i + \beta_0 \Delta Y_{it} + \beta_1 \Delta Y_{it-1} + \gamma_i \text{ year}_t + e_{it} \qquad \text{and} \qquad (2a)$$
$$\Delta IQ_{it} = \alpha_i + \beta_0 \Delta Y_{it} + \beta_1 \Delta Y_{it-1} + \gamma_i \text{ year}_t + e_{it}. \qquad (2b)$$

Depending on the regression we used combinations of country-specific time trends, fixed effects, or a vector of control variables (the results are robust with respect to which controls we include).

4. Main results: 2SLS regressions

Tables 3a and 3b contain our main results. They present OLS and IV-2SLS estimates of the impact of income growth on governance quality, controlling for a variety of other variables.

Table 3a. Income and quality of governance.

Explanatory variables	Dependent variable: governance quality					
	OLS (1)	OLS (2)	OLS (3)	IV-2SLS (4)	IV-2SLS (5)	IV-2SLS (6)
Economic growth, t	0.86***	0.8	0.268	1.155	8.2	15.2
	(2.91)	(1.05)	(0.82)	(1.38)	(17.2)	(27.0)
Economic growth, t-1		-0.09	0.31	1.40	-6.1	-3.08
		(1.4)	(0.74)	(1.22)	(13.3)	(14.8)
British colony		0.29			0.17	-0.43
		(1.15)			(0.7)	(13.7)
Instability		-0.03*			-0.015*	-0.16
		(0.29)			(0.31)	(0.29)
Aid per capita		0.006*			0.007*	0.009*
		(0.00)			(0.003)	(0.004)
Ethnic. Fraction.		-6.55**			-6.5**	(-3.8
		(2.1)			(1.1)	(33.9)
Population density		-0.05**			-0.05**	0.04*
		(0.02)			(0.022)	(0.02)
Fixed effects	No	No	Yes	Yes	No	No
Country time trends	No	Yes	Yes	Yes	Yes	No
R^2	0.17	0.65	0.01	0.54	0.598	0.007
Observations	413	405	413	413	405	405

Hubert robust standard errors in parenthesis for OLS estimates. Regression disturbances are clustered at country level. * denotes statistical significance at the 10% level, ** denotes statistical significance at the 5% level, and *** denotes significance at the 1% level. Instruments for income growth are growth in rainfall and one period lagged growth in rainfall.

Table 3b. Income and change in quality of governance.

Explanatory variables	Dependent variable: change in governance quality				
	OLS	OLS	OLS	IV-2SLS	IV-2SLS
	(1)	(2)	(3)	(4)	(5)
Economic growth, t	-0.262	-0.26	-0.45	-19.1	-0.141
	(0.53)	(0.53)	(0.67)	(19.3)	(0.544)
Economic growth, t-1	-0.45	-0.45	-0.5	-15	-0.3
	(0.32)	(0.32)	(0.63)	-28	(0.423)
Population (lagged)	0.0067	0.006			0.33**
	(0.02))	(0.02))			(0.08))
Wars	-0.06	-0.063*			-0.03*
	(0.06)	(0.006)			(0.29)
Democracy (lagged Polity IV measure)	0.07*	0.07			0.669**
	(0.01)	(0.1)			(0.16)
Fixed effects	No	No	Yes	Yes	No
Country time trends	No	Yes	Yes	Yes	Yes
R^2	0.004	0.2	0.02	0.004	0.2
Observations	383	383	383	383	383

Hubert robust standard errors in parenthesis for OLS estimates. Regression disturbances are clustered at country level. * denotes statistical significance at the 10% level, ** denotes statistical significance at the 5% level, and *** denotes significance at the 1% level. Instruments for income growth are growth in rainfall and one period lagged growth in rainfall.

In the first column of Table 3a we run a 'naïve' regression – searching for correlation between income growth and governance. We find a positive relationship between these variables. But, as is well known, such regressions likely suffer from omitted variable bias and endogeneity bias. In the other columns we try to control for this by adding additional control variables, or by instrumenting for income (using the OLS results reported in Table 2 as the first stage of the 2SLS procedure).

Our results highlight the potential dangers of estimating ill-specified models: the positive association between income growth and institutions disappears in columns (2)–(3). Colonial heritage, political instability, ethnic fractionalization, population density and aid[63] are all potential determinants of institutional quality. To the extent that income shocks affect any of these variables, they may impact on governance indirectly. But the analysis suggests that income growth, when controlling for these other variables, has no statistically significant direct effect.

Our 2SLS approach summarized in columns (4)-(6) suggests similar results. When instrumenting for income shocks to address omitted variables or reverser causality, income shocks are not significantly associated with governance changes. These results do not change when we add a measure of rainfall volatility at the country level, and is also robust to changes in specification and explanatory variables.

We believe these results lend support to the view that the causality between income and institutions runs from the latter to the former (at least in the short run). Since income does not directly feed back into governance, there is no evidence of a virtuous cycle of institutional and economic development. In the absence of an immediate system response, it appears as if institutional reform should be a conscious choice of policy makers. If such policy responses to beef up governance measures are lacking, income shocks are perhaps hard to sustain. This appears consistent with earlier empirical work on (African) growth accelerations – they do exist, but tend to fizzle out (e.g. Hausmann *et al.*, 2005; Rodrik, 2006).

In Table 3b we report similar results for the case where the dependent variable is changes in governance quality. Variations in income (current and lagged) have no statistically significant impact on this variable, and this finding is robust to a variety of specifications and alternative estimation techniques.

5. Discussion and conclusion

We have tried to shed light on one of the ongoing controversies dividing the literature on institutions and income. There exists a strong and positive correlation between per capita income and institutional quality – but what does this mean? While earlier work has sought to instrument for institutions when explaining incomes, we have chosen the opposite route of instrumenting for income when explaining institutional quality (or rather: the quality of governance). Our results indicate that there is no mutually reinforcing relationship – or *virtuous cycle* – between income growth and institutions in the short run: we find no evidence of income shocks affecting institutions. Since the two variables are positively

[63] There remains the possibility that the causality with respect to aid is reversed if donors provide more aid to countries with strong institutions. To test for this we used lagged values for aid in the regression – which remained positive and significant at the 10% level. As a further check we estimated a specification with *future* values of aid as an explanatory variable, and this turned out to be statistically insignificant. Though the direction of causality is not necessarily established, this provides at least partial evidence that aid has been associated with institutional strengthening in this sample.

correlated, we are left with the result that causality must run the other way; from institutions to income (see also Bockstette *et al.*, 2002). We have not explored the magnitude of this effect, but this has been done extensively and convincingly by others (e.g. Acemoglu *et al.*, 2005a; Easterly and Levine, 2003; Rodrik *et al.*, 2004).

Our findings are consistent with new results presented by Acemoglu *et al.* (2005b). They analyze the empirical foundations of the so-called 'modernization theory,' arguing the existence of a causal relationship from high incomes to democracy (e.g. Barro, 1999). Controlling for country fixed effects, however, removes the statistical association. In other words, Acemoglu finds that income does *not* seem to be a driver of democratic change, just as we find, using analogous techniques, that income shocks are no driver of institutional quality in general.

We are therefore left with the result that governance measures are variable and subject to constant change, but that income growth *per se* is not one of the drivers of change. While institutional reform may be facilitated by economic development (Rodrik, 2006, World bank 2005), our results suggest that this process may not be spontaneous, and does not happen immediately in response to income shocks. It appears that to sustain economic growth over longer periods, and prevent it from fizzling out, policy makers should consciously work to improve governance (Hausmann *et al.*, 2005).

An important caveat to our conclusion exists. We use data on annual income variation to explain changes and levels in governance quality – over a relatively short time horizon from 1981-1999. While this (barely) fits within the timeframe proposed by Williamson (2000) for measuring governance changes, there may be a potential mismatch of temporal scales if institutional change is a slow moving process. We are aware of the limitations of our data, and clearly cannot (and indeed do not want to) reject the idea that there can be long-term feedback effects from income to institutions.[64]

Finally, we turn to policy-making – what lessons for policy makers can be distilled from our findings? To explore this issue further one needs to dig a little deeper and distinguish between different sources of income. We believe our results nicely complement existing work that has shown, through case studies and cross section regression analysis, that certain forms of economic activity tend to adversely affect institutional quality while other forms leave institutions unaffected. It matters whether (windfall) rents ends up in the hands of a broad base of the population, or instead is controlled by privileged elites. In the latter case, rent capture often provides incentives to distort institutions and policies. For example, foreign aid

---

[64] Demand for civil liberties and democracy is likely a luxury good, and developments in Southeast Asia clearly support the idea that institutional change can eventually follow economic development (e.g. Durham 1999 and others). Such a link might also exist through improvements in education, translating into more democratic and accountable governments. (But see Acemoglu *et al.* (2005b) for a critical assessment of the view that educational advances are a prerequisite for democracy.)

channeled through the government has the ability to adversely affect institutions (Brautigam and Knack 2004). Similarly, resource rents derived from 'pointy' natural resources (resources clustered in space and easily controllable by force – think of diamond mines and oil fields) tend to undermine institutional quality (Bulte *et al.*, 2005).[65] In contrast, rents from 'diffuse' resources like household agriculture and food production perhaps have no effect on institutions. Note that rainfall in Africa, through its impact on agricultural production, is closely aligned with the latter form of income.

Future work should be geared towards identifying and separating the opposing effects implied by income shocks. This implies augmenting our empirical analyses by more detailed measures of income, capital and institutions. Moreover, it would be useful to extend the time frame within which governance may respond to changes in income. This would require linking governance measures to more structural income shocks (e.g. productivity shocks in agriculture). This is left for future work.

References

Acemoglu, D., S. Johnson and J.A. Robinson, 2001). The Colonial Origins of Comparative Development: An Empirical Investigation. American Economic Review 91: 1369-1401.

Acemoglu, D., S. Johnson, J.A. Robinson and P. Yared, 2005a. From Education to Democracy? American Economic Review (Papers & Proceedings) 95: 44-49.

Acemoglu, D., S. Johnson, J.A. Robinson and P. Yared, 2005b. Income and Democracy. Department of Economics, Massachusetts Institute of Technology, Working Paper.

Aidt, T., 2003. Economic Analysis of Corruption: A Survey. Economic Journal 113: F632-652.

Barro, R.J., 1999. Determinants of Democracy, Journal of Political Economy, University of Chicago Press, vol. 107(S6), S158-29.

Bockstette, V., A. Chandra and L. Putterman, 2002. States and Markets: The Advantage of an Early Start. Journal of Economic Growth 7: 347-369.

Brautigam, D. and S. Knack, 2004. Foreign Aid, Institutions and Governance in Sub-Saharan Africa. Economic Development and Cultural Change 52: 255-285.

Bulte, E.H., R. Damania and R.T. Deacon, 2005. Resource Intensity, Institutions and Development. World Development 33: 1029-1044.

Collier, P. and A. Hoeffler, 2002. On The Incidence of Civil War in Africa. Journal of Conflict Resolution 46: 13-28.

Durham, J.B., 1999. Economic Growth and Political Regimes. Journal of Economic Growth 4: 81-111.

---

[65] For example, the political scientist Ross (2001) introduced the term 'rent seizing' to describe deliberate attempts by politicians to destroy the institutions that constrain them to gain access to resource rents. In the context of tropical deforestation Ross shows that 'timber' booms have induced Asian political leaders to dismantle the institutions that promote sustainable forest management to gain better access to the allocation of timber rents.

Easterly, W. and R. Levine, 2003. Tropics, germs, and crops: How endowments influence economic development. Journal of Monetary Economics 50: 3-39.

Fearon, J. and D. Laitin, 2003. Ethnicity, Insurgency, and Civil War. American Political Science Review 97: 75-90.

Forbes, K., 2000. A Reassessment of the Relation between Inequality and Growth. American Economic Review 90: 869-887.

Glaeser, E.L., R. La Porta, F. Lopez-De-Silanes and A. Schleifer, 2004. Do Institutions Cause Growth? Journal of Economic Growth 9: 271-303.

Hall, R and C. Jones, 1999. Why do some Countries Produce so Much More output than Others? Quarterly Journal of Economics 114, 83-116.

Hausmann, R. L. Pritchett and D. Rodrik, 2005. Growth Accelerations. Journal of Economic Growth 10: 303-329.

Knack, S. and P. Keefer, 1995. Institutions and Economic Performance: Cross Country Tests using Alternative Institutional Measures. Economics and Politics 7, 207-227.

Mauro, P., 2004. The Persistence of Corruption and Slow Economic Growth. IMF Staff Papers 41: 1-18.

Miguel, E., S. Satyanath and E. Sergenti, 2004. Economic Shocks and Civil Conflict: An Instrumental Variables Approach. Journal of Political Economy 112: 725-753.

Paxson, C.H., 1992. Using Weather Variability To Estimate the Response of Savings to Transitory Income in Thailand. American Economic Review 82: 15-33.

Rodrik, D., 2006. Goodbye Washington Consensus, Hello Washington Confusion. Journal of Economic Literature, In Press.

Rodrik, D., A. Subramanian and F. Trebbi, 2004. Institutions Rule: The Primacy of Institutions over Geography and Integration in Economic Development. Journal of Economic Growth 9: 131-165.

Temple, J. and P. Johnson, 1998. Social Capability and Economic Growth. Quarterly Journal of Economics 113: 965-990.

Williamson, O.E., 2000. The New Institutional Economics: Taking Stock, Looking Ahead. Journal of Economic Literature 38: 595-613.

World Bank, 2005. Economic Growth in the 1990s: Learning from a Decade of Reform. Washington DC: World Bank.

About the contributors

Peter **Arellanes** is a former graduate student at Cornell University, USA

Chris **Barrett** is Professor at the Department of Applied Economics and Management, Cornell University, 315 Warren Hall, Ithaca, NY 14853-7801 USA; cbb2@cornell.edu

Michel **Benoit-Cattin** is Director Socio-Economic Research at CIRAD, 34830 Montpellier, France; michel.benoit-cattin@cirad.fr

Erwin **Bulte** is Professor of Development Economics at Wageningen University and Tilburg University, the Netherlands; erwin.bulte@wur.nl

Richard **Damania** is Senior Researcher at the World Bank; rdamania@worldbank.org

Javier **Escobal** is Research Director with the Group for the Analysis of Development (GRADE) in Lima, Peru; jescobal@grade.org.pe

Ramon **Espinel** is Professor of Development Economics at Escuela Politécnica del Litoral (ESPOL), Ecuador, Adjunct Professor at the University of Florida, Gainesville, USA and Visiting Professor at Ghent University, Belgium; respinel@goliat.espol.edu.ec

Peter **Hazell** is visiting professor at Imperial College London, United Kingdom

Nico **Heerink** is Associate Professor at the Development Economics Group, Department of Social Sciences, Wageningen University, Wageningen, the Netherlands; Nico.Heerink@wur.nl

Stein **Holden** is Professor at the Norwegian University of Life Sciences; stein.holden@umb.no.

Hans **Jansen** is Senior Research Fellow and Coordinator for Central America for the International Food Policy Research Institute (IFPRI); hjansen@cgiar.org

Andrew **Karana** is an Agricultural Economist at the World Bank (Kenya Country Office); akaranja@worldbank.org

Michiel A. **Keyzer** is Director of the World Food Studies of the Vrije Universiteit (SOW-VU), Amsterdam, the Netherlands; M.A.Keyzer@sow.vu.nl

Froukje Kruijssen is researcher at the International Plant Genetic Resources Institute (IPGRI), Malaysia; F.Kruijssen@cgiar.org

Gideon **Kruseman** is Senior Research Fellow at the LEI-Wageningen UR, the Netherlands; Gideon.kruseman@wur.nl

Arie **Kuyvenhoven** was Professor at the Wageningen University and Research Centre, the Netherlands; Arie.Kuyvenhoven@wur.nl

David **Lee** is professor at at the Department of Applied Economics and Management,Cornell University, USA; drl5@cornell.edu.

Henk A.J. **Moll** is Associate Professor at Wageningen University, the Netherlands; henk.moll@wur.nl

Sam **Morley** is Visiting Research Fellow, IFPRI, Washington DC, USA

Tolulope **Olofinbiyi** is senior research assistant at the International Food Policy Research Institute (IFPRI); t.olofinbiyi@cgiar.org

Prabhu **Pingali** is Director of the Agriculture and Development Economics Division of the Food and Agriculture Organization of the United Nations, Rome, Italy; prabhu.pingali@fao.org

Per **Pinstrup-Andersen** is Professor of Food, Nutrition and Public Policy and J. Thomas Clark Professor of Entrepreneurship, Division of Nutritional Sciences, Cornell University, Ithaca, New York. USA; pp94@cornell.edu

Jan **Pronk** is former Dutch Minister for Development Cooperation (1973-77 and 1989-1998), Special Representative Secretary General of the United Nations in Sudan, Khartoum (2004 - 2006) and Professor at the Institute of Social Studies (ISS) in The Hague, the Netherlands

Futian **Qu** is Vice President and Professor at the Nanjing Agricultural University, Nanjing, P.R. China

Terri **Raney** is Senior Economist of the Agriculture and Development Economics Division of the Food and Agriculture Organization of the United Nations, Rome, Italy; terri.raney@fao.org

Teunis **van Rheenen** is special assistant to the DG at the International Food Policy Research Institute (IFPRI); t.vanrheenen@cgiar.org

Johannes **Roseboom** is an Innovation Policy Consultant; j.roseboom@planet.nl

Ruerd **Ruben** is Professor in Development Studies and Director of the Centre for International Development Issues (CIDIN), Radboud University Nijmegen, the Netherlands; R.Ruben@maw.ru.nl

Tan **Shuhao** is lecturer with Nanjing Agricultural University, Nanjing, P.R. China

Johan F.M. **Swinnen** is Professor at the LICOS Centre for Institutions and Economic Performance and the Department of Economics, University of Leuven (KUL), Belgium; Jo.Swinnen@econ.kuleuven.be

Guido **van Huylenbroeck** is Professor of Agricultural and Rural Environmental Economics at Ghent University, Belgium; Guido.VanHuylenbroeck@UGent.be

Lia **van Wesenbeeck** is researcher with the World Food Studies of the Vrije Universiteit (SOW-VU), Amsterdam, the Netherlands; C.F.A.vanWesenbeeck@sow.vu.nl

Joachim **von Braun** is Director General of the International Food Policy Research Institute (IFPRI), Washington DC, USA; j.von-braun@cgiar.org

Ruerd Ruben is Professor in Development Studies and Director of the Center for International Development Issues (CIDIN), Radboud University, Nijmegen, the Netherlands. R.Ruben@maw.ru.nl

Tan Shuhao is lecturer with Nanjing Agricultural University, Nanjing, PR China.

Johan F.M. Swinnen is Professor at the LICOS Centre for Institutions and Economic Performance and the Department of Economics, University of Leuven (KUL), Belgium. Jo.Swinnen@econ.kuleuven.be

Guido van Huylenbroeck is Professor of Agricultural and Rural Environmental Economics at Ghent University, Belgium. Guido.VanHuylenbroeck@UGent.be

Lia van Wesenbeeck is researcher with the World Food Studies of the Vrije Universiteit (SOW-VU), Amsterdam, the Netherlands. CPA.vanWesenbeeck@sow.vu.nl

Joachim von Braun is Director General of the International Food Policy Research Institute (IFPRI), Washington DC, USA. j.von-braun@cgiar.org

Keyword index

Development economics between markets and institutions